Advanced Welding Engineering and Technology for Marine Applications

This book examines the radical technological changes occurring in welding engineering and technology, with a focus on offering a better understanding of the latest developments for marine applications. It presents a detailed analysis of interdisciplinary knowledge, numerical modeling and simulation, and the various applications of welding systems. The theoretical results, practical solutions, and guidelines presented are valuable for both researchers working in the area of marine engineering and practitioners looking for solutions to welding-related problems in the maritime industry.

- Focuses on metallurgical and material interactions associated with joining.
- Presents the latest developments in the quality assessment of welded joints.
- Examines health and safety, as well as the related environmental aspects of marine welding.

Maritime Engineering
Science, Technology, and Applications Series
Edited by Kaushik Kumar

This new series will target and provide a collection of textbooks and research books on Maritime Engineering, Science, and Technology (a discipline of its own offering UG, PG, and Ph.D. degrees at various universities, of US, Europe, Australia, Asia, etc.), along with applications applied under Manufacturing, Mechanical, and Industrial Engineering. The series will provide an understanding of the Maritime domain using language of common man and capabilities to applications. The series will incorporate tools and applications to assist in the learning process. The series will contain books on various aspects of Naval Manufacturing and Engineering, Naval Architecture, Ocean Engineering, Marine Engineering, Autonomous Underwater and Sea Surface Vehicles, etc.

Advanced Welding Engineering and Technology for Marine Applications
Kaushik Kumar, Chikesh Ranjan, Saurav Datta, and Sanjiv Kumar Tiwari

Advanced Welding Engineering and Technology for Marine Applications

Kaushik Kumar, Chikesh Ranjan,
Saurav Datta, and Sanjiv Kumar Tiwari

CRC Press
Taylor & Francis Group
Boca Raton London New York

CRC Press is an imprint of the
Taylor & Francis Group, an **informa** business

Designed cover image: Shutterstock

First edition published 2026
by CRC Press
2385 NW Executive Center Drive, Suite 320, Boca Raton FL 33431

and by CRC Press
4 Park Square, Milton Park, Abingdon, Oxon, OX14 4RN

CRC Press is an imprint of Taylor & Francis Group, LLC

ISBN: 978-1-032-10849-0 (hbk)
ISBN: 978-1-032-10850-6 (pbk)
ISBN: 978-1-003-21735-0 (ebk)

DOI: 10.1201/9781003217350

Typeset in Times
by codeMantra

Contents

Acknowledgments xix
Preface xxi
Author Biographies xxiii

Chapter 1 Introduction to Marine Welding Engineering 1

1.1 Introduction 1
1.2 Overview of Marine Welding 1
1.3 Historical Development of Marine Welding 2
1.4 Role of Welding in Modern Marine Industries 4
1.4.1 Shipbuilding and Hull Fabrication 4
1.4.2 Offshore Structures and Oil Platforms 4
1.4.3 Repair and Maintenance 4
1.4.4 Lightweight Marine Vessels 5
1.4.5 Integration of Advanced Materials 6
1.4.6 Automation and Digitalization 6
1.4.7 Sustainability and Green Shipbuilding 6
1.5 Importance of Welding in Shipbuilding and Offshore Structures 6
1.5.1 Structural Integrity and Safety Considerations 6
1.5.2 Cost and Time Efficiency in Marine Fabrication 7
1.5.3 Contribution to Sustainability and Life Cycle Performance 7
1.6 Unique Challenges in Marine Welding 7
1.6.1 Harsh Environmental Conditions 8
1.6.2 Corrosion Resistance and Material Selection 9
1.6.3 Underwater Welding Complexities 9
1.6.4 Welding of Dissimilar Metals 10
1.6.5 Impact of Regulations and Classification Society Standards 10
1.7 Future Trends and Innovations in Marine Welding 11
1.7.1 Automation and Robotics in Shipyard Welding 11
1.7.2 Advances in Welding Consumables for Marine Environments 12
1.7.3 Non-Destructive Testing (NDT) in Harsh Conditions 13
1.8 Summary 14
1.9 Review Questions 15

Chapter 2 Materials Used in Marine Welding 17

2.1 Introduction 17
2.1.1 Role of Material Selection in Marine Welding 17

2.1.2 Influence of Marine Environment on Material Performance 18
2.2 Steel Alloys for Shipbuilding 20
2.2.1 Carbon Steel Grades and Applications 20
2.2.2 High-Strength Low-Alloy (HSLA) Steels 20
2.2.3 Welding Characteristics of Marine Steels 21
2.2.4 Standards and Classifications for Marine Steel 21
2.3 Aluminum Alloys for Shipbuilding 21
2.3.1 Common Marine-Grade Aluminum Alloys (5xxx, 6xxx Series) 22
2.3.2 Advantages of Aluminum in Ship Structures 22
2.3.3 Welding Methods for Aluminum in Marine Applications 23
2.3.4 Challenges in Aluminum Welding (Distortion, Porosity) 23
2.4 Corrosion-Resistant Materials for Marine Application 23
2.4.1 Stainless Steel Types for Marine Use (Austenitic, Duplex, Super Duplex) 24
2.4.2 Copper-Nickel Alloys for Seawater Systems 24
2.4.3 Titanium and Exotic Alloys in Offshore Structures 25
2.4.4 Coatings and Cladding Techniques for Corrosion Resistance 25
2.5 Material Compatibility and Dissimilar Metal Welding 25
2.5.1 Issues in Welding Steel to Aluminum and Other Metals 26
2.5.2 Techniques to Minimize Galvanic Corrosion 27
2.6 Summary 27
2.7 Review Questions 28

Chapter 3 Welding Processes in the Marine Industry 30

3.1 Introduction 30
3.1.1 Role of Welding Processes in Marine Fabrication 30
3.1.2 Factors Affecting Process Selection in Shipyards 31
3.1.3 Influence of Marine Environment on Welding Techniques 31
3.2 Arc Welding Processes 31
3.2.1 Shielded Metal Arc Welding (SMAW) 31
3.2.2 Gas Metal Arc Welding (GMAW) 33
3.2.3 Flux-Cored Arc Welding (FCAW) 35
3.2.4 Gas Tungsten Arc Welding (GTAW) 37
3.3 Submerged Arc Welding (SAW) 39
3.3.1 Principle and Equipment 39
3.3.2 Advantages for Large Ship Panels and Offshore Platforms 40

3.3.3 Limitations and Quality Considerations 41
3.4 Resistance Welding 41
3.4.1 Working Principle of Resistance Welding 43
3.4.2 Types of Resistance Welding...................................... 44
3.4.3 Spot Welding 44
3.4.4 Seam Welding 46
3.4.5 Projection Welding...47
3.4.6 Flash Butt Welding.. 48
3.5 Friction Stir Welding (FSW) 51
3.5.1 Process Fundamentals...51
3.5.2 Applications in Aluminum Ship Structures................52
3.5.3 Benefits for Strength and Corrosion Resistance..........53
3.5.4 Challenges and Future Potential 53
3.6 Process Comparison and Selection Guidelines 54
3.6.1 Criteria for Choosing a Welding Process 54
3.6.2 Cost, Speed, and Quality Considerations.................. 54
3.6.3 Case Studies from Shipbuilding and Offshore Projects 54
3.7 Summary 54
3.8 Review Questions 55

Chapter 4 Metallurgical Aspects of Marine Welding... 56

4.1 Introduction 56
4.1.1 Importance of Metallurgy in Marine Welding........... 56
4.1.2 Effect of Marine Environment on Welded Joints........57
4.1.3 Overview of Common Marine Welding Materials 57
4.2 Heat-Affected Zone (HAZ) in Marine Alloys 58
4.2.1 Definition and Metallurgical Changes in HAZ...........58
4.2.2 HAZ in Carbon and Low-Alloy Steels........................58
4.2.3 HAZ in Aluminum and Stainless Steels 59
4.2.4 Minimizing HAZ Degradation in Marine Fabrication 59
4.3 Weldability of High-Strength Steels....................................... 60
4.3.1 Types of High-Strength Steels in Marine Applications.. 60
4.3.2 Factors Influencing Weldability (Preheat, Cooling Rate)... 60
4.3.3 Controlling Microstructural Changes for Toughness 61
4.4 Weldability of Corrosion-Resistant Alloys..............................61
4.4.1 Stainless Steels (Austenitic, Duplex, Super Duplex) 61
4.4.2 Copper-Nickel and Titanium Alloys 62
4.4.3 Special Considerations in Welding Non-Ferrous Marine Alloys...62

4.5 Hydrogen-Induced Cracking (HIC) 63
4.5.1 Mechanism of Hydrogen Embrittlement in Marine Steels 63
4.5.2 Sources of Hydrogen During Welding 63
4.5.3 Prevention and Control Techniques 64
4.6 Stress Corrosion Cracking (SCC) in Marine Welds 64
4.6.1 Mechanism and Contributing Factors 64
4.6.2 SCC in Stainless Steels and Aluminum Alloys 65
4.6.3 Mitigation Strategies for Marine Structures 65
4.7 Metallurgical Testing and Evaluation 66
4.7.1 Microstructural Analysis of Weld Zones 66
4.7.2 Hardness and Toughness Testing in Marine Welds 66
4.7.3 Corrosion Testing in Simulated Seawater 67
4.8 Summary 67
4.9 Review Questions 68

Chapter 5 Robotic and Automated Welding in Shipbuilding 69

5.1 Introduction 69
5.1.1 Evolution of Welding Automation in Shipbuilding 69
5.1.2 Benefits and Drivers for Automation in Marine Fabrication 70
5.1.3 Overview of Global Trends in Shipyard Robotics 70
5.2 Role of Automation in Shipyard Welding 71
5.2.1 Process Standardization and Repeatability 71
5.2.2 Enhancing Weld Quality and Reducing Human Error 72
5.2.3 Productivity Gains in Large-Scale Ship Fabrication 72
5.2.4 Examples of Automated Welding Systems in Leading Shipyards 73
5.3 Types of Robotic Welding Systems in Shipbuilding 73
5.3.1 Fixed Welding Robots 73
5.3.2 Mobile Welding Robots 74
5.3.3 Gantry and Track-Mounted Systems for Hull Construction 75
5.3.4 Collaborative Robots (Cobots) in Ship Sections 77
5.3.5 Underwater and Remote-Controlled Welding Robots 78
5.4 AI and Machine Learning in Welding Automation 79
5.4.1 Role of AI in Real-Time Weld Parameter Adjustment 80
5.4.2 Machine Vision for Seam Tracking and Defect Detection 80

5.4.3 Predictive Maintenance of Welding Equipment 81
5.4.4 Intelligent Quality Control Systems in Shipyards 81
5.5 Integration of Automation with Digital Shipbuilding 82
5.5.1 CAD/CAM-to-Robot Interface 82
5.5.2 Digital Twin Applications for Welding Planning......... 83
5.5.3 Data Analytics for Continuous Improvement 84
5.6 Challenges in Implementing Robotic Welding......... 85
5.6.1 High Initial Investment and Return on Investment (ROI) Considerations......... 86
5.6.2 Workforce Training and Skill Upgradation 86
5.6.3 Adaptation to Complex Marine Structures 86
5.6.4 Integration with Digital Shipbuilding Systems......... 86
5.6.5 Real-Time Quality Assurance......... 86
5.6.6 Maintenance and Reliability Issues......... 86
5.6.7 Flexibility for Low-Volume or Customized Production 86
5.6.8 Safety and Workspace Constraints......... 87
5.7 Future Trends in Marine Welding Automation 87
5.8 Summary 87
5.9 Review Questions 88

Chapter 6 Hybrid Welding Technologies for Marine Applications 90

6.1 Introduction 90
6.1.1 Concept and Importance of Hybrid Welding in the Marine Industry......... 90
6.1.2 Advantages of Hybrid Welding over Conventional Processes.........91
6.1.3 Scope of Hybrid Techniques in Shipbuilding and Offshore Structures 92
6.2 Hybrid Laser-Arc Welding (HLAW)......... 92
6.2.1 Principle and Process Mechanism 93
6.3 Underwater Welding Techniques......... 94
6.3.1 Wet Welding......... 96
6.3.2 Dry Welding (Habitat Welding) 97
6.4 Metallurgical Considerations in Hybrid and Underwater Welding......... 99
6.4.1 Heat-Affected Zone (HAZ) Characteristics......... 99
6.4.2 Microstructural Changes in Marine Alloys 100
6.4.3 Corrosion Resistance and Protective Measures 100
6.5 Quality Assurance and Testing in Hybrid Marine Welding.........101
6.5.1 Inspection Standards for Hybrid and Underwater Welds 101

6.5.2 Case Studies from Shipbuilding and Offshore Repair ... 102
6.6 Future Trends in Hybrid Marine Welding ... 103
6.7 Summary ... 104
6.8 Review Questions ... 105

Chapter 7 Wire Arc Additive Manufacturing (WAAM) in Marine Applications ... 107

7.1 Introduction ... 107
7.1.1 Evolution of Additive Manufacturing in the Marine Sector ... 107
7.1.2 Advantages of WAAM over Conventional Marine Fabrication ... 108
7.1.3 Scope of WAAM in Shipbuilding and Offshore Engineering ... 110
7.2 Fundamentals of WAAM for Shipbuilding ... 110
7.2.1 Principle and Working Mechanism of WAAM ... 111
7.2.2 Comparison with Other Additive Manufacturing Processes ... 113
7.2.3 WAAM Equipment and Setup in Marine Fabrication ... 113
7.2.4 Layer-by-Layer Deposition in Large Marine Structures ... 116
7.3 WAAM Process Parameters and Material Considerations ... 117
7.4 Challenges of WAAM in the Marine Industry ... 120
7.5 Future Trends of WAAM in the Marine Industry ... 122
7.6 Case Study: Wire Arc Additive Manufacturing of SS308L Using CMT-Based Robotic Metal Inert Gas (MIG) Welding ... 124
7.7 Summary ... 126
7.8 Review Questions ... 126

Chapter 8 Welding for Offshore and Subsea Structures ... 128

8.1 Introduction ... 128
8.1.1 Importance of Welding in Offshore and Subsea Engineering ... 129
8.1.2 Types of Offshore Structures (Fixed, Floating, Subsea) ... 130
8.1.3 Role of Welding in Structural Integrity and Safety ... 134
8.2 Challenges in Welding Offshore Rigs and Platforms ... 135
8.2.1 Environmental Challenges ... 135
8.2.2 Metallurgical and Material-Related Challenges ... 135

8.2.3 Accessibility and Positional Welding........................136
8.2.4 Safety and Human Factors..136
8.2.5 Equipment and Process Control Limitations............136
8.2.6 Inspection and Quality Assurance Challenges.........136
8.2.7 Economic and Logistical Constraints.......................137
8.2.8 Mitigation Strategies and Technological Advancements in Offshore Welding.........................137
8.3 Welding Materials for Offshore Applications........................139
8.3.1 Base Materials Used in Offshore Structures.............140
8.3.2 Filler Metals and Welding Consumables..................140
8.3.3 Corrosion and Hydrogen Control.............................141
8.3.4 Metallurgical Compatibility and Heat-Affected Zone (HAZ) Control...141
8.3.5 Material Standards and Qualification Requirements..141
8.3.6 Emerging Trends in Offshore Welding Materials.....141
8.4 Hyperbaric Welding..142
8.4.1 Principles of Hyperbaric Welding............................144
8.4.2 Wet Hyperbaric Welding (Direct Seawater Environment)..145
8.4.3 Dry Hyperbaric Welding (Habitat Welding).............147
8.5 Deep-Sea Welding Techniques..149
8.5.1 Challenges in Deep-Sea Welding.............................150
8.5.2 Robotic and Remotely Operated Welding Systems...150
8.5.3 Deepwater Habitat and Pressure-Controlled Systems...151
8.5.4 Laser and Hybrid Welding for Deep-Sea Applications..151
8.6 Quality Control and Inspection in Offshore Welding............151
8.6.1 Standards and Codes for Offshore Welding Quality...152
8.6.2 Welding Process Control and Monitoring................152
8.6.3 Non-Destructive Testing (NDT) and Inspection Methods...153
8.6.4 Metallurgical and Mechanical Testing.....................153
8.6.5 Corrosion and Fatigue Inspection............................153
8.6.6 Digital Quality Assurance and Predictive Maintenance..153
8.7 Future Trends in Offshore and Subsea Welding....................155
8.7.1 Robotic and Autonomous Welding Systems............156
8.7.2 Integration of Artificial Intelligence (AI) and Digital Twin Technology...156
8.7.3 Hybrid and Laser-Based Welding Advancements...157
8.7.4 Advanced Materials and Smart Consumables..........158

8.7.5 Sustainable and Energy-Efficient Welding Practices 158
8.7.6 Integration with Offshore Renewable Energy and Deepwater Infrastructure 159
8.8 Summary 159
8.9 Review Questions 159

Chapter 9 Non-Destructive Testing (NDT) for Marine Welds 161

9.1 Introduction 161
9.1.1 Importance of NDT in Marine Welding Quality Assurance 162
9.1.2 Role of NDT in Safety, Reliability, and Regulatory Compliance 162
9.1.3 Challenges of NDT in Marine Environments 163
9.2 Overview of NDT Methods for Marine Applications 164
9.2.1 Surface vs Volumetric Testing Techniques 165
9.2.2 Advanced and Specialized NDT Methods 166
9.2.3 Underwater and Remote NDT Techniques 166
9.3 Ultrasonic Testing (UT) 168
9.3.1 Principles of Ultrasonic Inspection 168
9.3.2 Conventional UT vs. Phased Array UT in Marine Welds 170
9.3.3 Advantages, Limitations, and Applications in Shipbuilding 171
9.4 Radiographic Testing (RT) 171
9.4.1 Principle of Radiographic Testing (RT) 172
9.4.2 Applications in Shipyard and Offshore Use 173
9.4.3 Advantages of Radiographic Testing (RT) 173
9.4.4 Limitations of Radiographic Testing (RT) 174
9.5 Magnetic Particle Testing (MT) 174
9.5.1 Principle of Magnetic Flux Leakage 175
9.5.2 Application in Ferromagnetic Marine Materials 176
9.5.3 Advantages and Limitations in Shipyard and Offshore Use 176
9.6 Other Relevant NDT Methods in Marine Welding 177
9.6.1 Dye Penetrant Testing (PT) for Surface Crack Detection 177
9.6.2 Eddy Current Testing (ECT) for Non-Ferrous Alloys 178
9.6.3 Acoustic Emission Monitoring (AEM) for Structural Health 180
9.7 Inspection Techniques for Welded Marine Structures 181
9.7.1 Hull and Deck Weld Inspection Procedures 182
9.7.2 Offshore Platform Weld Integrity Assessment 183

9.7.3 Underwater Weld Inspection Techniques (ROV and Diver-Assisted) ... 183
9.8 Standards and Certification for Marine Weld NDT ... 184
9.8.1 Classification Society Requirements (ABS, DNV, LR) ... 184
9.8.2 International Standards (ISO, ASME, AWS) ... 185
9.8.3 Personnel Qualification and Training ... 186
9.9 Future Trends in Marine Weld Inspection ... 187
9.9.1 Automated and Robotic NDT Systems ... 187
9.9.2 AI-Based Defect Recognition ... 188
9.9.3 Integration with Digital Twins for Continuous Monitoring ... 188
9.10 Summary ... 189
9.11 Review Questions ... 189

Chapter 10 Welding Standards and Regulations in the Marine Industry ... 192

10.1 Introduction ... 192
10.1.1 Importance of Welding Standards in Marine Fabrication ... 193
10.1.2 Role of Regulations in Safety, Quality, and Compliance ... 193
10.1.3 Global vs. Regional Standardization in the Marine Sector ... 194
10.2 International Maritime Organization (IMO) Welding Standards ... 195
10.2.1 Overview of IMO's Role in Maritime Safety ... 195
10.2.2 Relevant IMO Conventions and Guidelines for Welding ... 195
10.2.3 Integration of IMO Standards into Shipbuilding Practices ... 197
10.3 ABS Welding Standards ... 197
10.3.1 ABS Rules for Materials and Welding Procedures ... 198
10.3.2 Approval Processes for Welders and Welding Consumables ... 199
10.3.3 Case Examples of ABS Compliance in Shipyards ... 199
10.4 Det Norske Veritas (DNV) Welding Standards ... 200
10.4.1 DNV Rules for Ship and Offshore Structure Welding ... 200
10.4.2 Welding Procedure Qualification under DNV ... 201
10.4.3 DNV's Approach to Quality Assurance and Inspection ... 202
10.5 Lloyd's Register (LR) Welding Standards ... 203

10.5.1 LR Requirements for Marine Welding Procedures and Materials.. 203
10.5.2 Certification and Welder Qualification Processes 204
10.5.3 LR Inspection and Audit Practices.......................... 205
10.6 Shipbuilding and Offshore Structural Welding Codes.......... 205
10.6.1 Overview of AWS D1.1, AWS D3.6M, and Related Codes.. 206
10.6.2 ISO and EN Standards Relevant to Marine Welding 206
10.6.3 Offshore Structural Welding Requirements (API, NORSOK)... 207
10.6.4 Harmonization of Multiple Codes in Global Shipyards... 208
10.7 Compliance and Quality Control in Marine Welding........... 208
10.7.1 Documentation and Traceability Requirements....... 209
10.7.2 Role of Third-Party Inspection and Classification Societies.. 209
10.7.3 Audit and Certification Processes in Marine Projects... 210
10.8 Future Trends in Welding Standards..................................... 211
10.8.1 Digitalization and Real-Time Welding Data Compliance... 211
10.8.2 AI-Driven Quality Monitoring and Standardization 212
10.8.3 Sustainability and Environmental Compliance in Marine Welding 212
10.9 Summary 213
10.10 Review Questions 214

Chapter 11 Corrosion and Fatigue in Marine Welds 216

11.1 Introduction 216
11.1.1 Importance of Corrosion and Fatigue Considerations in Marine Welding........................... 217
11.1.2 Influence of Marine Environment on Weld Durability 217
11.1.3 Overview of Common Degradation Mechanisms in Marine Structures 218
11.2 Types of Corrosion in Marine Welds 219
11.2.1 General Corrosion in Welded Joints.......................... 219
11.2.2 Galvanic Corrosion between Dissimilar Metals 221
11.2.3 Crevice and Pitting Corrosion in Weld Zones.......... 222
11.2.4 Stress Corrosion Cracking (SCC) in Marine Applications... 224

11.3 Corrosion Prevention Techniques in Marine Welded Structures 225
11.3.1 Material Selection for Corrosion Resistance 226
11.3.2 Protective Coatings and Cladding in Weld Zones 227
11.3.3 Cathodic Protection (Sacrificial Anode and Impressed Current) 228
11.3.4 Weld Design Modifications to Minimize Corrosion Risk 228
11.3.5 Post-Weld Treatments and Surface Finishing 229
11.4 Fatigue in Marine Welds 229
11.4.1 Mechanism of Fatigue Failure in Welded Joints 230
11.4.2 Factors Affecting Fatigue Life in Marine Structures 231
11.4.3 Stress Concentration Effects at Weld Toes and Roots 231
11.5 Fatigue Life Assessment of Welded Joints in Ships and Offshore Platforms 232
11.5.1 ***S-N*** Curve Approach for Fatigue Evaluation 232
11.5.2 Fracture Mechanics Approach to Fatigue Analysis 233
11.5.3 Influence of Welding Defects on Fatigue Performance 234
11.6 Testing and Monitoring of Corrosion and Fatigue 234
11.6.1 Laboratory Corrosion Testing for Marine Welds 235
11.6.2 Full-Scale Fatigue Testing of Welded Components 236
11.6.3 In-Service Monitoring Techniques for Offshore and Ship Structures 237
11.7 Standards and Guidelines for Corrosion and Fatigue Design 238
11.7.1 Classification Society Requirements (ABS, DNV, LR) 238
11.7.2 ISO, AWS, and API Standards for Marine Weld Durability 239
11.8 Future Trends in Corrosion and Fatigue Mitigation 240
11.8.1 Advanced Coatings and Nanomaterials for Weld Protection 240
11.8.2 AI-Based Predictive Maintenance for Fatigue and Corrosion 241
11.8.3 Integration of Digital Twins for Weld Life Prediction 242
11.9 Summary 242
11.10 Review Questions 243

Chapter 12 Sustainability and Environmental Considerations in Marine Welding 245

12.1 Introduction 245
12.1.1 Importance of Sustainability in the Marine Welding Sector 245
12.1.2 Environmental Impacts of Conventional Shipyard Welding 246
12.1.3 Global Regulations and Maritime Environmental Goals (IMO, EU Green Deal) 246
12.2 Reducing Carbon Footprint in Shipyard Welding 247
12.2.1 Energy-Efficient Welding Processes (GMAW, FCAW, HLAW) 248
12.2.2 Use of Renewable Energy in Welding Operations 249
12.2.3 Low-Emission Welding Consumables and Shielding Gases 249
12.2.4 Process Optimization to Reduce Energy Consumption 249
12.2.5 Carbon Accounting and Life Cycle Assessment (LCA) for Marine Welding 250
12.3 Recycling and Eco-Friendly Welding Technologies 250
12.3.1 Recycling of Scrap Metal from Shipbuilding and Repair 251
12.3.2 Reuse of Weld Consumables and By-Products 252
12.3.3 Water-Based and Low-VOC Welding Consumables 252
12.3.4 Solid-State Welding for Reduced Environmental Impact (FSW, WAAM) 253
12.3.5 Hybrid and Automated Welding for Material Efficiency 255
12.4 Waste Management in Marine Welding Operations 256
12.4.1 Handling and Disposal of Slag, Fumes, and Metal Dust 257
12.4.2 Hazardous Waste Minimization Techniques 257
12.4.3 Compliance with Marine Pollution Prevention Standards (MARPOL) 258
12.5 Health and Safety in Sustainable Welding Practices 258
12.5.1 Reducing Weld Fume Exposure with Green Technologies 259
12.5.2 Eco-Friendly Personal Protective Equipment (PPE) 259
12.5.3 Worker Training for Sustainable Welding Practices 260
12.6 Summary 260
12.7 Review Questions 261

Chapter 13 Future Trends in Marine Welding Engineering 263

13.1 Introduction 263
13.1.1 Evolution of Welding Practices in the Marine Industry 263
13.1.2 Drivers for Innovation in Marine Welding 265
13.1.3 Role of Emerging Technologies in Shipbuilding and Offshore Structures 265
13.2 Digital Twin Technology in Marine Welding 266
13.2.1 Concept and Principles of Digital Twin Technology 266
13.2.2 Integration with Welding Process Simulation 267
13.2.3 Predictive Maintenance of Welded Structures 268
13.2.4 Application in Shipyard Workflow Optimization 268
13.3 Smart Sensors and Real-Time Monitoring in Marine Welding 269
13.3.1 Types of Sensors for Weld Quality Monitoring (Thermal, Acoustic, Optical) 270
13.3.2 Real-Time Data Acquisition during Welding Operations 270
13.3.3 Defect Detection and Automated Correction Systems 271
13.3.4 Remote and Underwater Weld Monitoring Solutions 271
13.3.5 Integration with AI for Intelligent Weld Control 272
13.4 Artificial Intelligence and Machine Learning in Marine Welding 272
13.4.1 AI-Based Weld Parameter Optimization 273
13.4.2 Automated Quality Assessment and Defect Prediction 274
13.4.3 Adaptive Welding Systems for Complex Marine Structures 274
13.5 Automation and Robotics in Next-Generation Marine Welding 275
13.5.1 Autonomous Welding Robots in Shipbuilding 276
13.5.2 Hybrid Human–Robot Collaboration Models 276
13.5.3 Integration with CAD/CAM for Fully Automated Welding Lines 277
13.6 Sustainable and Low-Carbon Welding Innovations 277
13.6.1 Energy-Efficient Welding Processes for the Future 278
13.6.2 Green Consumables and Environmentally Friendly Practices 279
13.6.3 Role of Digital Tools in Sustainability Tracking 279
13.7 Challenges in Adopting Future Welding Technologies 280
13.7.1 High Investment and Technology Integration Costs 281

13.7.2 Training and Skill Requirements for Shipyard Workforce 281
13.7.3 Compatibility with Existing Marine Welding Standards 282
13.8 Outlook and Long-Term Trends 282
13.8.1 Fully Autonomous Shipyard Welding Systems 283
13.8.2 Integration with Industry 5.0 and Human-Centric Manufacturing 284
13.8.3 Global Collaboration on Smart Marine Welding Technologies 284
13.9 Summary 285
13.10 Review Questions 285

Bibliography 287

Index 291

Acknowledgments

The completion of this book, *Advanced Welding Engineering and Technology for Marine Applications*, has been a deeply rewarding project, made possible through the generous support, guidance, and encouragement of numerous individuals and institutions. We would like to express our profound gratitude to all those who, in various ways, contributed to the successful realization of this work.

First and foremost, we would like to thank God for providing the opportunity, believe in passion and hard work, and pursue dreams. In the process of putting this book together, it was realized how true this gift of writing is for anyone. This could never have been done without the power provided by you and, of course, the faith in You, the Almighty.

Our deepest gratitude goes to our families for their unending encouragement and understanding. Your patience during the long hours of writing and your unwavering belief in us provided the emotional support necessary to see this project through to completion. Your words of encouragement and your ability to remind us of the bigger picture have been invaluable, which also helped us to take yet another challenge which decreases the amount of time we could spend with them. They were our inspiration and motivation, and their constant encouragement and late-night brainstorming sessions were invaluable. The unwavering support and insightful feedback helped to shape the direction of this book, and to say the least, their belief in our vision made all the difference. We hope that one day they will read this book and understand why we spent more time in front of our laptops and computers than with them. We would like to thank our parents and grandparents for allowing us to follow our ambitions.

We extend our sincere appreciation to the academic and research mentors whose insights and constructive discussions shaped my understanding of welding science and marine engineering. Their invaluable advice and scholarly rigor have continually inspired me to maintain the highest standards of technical accuracy and clarity throughout this manuscript. Our heartfelt thanks are also due to colleagues and peers in the fields of materials science, mechanical engineering, and naval architecture, whose suggestions and exchange of ideas enriched several aspects of this book.

We are particularly indebted to the industry professionals, shipyard engineers, and classification society experts who shared their practical knowledge and experience regarding modern welding technologies, automation, and inspection standards. Their contributions ensured that the content remains grounded in real-world applications and aligned with the latest industrial trends. I also acknowledge the technical reviewers and subject specialists who provided critical feedback on early drafts of the chapters, leading to substantial improvements in scope, depth, and readability.

Special thanks are due to research scholars and postgraduate students, whose questions, discussions, and enthusiasm for learning continually motivated me to refine explanations and expand this book's educational value. Their curiosity and dedication represent the very spirit of inquiry that this work seeks to encourage. I am equally grateful to professional societies, standardization organizations, and academic institutions that have advanced the discipline of marine welding engineering through publications, conferences, and collaborative initiatives.

We wish to record my appreciation for the publishers and editorial team for their professionalism, patience, and meticulous attention to detail during the production process. Their efforts have been instrumental in transforming this manuscript into a polished and accessible reference for readers across academia and industry.

Finally, I dedicate this work to all researchers, engineers, educators, and students striving to advance marine welding technology toward safer, smarter, and more sustainable maritime practices. It is our hope that this book will serve as a lasting contribution to their learning, research, and professional growth and that it may inspire further innovation in the fascinating and ever-evolving field of marine welding engineering.

Kaushik Kumar
Chikesh Ranjan
Saurav Datta
Sanjiv Kumar Tiwari

Preface

The authors are pleased to present the book *Advanced Welding Engineering and Technology for Marine Applications*. The book title has been conceived as a comprehensive and structured resource that brings together the scientific principles, technological practices, and emerging innovations governing welding applications in the marine environment. The objective is to bridge the knowledge gap between classical metallurgical and process-based understanding and the contemporary advancements that are redefining marine fabrication. It seeks to serve as an authoritative reference for undergraduate and postgraduate students of marine, mechanical, and materials engineering, as well as for researchers, academicians, design engineers, classification societies, and shipyard professionals engaged in welding technology and quality assurance.

The marine industry represents one of the most technologically demanding sectors of engineering, where structures and systems must operate reliably under complex environmental and mechanical stresses. Among the multifaceted engineering disciplines supporting this industry, **marine welding engineering** holds a position of paramount importance. It forms the foundation for the construction, maintenance, and modernization of ships, offshore structures, underwater systems, and other marine assets that constitute the global maritime infrastructure. The evolution of marine welding—from traditional manual operations to intelligent, automated, and digitally integrated processes—reflects the broader transformation of the maritime sector toward enhanced safety, efficiency, and sustainability.

The entire book is segregated into **13 chapters**. The organization of this volume reflects both a pedagogical and an application-oriented approach. The opening chapters provide a **fundamental introduction to marine welding engineering**, including its historical development, role in modern shipbuilding and offshore industries, and the challenges arising from harsh marine conditions. Readers are introduced to the interplay between welding processes, material behavior, and structural design considerations essential for ensuring the safety and longevity of marine assets. The subsequent chapters offer an in-depth discussion of **materials used in marine welding**, covering ferrous and non-ferrous alloys, corrosion-resistant systems, and dissimilar metal joining techniques that address the unique environmental and mechanical requirements of marine structures. This book then progresses to a detailed exposition of **welding processes and their metallurgical implications**, emphasizing process selection, heat-affected zone (HAZ) phenomena, and the prevention of weld defects in marine-grade materials. Special attention is devoted to **advanced and hybrid processes**, such as laser-arc hybrid welding, underwater welding, and friction stir welding, which have gained significant industrial relevance in shipbuilding and offshore fabrication. The inclusion of chapters on **robotic and automated welding systems, artificial intelligence (AI)-enabled process control, and wire arc additive manufacturing (WAAM)** underscores the growing impact of digital technologies and Industry 5.0 paradigms in transforming traditional shipyard operations into intelligent, adaptive, and data-driven manufacturing ecosystems. Recognizing the critical importance of safety and reliability, this book also covers **non-destructive testing (NDT)** techniques, standards, and regulations applicable to marine welds, referencing global practices of leading classification societies such as ABS, DNV, and Lloyd's Register. Furthermore, dedicated chapters address the challenges of **corrosion and fatigue in marine welds**, elucidating degradation mechanisms and mitigation strategies that are vital for ensuring life cycle performance and structural integrity. The integration of **sustainability and environmental considerations** in welding operations—such as low-carbon processes, eco-friendly consumables, and waste minimization—reflects the global commitment of the maritime sector toward greener and more energy-efficient manufacturing. Finally, this book provides a forward-looking perspective on **future trends in marine welding engineering**, encompassing smart sensors, digital twin technology, predictive maintenance, and autonomous welding systems. These emerging domains signal a paradigm shift in marine fabrication, where human expertise and AI converge to achieve precision, repeatability, and environmental responsibility.

Throughout the text, theoretical discussions are complemented by illustrative examples, industrial case studies, and review questions designed to reinforce conceptual understanding and promote independent learning. The emphasis on both fundamentals and innovation ensures that this book can function as an academic textbook, a research reference, and a practical guide for professionals seeking to upgrade their knowledge in marine welding science and technology.

The preparation of this book has been guided by the conviction that welding is not merely a joining process but a strategic enabler of structural integrity, performance, and sustainability in the marine domain. By integrating classical engineering principles with contemporary research and industrial practices, this work aspires to contribute meaningfully to the education, professional development, and technological advancement of the global maritime community.

It is sincerely hoped that this volume will serve as a valuable companion for students, educators, researchers, and practicing engineers—supporting both foundational learning and advanced exploration of marine welding engineering as it evolves toward the digital and sustainable era of modern shipbuilding and offshore engineering.

In conclusion, this book is a culmination of years of exploration, research, and personal growth. It is our hope that the insights and knowledge shared within these pages will inspire, challenge, and guide you in your own journey. Whether you are a seasoned professional or a curious learner, we believe that the ideas presented here will resonate and offer value. As you turn the pages, we invite you to engage with the material, question assumptions, and explore new perspectives. Thank you for allowing us to share this work with you—we hope it serves as a meaningful contribution to your endeavors.

Kaushik Kumar
Chikesh Ranjan
Saurav Datta
Sanjiv Kumar Tiwari

Author Biographies

Kaushik Kumar holds a **Ph.D.** in **Engineering** from **Jadavpur University, India**; an **MBA** in **Marketing Management** from **Indira Gandhi National Open University, India**; and a **Bachelor of Technology** from **Regional Engineering College (Now National Institute of Technology), Warangal, India**. For 11 years, he worked in a manufacturing unit of global repute. He is currently working as a Professor in the Department of Mechanical Engineering, Birla Institute of Technology, Mesra, Ranchi, India. He has 24 years of teaching and research experience. His research interests include composites, optimization, non-conventional machining, CAD/CAM, rapid prototyping, and quality management systems toward product development for societal and industrial usage, and he has received **50+ patents** for them, most of which are donated to start-ups and industrial fraternity through **Jharkhand Small Industries Association (JSIA)**. He has published **60+ books** (which are referred to as text books and reference books by **40+ universities/institutes** in their academic curriculum), **100+ book chapters,** and **200+ research papers** in peer-reviewed reputed national and international journals. Kaushik has also served as **editor-in-chief, series editor, guest editor, editor, editorial board member,** and **reviewer** for international and national journals. He has been felicitated with many awards and honors including **Distinguished Alumnus Award for Professional Excellence 2023 under Academic and Research** from the **National Institute of Technology, Warangal, India**. He has also received sponsored research and consultancy projects of **more than 1 crore** from the Government of India and abroad. Kaushik has delivered expert lectures as a keynote speaker at international and national conferences, and as a resource person at various workshops, faculty development programmes (FDPs), and short-term courses. He has guided many students of doctoral, master's, and undergraduate programs of his home and other institutions in India and abroad. He also has served as reviewer and examiner of doctoral and master's dissertations for institutes in India and abroad.

Chikesh Ranjan is currently serving as a **Project Engineer for Project SwaYaan—Capacity Building in Drone/UAS** at the Department of Mechanical Engineering, National Institute of Technology (NIT) Rourkela, Rourkela (Odisha), India. He holds a **PhD** in Mechanical Engineering from the Department of Mechanical Engineering, Birla Institute of Technology, Mesra, Ranchi, Jharkhand, India. Chikesh also possesses a **Master of Engineering** degree in Mechanical Engineering, specializing in the Design of Mechanical Equipment, from the same institution.

His academic journey began with a **Bachelor of Engineering** in Mechanical Engineering from Marathwada Institute of Technology, Aurangabad, Maharashtra, India, where he graduated with first class with distinction. Chikesh's innovative contributions have been recognized through the publication of **15** journal papers, **10** authored books, and **13** book chapters. He has actively organized **ten workshops**, coordinated **13 technical events**, and been involved in **18 technical events** overall.

Additionally, he has participated in **31 training/FDP programs** and attended **34 national/international workshops, seminars, and webinars**.

Chikesh has been granted **10 design patents** and **07 foreign patents**, demonstrating his dedication to advancing mechanical design and technology. With 9 years of teaching and research experience of global repute, his areas of teaching and research interest include composites, non-conventional machining, CAD/CAM, and robotics. He is a proud member of the Institution of Engineers, India, and has received **07 prestigious awards and honors** for his remarkable contributions to the field of Mechanical Engineering.

Saurav Datta is presently working as a Professor in the Department of Mechanical Engineering, National Institute of Technology, Rourkela (Odisha). After receiving B.E. (Honors) in Mechanical Engineering from Jadavpur University, Kolkata, in 2003, he was awarded Ph.D. (Engineering) by Jadavpur University in December 2008. He joined the Department of Mechanical Engineering of NIT Rourkela in August 2008. His domain of research includes manufacturing, quality optimization, statistical modeling/optimization of manufacturing processes, industrial decision-making, traditional and non-traditional machining, and metal additive manufacturing. He is a life associate member of the Indian Institute of Welding. He has published more than 200 articles in international journals, national journals of repute, and proceedings of different international- and national-level conferences. To date, he has mentored 45 M.Tech. projects and supervised a total number of 11 Ph.D.s with the capacity of sole/principal supervisor. He executed two sponsored projects funded by Science and Engineering Research Board (SERB) and Council of Scientific & Industrial Research (CSIR), with the capacity of principal investigator and co-investigator, respectively. He is currently executing two more sponsored projects on metal additive manufacturing funded by the Science and Technology Department, Government of Odisha, and CSIR, respectively. Recently, his name has been included within the top 2% of the most influential scientists—2023 by the Stanford University list.

Sanjiv Kumar Tiwari is an Assistant Professor from Birla Institute of Technology, Ranchi, India. His Ph.D. dissertation is about continuous improvement using the lean Six Sigma philosophy in manufacturing industries. His research interests include supply chain, quality engineering, decision support systems, and continuous improvement philosophies like lean manufacturing, Six Sigma, etc. He has authored nearly 20 research articles in various journals and conference proceedings.

1 Introduction to Marine Welding Engineering

1.1 INTRODUCTION

Welding is one of the most critical fabrication processes in the marine industry, forming the backbone of modern shipbuilding, offshore platform construction, and subsea repair. Unlike other manufacturing sectors, the marine environment imposes unique technical, safety, and durability requirements on welded structures, making marine welding engineering a highly specialized discipline.

The shipbuilding industry has transitioned from riveting in the early 20th century to fully welded structures today, enabling the construction of larger, lighter, and more efficient vessels. Similarly, offshore oil and gas platforms, floating production systems, and renewable energy structures such as offshore wind turbines rely heavily on advanced welding processes to ensure structural integrity in harsh operating conditions.

Marine welding engineering encompasses the selection of appropriate materials, welding processes, consumables, and inspection techniques tailored to withstand the corrosive effects of seawater, variable loading conditions, and extended service lifetimes. In particular, welded joints in ships and offshore structures must not only bear static and dynamic loads but also resist corrosion, fatigue, and stress cracking, making welding both a technical and safety-critical task (Pemberton 1953).

The discipline also extends to specialized techniques such as underwater welding, hyperbaric welding, and hybrid welding methods, which are indispensable for subsea maintenance and deepwater applications. Furthermore, the increasing use of automation, robotics, and digital technologies is reshaping marine welding practices, improving consistency, reducing human risk, and enhancing productivity.

This chapter introduces the fundamental aspects of marine welding, emphasizing its role in shipbuilding and offshore structures and outlining the unique challenges faced in this demanding field.

1.2 OVERVIEW OF MARINE WELDING

Marine welding refers to the specialized processes of joining metallic structures in maritime environments—including ship construction and repair, offshore platforms, underwater pipelines, and subsea installations—and it demands far more rigorous controls than typical land-based welding because of the harsh saltwater environment, large structural loads, tight spaces, variable welding positions, and the need for watertight, long-lasting welds. Over time, shipbuilding evolved from riveting to welded assemblies to improve strength, weight, water tightness, and speed of construction.

DOI: 10.1201/9781003217350-1

Typical marine welding operations use arc-based methods (such as shielded metal arc welding [SMAW], gas metal arc welding [GMAW], flux-cored welding, submerged arc welding [SAW], and gas tungsten arc welding [GTAW]) and sometimes more exotic techniques (e.g., hyperbaric or underwater welding, laser or hybrid welding) depending on access, material thickness, environment, and certification requirements. Because marine structures are exposed to corrosive saltwater, wave action, thermal cycling, vibration, and sometimes elevated pressures (in the case of subsea repair or hyperbaric welding), welders must ensure proper joint design, pre- and post-weld heat treatments, correct filler materials and electrodes, and rigorous non-destructive testing (NDT) and quality assurance to avoid cracks, porosity, warping, insufficient fusion, or other defects that could lead to structural failure. In addition, welders must often work in challenging positions (vertical, overhead, confined spaces), with exposure to wind, moisture, and sometimes underwater conditions, and must comply with marine classification society standards (e.g., American Bureau of Shipping [ABS], Det Norske Veritas [DNV]/Lloyd's Register [LR]) and certification requirements (e.g., diver welders for underwater repair). Recent trends in marine welding include increased automation and robotics (especially for repetitive or hazardous tasks), hybrid welding technologies (combining laser, plasma, or arc methods), advanced corrosion-resistant alloys (stainless steels, aluminum alloys, nickel-based alloys), and stricter integration of environmental, safety, and sustainability considerations (less waste, lower distortion, better inspection). Because of all these constraints, marine welding is often more expensive, more tightly regulated, and more reliant on experienced welders and inspectors than "ordinary" fabrication welding, but the payoff is ships, platforms, and subsea structures that are safe, durable, and fit for harsh marine service.

1.3 HISTORICAL DEVELOPMENT OF MARINE WELDING

The development of welding in the marine industry has been closely tied to advancements in materials, ship design, and global industrial demands. Early ship construction relied heavily on **mechanical fastening techniques**, such as riveting and bolting, to assemble steel plates and structural members. While reliable, riveted joints were heavy, time-consuming to install, and prone to leakage under harsh marine conditions.

The **transition from riveting to welding** began in the early 20th century. Welding offered significant benefits, including weight reduction, improved watertight integrity, and enhanced structural strength. The first recorded applications of welding in shipbuilding occurred during **World War I**, when experimental welded vessels were constructed to save time and materials. However, widespread adoption was limited due to concerns about weld quality, lack of skilled labor, and insufficient understanding of metallurgical effects (Pogrebisky 2016).

During **World War II**, the urgency of mass-producing ships accelerated the adoption of welding. Notably, the **Liberty Ships** built in the United States were among the first large-scale, fully welded merchant vessels. Welding not only enabled faster production but also exposed challenges such as brittle fracture in cold waters, leading to greater emphasis on weld metallurgy, heat treatment, and inspection techniques. These lessons laid the foundation for modern welding standards in shipbuilding.

By the **post-war era**, welding had become the dominant method of ship construction worldwide. Continuous improvements in **arc welding processes**, including SMAW, SAW, and later GMAW, enabled more reliable fabrication of hulls, decks, and offshore structures. Advances in **NDT** further increased confidence in welded joints, allowing classification societies to incorporate welding into international shipbuilding codes.

In recent decades, the marine industry has witnessed the integration **of high-efficiency and specialized techniques**, such as flux-cored arc welding (FCAW), GTAW, and hybrid laser-arc welding (HLAW), alongside **automation and robotics** in modern shipyards. The development of **underwater and hyperbaric welding** has further expanded the scope of welding to subsea maintenance and offshore energy applications.

Today, marine welding continues to evolve in response to demands for **larger, safer, and more sustainable vessels and offshore structures**. The ongoing shift toward **digital twin technology, smart sensors, and eco-friendly welding practices** represents the latest stage in a century-long progression, marking welding as an indispensable enabler of maritime engineering.

Figure 1.1 illustrates the evolution of marine welding technologies across more than a century, marking key milestones that have shaped modern shipbuilding and offshore fabrication. The timeline begins with the introduction of oxy-fuel welding in the early 1900s, used for early steel hull construction. By the 1950s, the adoption of SMAW revolutionized ship fabrication with greater versatility and strength. The 1980s saw the rise of SAW, enabling large-scale, high-efficiency welding of hull and deck structures. In the present era, advanced techniques like HLAW and artificial intelligence (AI)-assisted automation have become central to achieving precision,

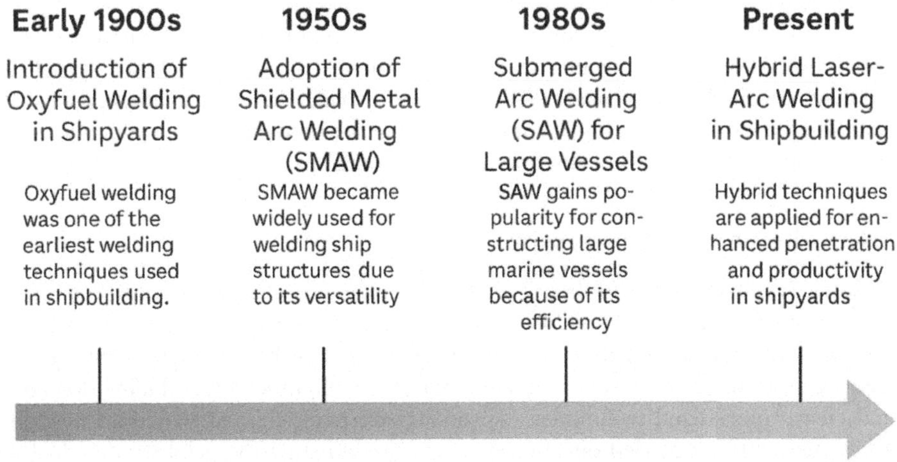

FIGURE 1.1 Historical development of marine welding.

productivity, and sustainability in shipyards. This progressive timeline highlights the continuous transition from manual craftsmanship to intelligent, digitally integrated marine welding systems, reflecting the industry's evolution toward efficiency, safety, and innovation.

1.4 ROLE OF WELDING IN MODERN MARINE INDUSTRIES

Welding plays a critical role in the construction, maintenance, and repair of ships and offshore structures. Modern marine industries rely heavily on welding technologies to ensure safety, durability, and efficiency in demanding ocean environments. The following key areas highlight its role (Moore 2009).

1.4.1 Shipbuilding and Hull Fabrication

Welding serves as the backbone of shipbuilding, particularly in hull fabrication, where large steel and aluminum plates must be joined with precision and strength. Unlike traditional riveting, welding provides seamless connections that not only enhance the overall structural integrity but also ensure watertight performance, which is critical for marine vessels. By creating stronger and more reliable joints, welding has revolutionized modern ship construction, allowing for larger, safer, and more efficient ships to be built for both commercial and defense applications.

1.4.2 Offshore Structures and Oil Platforms

Welding plays a pivotal role in the construction and maintenance of offshore structures such as oil rigs, drilling platforms, subsea pipelines, and floating production storage and offloading (FPSO) units. These installations are exposed to extreme marine environments, including high-pressure waves, strong currents, and corrosive seawater, which demand exceptional structural strength and reliability. Welded joints provide the necessary toughness and durability to withstand dynamic loads and harsh operating conditions. Additionally, advanced welding techniques, including SAW and hyperbaric welding, are employed to ensure safety and performance in deep-sea applications. By enabling the fabrication of robust and long-lasting offshore platforms, welding directly supports the global energy sector and marine resource exploration.

1.4.3 Repair and Maintenance

Welding is indispensable in the repair and maintenance of ships and offshore structures, ensuring operational reliability and extended service life. Damaged components such as hull sections, pipelines, or structural joints can be restored through in situ welding without the need for complete replacement, significantly reducing downtime and costs. Specialized techniques like underwater welding and hyperbaric welding allow repairs to be carried out directly at sea, eliminating the need for dry docking in many cases. This adaptability makes welding a vital tool for addressing corrosion, fatigue cracks, and accidental damage, thereby maintaining the safety, efficiency, and seaworthiness of marine assets in challenging environments. Figure 1.2 illustrates an

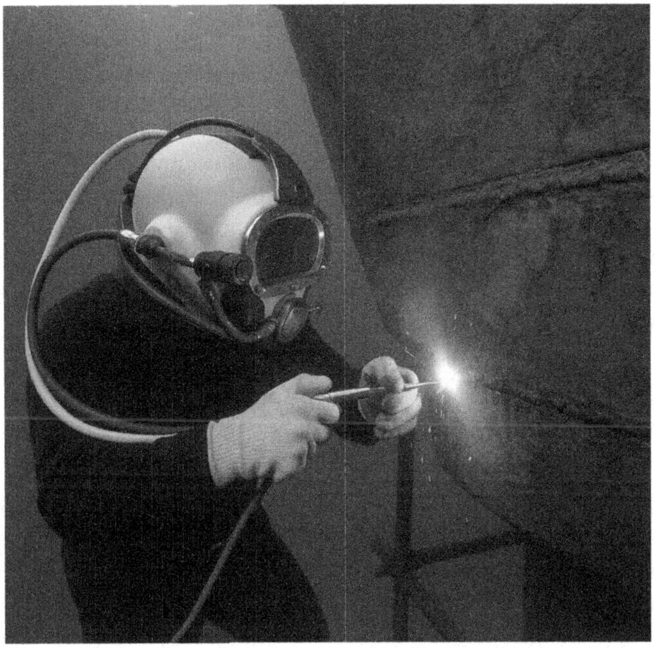

FIGURE 1.2 Underwater repair welding on ship hull/offshore platform.

underwater repair welding operation performed by a professional diver on a ship hull or offshore platform. The diver, equipped with a full-face helmet, insulated gloves, and air and power supply hoses, uses an SMAW process to repair corroded or damaged steel surfaces. The electrode holder is connected to a waterproof cable that delivers controlled current to maintain a stable arc, even in a submerged environment. The welding zone produces visible gas bubbles as the arc generates intense heat to fuse the metal, forming a protective film around the molten pool. The safety tether and communication line ensure diver stability and real-time coordination with surface operators. This figure highlights the precision, technical skill, and safety requirements involved in underwater repair welding, which is essential for maintaining the structural integrity of marine vessels, subsea pipelines, and offshore installations.

1.4.4 Lightweight Marine Vessels

Welding plays a critical role in the construction of lightweight marine vessels, where aluminum alloys and advanced composites are increasingly used to reduce weight and improve fuel efficiency. Precise welding techniques ensure strong and durable joints while maintaining the lightweight characteristics essential for high-speed crafts, ferries, and naval patrol vessels. By minimizing weight without compromising strength, welding contributes to improved maneuverability, higher payload capacity, and lower operational costs. Furthermore, specialized processes such as GMAW and friction stir welding (FSW) are widely employed to achieve defect-free welds in thin aluminum plates, making them ideal for modern lightweight shipbuilding.

1.4.5 Integration of Advanced Materials

The integration of advanced materials in marine structures has significantly expanded the role of welding in shipbuilding and offshore engineering. High-strength steels, duplex and super duplex stainless steels, and corrosion-resistant alloys are increasingly used to enhance durability, reduce maintenance, and improve performance in harsh marine environments. Welding enables the effective joining of these materials while maintaining their mechanical properties and resistance to seawater-induced corrosion. Advanced welding processes, such as GTAW, HLAW, and FSW, are particularly suited for handling the challenges of these specialized materials. By facilitating the use of advanced alloys and composites, welding supports the development of lighter, stronger, and longer-lasting marine vessels and offshore structures.

1.4.6 Automation and Digitalization

Automation and digitalization have transformed marine welding by improving efficiency, precision, and consistency in shipbuilding and offshore fabrication. Robotic welding systems are increasingly employed in shipyards to handle repetitive and large-scale welds with greater accuracy, reducing human error and labor-intensive work. Digital tools such as computer-aided welding design, real-time monitoring, and predictive analytics enable better control over weld quality and process optimization. The integration of AI and digital twin technologies further enhances defect detection, maintenance planning, and life cycle management of welded structures. Together, automation and digitalization not only increase productivity but also ensure safer working environments and higher-quality outputs in modern marine welding practices.

1.4.7 Sustainability and Green Shipbuilding

Sustainability has become a central focus in modern marine welding, particularly with the rise of green shipbuilding initiatives aimed at reducing environmental impact. Welding supports the construction of energy-efficient vessels by enabling the use of lightweight materials and advanced designs that lower fuel consumption and greenhouse gas emissions. Eco-friendly welding technologies, such as low-fume processes and recyclable filler materials, are increasingly adopted to minimize pollution and waste in shipyards. Additionally, precise welding techniques reduce material usage and improve the longevity of ships, thereby lowering the overall environmental footprint across their life cycle. By aligning with global regulations and sustainability goals, welding plays a crucial role in advancing greener and more responsible marine industries.

1.5 IMPORTANCE OF WELDING IN SHIPBUILDING AND OFFSHORE STRUCTURES

1.5.1 Structural Integrity and Safety Considerations

In shipbuilding and offshore engineering, welding is the foundation of structural integrity, as it provides the strength and reliability necessary for vessels and platforms to withstand extreme marine conditions. Welded joints ensure watertight and airtight construction, which is vital for buoyancy and overall seaworthiness. Properly executed welds distribute loads evenly across structural components, reducing stress concentrations that could otherwise lead to cracks or catastrophic failures. In offshore platforms, where structures are exposed to high waves, wind, and corrosive seawater, the safety of workers and assets depends heavily on weld quality. International classification societies and marine standards set strict guidelines for welding practices, inspection, and certification to guarantee safety and compliance. Ultimately, welding not only enables the construction of massive hulls, decks, and subsea frameworks but also safeguards human lives and marine assets by ensuring structural robustness and long-term reliability.

1.5.2 Cost and Time Efficiency in Marine Fabrication

Welding provides significant advantages in cost and time efficiency for shipbuilding and offshore fabrication compared to older methods such as riveting. By enabling the direct fusion of metal plates and structural members, welding eliminates the need for additional fastening components, thereby reducing material costs. Automated and semi-automated welding technologies further accelerate the fabrication process, allowing large vessels and offshore platforms to be constructed within shorter time frames while maintaining consistent quality. The ability to perform on-site repairs through in situ welding also minimizes downtime, avoiding the expense and delays associated with dry docking or component replacement. Furthermore, modern welding techniques optimize resource utilization by reducing scrap and rework, ensuring that shipyards achieve higher productivity with lower labor intensity. Together, these factors make welding an essential driver of cost-effective and time-efficient marine engineering projects.

1.5.3 Contribution to Sustainability and Life Cycle Performance

Welding plays a vital role in advancing sustainability and enhancing the life cycle performance of marine structures. By enabling the integration of lightweight alloys and corrosion-resistant materials, welding contributes to building energy-efficient vessels that consume less fuel and release fewer emissions. High-quality welds improve durability and reduce the frequency of repairs, thereby minimizing material waste, energy consumption, and operational downtime over a ship's service life. Additionally, modern eco-friendly welding technologies, such as low-fume processes and recyclable filler materials, align with green shipbuilding initiatives and international environmental standards. Through precise fabrication and long-lasting performance, welding ensures that ships and offshore platforms not only meet operational demands but also achieve extended service lives with reduced environmental impact, supporting global goals of sustainable and responsible marine engineering.

1.6 UNIQUE CHALLENGES IN MARINE WELDING

Marine welding involves several unique challenges compared to conventional land-based applications due to the harsh and highly demanding operational environment.

Factors such as saltwater exposure, high humidity, fluctuating temperatures, and corrosive conditions necessitate specialized welding techniques and material selection. Moreover, the integration of new alloys, underwater operations, and strict adherence to international standards add complexity to the process. These challenges must be carefully addressed to ensure structural safety, long-term performance, and compliance with industry requirements.

1.6.1 Harsh Environmental Conditions

Marine welding operations and welded joints are constantly challenged by harsh environmental conditions. Saltwater exposure is one of the most aggressive factors, as it accelerates corrosion, leading to pitting, crevice attack, and material degradation around weld zones. High humidity in shipyards and offshore locations further promotes oxidation and moisture entrapment, affecting weld quality and increasing the risk of hydrogen-induced cracking. Temperature variations ranging from extreme cold in arctic operations to high heat in tropical regions cause thermal stresses, expansion–contraction cycles, and reduced toughness in welds. These combined environmental factors necessitate the use of corrosion-resistant materials, protective coatings, controlled welding procedures, and strict quality assurance measures to ensure long-term performance and safety of marine structures. Figure 1.3 illustrates a commercial diver conducting wet welding operations on a submerged marine structure using the SMAW process. The diver, fully equipped with a diving helmet, insulated gloves, and life support system, performs welding under direct contact with seawater. The electrode generates a high-temperature arc

FIGURE 1.3 Commercial diver performing wet welding on a marine structure.

that melts both the filler and base metal, creating a weld bead that solidifies rapidly despite the cooling effect of water.

The marine structure, typically part of an offshore platform or subsea pipeline, is repaired by applying a series of weld passes, while the diver maintains visibility using arc-generated illumination and underwater lighting. Gas bubbles released during the process indicate active arc combustion and heat dissipation. This technique, while highly specialized, remains essential for emergency underwater repairs, hull reinforcement, and maintenance of offshore assets, emphasizing the diver's expertise, environmental control, and precision in performing high-risk wet welding operations.

1.6.2 Corrosion Resistance and Material Selection

Corrosion resistance is a critical factor in marine welding because welded joints are often the most vulnerable regions in a structure. Improper material choice can accelerate localized corrosion, jeopardizing both safety and service life. To mitigate these risks, shipbuilding and offshore industries employ corrosion-resistant steels, stainless steels, and aluminum alloys that maintain durability in saltwater environments. Filler metals are carefully selected to match or exceed the corrosion resistance of base materials, ensuring weld integrity. In addition, protective coatings, cathodic protection, and post-weld treatments such as passivation further enhance corrosion resistance. The strategic selection of materials not only reduces maintenance costs but also improves reliability, extending the life cycle of ships and offshore platforms.

1.6.3 Underwater Welding Complexities

Underwater welding is one of the most demanding aspects of marine welding due to the challenging environment in which it is performed. In wet welding, the diver welder works directly in the water, where visibility is limited, arc stability is affected by water movement, and the risk of hydrogen embrittlement is high due to moisture decomposition. This often results in porosity, cracking, and reduced mechanical strength of the weld. In contrast, dry or hyperbaric welding is performed in a sealed chamber around the workpiece, which provides a controlled environment with better weld quality but at significantly higher cost and logistical complexity. Additionally, divers must contend with high water pressure at greater depths, which increases safety risks and limits working time. Specialized consumables, advanced training, and strict safety protocols are therefore essential for ensuring reliable welds in underwater repair and maintenance of ships, subsea pipelines, and offshore structures (Vijay and Radhika 2024). Figure 1.4 depicts a professional commercial diver engaged in underwater welding/inspection work around a marine structure. The diver is equipped with a surface-supplied diving helmet, protective gloves, and connected umbilical lines that provide breathing gas, communication, and power supply. Such underwater welding is critical in the maintenance and repair of offshore oil platforms, ship hulls, and subsea pipelines. It demonstrates the challenging conditions faced by marine welders, including

FIGURE 1.4 Underwater welding operation by a commercial diver.

limited visibility, high water pressure, and the need for specialized equipment to ensure both safety and weld quality.

1.6.4 Welding of Dissimilar Metals

Welding dissimilar metals in marine applications presents unique challenges due to differences in thermal conductivity, melting points, and coefficients of thermal expansion between the materials being joined. For instance, welding steel to aluminum or stainless steel to carbon steel can lead to uneven heat distribution, brittle intermetallic compounds, and residual stresses that compromise joint integrity. Corrosion issues are also more pronounced, as galvanic corrosion can occur when two dissimilar metals are in contact in a saltwater environment. To address these issues, engineers often use transition joints, buttering layers, or specialized filler materials that minimize metallurgical incompatibilities. Advanced techniques such as FSW, explosion bonding, and the use of nickel-based alloys are also employed to achieve durable connections. Proper selection of welding processes, consumables, and protective measures is crucial to ensure that dissimilar metal joints meet the structural and environmental demands of marine operations.

1.6.5 Impact of Regulations and Classification Society Standards

In marine welding, compliance with international regulations and classification society standards is essential to ensure the safety, reliability, and durability of ships and offshore structures. Organizations such as the International Maritime Organization (IMO), ABS, LR, and DNV establish guidelines for welding procedures, material selection, welder qualifications, and inspection practices. These standards aim to minimize risks of structural failure by ensuring that welding practices meet stringent quality and safety requirements, particularly in harsh marine environments.

Strict adherence to codes governs every stage from the approval of welding consumables and procedures to the inspection of finished welds using NDT methods. For offshore and subsea structures, additional requirements address issues like fatigue resistance, corrosion protection, and performance under high pressures. Regulations also mandate the documentation and traceability of welding processes to maintain accountability. While these standards can increase costs and impose operational constraints, they ultimately safeguard human lives, protect marine ecosystems, and enhance the life cycle performance of marine assets.

1.7 FUTURE TRENDS AND INNOVATIONS IN MARINE WELDING

The marine welding industry is rapidly evolving to meet the increasing demands of modern shipbuilding, offshore construction, and sustainability goals. Emerging technologies are reshaping traditional welding practices by enhancing efficiency, safety, and quality. Among these, automation and robotics, advances in welding consumables, and innovations in NDT stand out as key areas driving the future of marine welding.

1.7.1 Automation and Robotics in Shipyard Welding

Automation and robotics are transforming shipyard welding by addressing the need for higher productivity, precision, and safety in marine fabrication. Traditional manual welding, though skilled, often struggles with the scale and complexity of modern ships and offshore platforms. Robotic welding systems, equipped with programmable arms, sensors, and AI-based control, ensure consistent weld quality, even in large repetitive tasks such as hull plating, bulkhead joints, and pipe welding. These systems reduce errors caused by fatigue and enable continuous operation, significantly speeding up construction timelines.

In addition, robotic welding minimizes human exposure to hazardous environments like confined ship compartments, elevated sections, or areas with toxic fumes. The integration of machine learning and digital twin technologies allows predictive adjustments, ensuring optimal weld parameters in real time. Furthermore, shipyards are adopting hybrid human–robot collaboration, where robots handle repetitive, heavy-duty welding while skilled welders focus on complex, precision-demanding tasks. As the industry moves toward smart shipyards, automation and robotics are expected to become central to improving efficiency, reducing costs, and ensuring the long-term competitiveness of marine manufacturing. Figure 1.5 shows a commercial diver engaged in wet welding at the seabed. The diver is equipped with a surface-supplied helmet, protective gloves, and an umbilical cable that provides breathing gas and communication. The welding arc is clearly visible, producing light and bubbles as the electrode interacts with the submerged metal surface. This process is crucial for the fabrication, repair, and strengthening of marine structures, including offshore oil platforms, subsea pipelines, and ship hulls. The figure emphasizes the

FIGURE 1.5 Diver conducting wet welding in a submerged environment.

high-risk conditions associated with underwater welding, including restricted mobility, limited visibility, and the technical complexity of maintaining weld quality under water pressure.

1.7.2 Advances in Welding Consumables for Marine Environments

Welding consumables play a crucial role in determining the strength, durability, and corrosion resistance of marine welds, making them a key area of innovation in shipbuilding and offshore engineering. Modern ships and offshore platforms operate in highly aggressive environments characterized by saltwater exposure, humidity, and fluctuating temperatures, which demand specialized electrodes, filler wires, and fluxes designed for long-term performance.

Recent advances include the development of low-hydrogen electrodes and specially alloyed filler metals that minimize the risk of hydrogen-induced cracking and improve weld toughness in high-strength steels. For aluminum and dissimilar metal welding, consumables have been engineered to provide better metallurgical compatibility and reduce the chances of galvanic corrosion. Additionally, flux-cored wires and self-shielded consumables are gaining popularity in shipyards due to their efficiency, ease of use, and ability to deliver consistent weld quality even in outdoor or windy environments.

A growing focus is also placed on eco-friendly consumables, such as low-emission fluxes and recyclable welding materials, aligning with sustainability goals in green shipbuilding. These advancements not only improve the service life and fatigue resistance of marine welded joints but also contribute to cost savings by reducing maintenance and repair needs. By tailoring consumables to specific marine alloys and operating conditions, the welding industry is ensuring higher reliability and safety for modern vessels and offshore structures.

1.7.3 Non-Destructive Testing (NDT) in Harsh Conditions

NDT is an essential part of marine welding to ensure the integrity, safety, and reliability of welded joints without causing any damage to the structure. In the marine industry, however, NDT faces unique challenges because inspections are often carried out in harsh environments, including high humidity, saltwater exposure, limited accessibility, and extreme temperatures on offshore platforms and ship hulls. These conditions demand robust inspection methods that can operate effectively in difficult and sometimes submerged settings.

Advanced techniques such as ultrasonic testing (UT) and phased array UT (PAUT) are increasingly used to detect internal flaws and cracks in thick welds, even in high-strength steels. Radiographic testing (RT) provides precise imaging for weld defects, though it requires careful handling in confined shipyard spaces. Magnetic particle testing (MPT) and dye penetrant testing (DPT) are widely applied for surface flaw detection on critical weld joints exposed to fatigue and corrosion.

In offshore and subsea environments, specialized remote-controlled and robotic NDT systems are now being deployed to inspect welds in inaccessible or dangerous areas, including underwater structures. Integration of digital technologies and real-time monitoring allows inspectors to analyze weld conditions immediately, ensuring faster decision-making during fabrication and maintenance. These advancements not only improve safety and reduce repair costs but also enhance the long-term durability and performance of marine welded structures under harsh operating conditions. Figure 1.6 illustrates a diver performing NDT on a submerged marine structure. NDT techniques, such as UT, magnetic particle inspection, and radiography, are employed underwater to evaluate weld integrity without causing damage to the structure. In harsh marine conditions, divers must operate with specialized equipment designed to withstand high pressure, corrosion, and low visibility. This process is essential for ensuring structural reliability, safety, and compliance with classification society standards in

FIGURE 1.6 Non-destructive testing (NDT) in harsh marine conditions.

ship hulls, offshore platforms, and subsea pipelines. Figure 1.6 highlights the critical role of NDT in preventive maintenance and life cycle performance of marine assets.

1.8 SUMMARY

This chapter highlights the evolution and critical role of welding in shipbuilding and offshore industries, replacing traditional riveting to ensure structural integrity, watertight performance, and safety. It discusses applications across hull fabrication, offshore platforms, repairs, lightweight vessels, and the integration of advanced materials, alongside the growing impact of automation and digitalization in modern shipyards. Unique challenges such as harsh marine environments, corrosion resistance, underwater welding, dissimilar metal joining, and compliance with classification society standards are emphasized, underscoring the complexity of marine welding. Overall, welding is established as a cornerstone of marine engineering, contributing to cost efficiency, sustainability, and life cycle performance of ships and offshore structures while setting the stage for advanced processes and innovations explored in later chapters.

1.9 REVIEW QUESTIONS

1. What factors led to the shift from riveting to welding in shipbuilding?
2. How has the historical development of welding influenced the marine industry?
3. Explain the role of welding in ensuring watertight integrity of ships.
4. Why is welding considered essential for structural integrity in marine applications?
5. Discuss the advantages of welding over traditional joining techniques in shipyards.
6. How does welding contribute to cost and time efficiency in ship fabrication?
7. What role does welding play in life cycle performance and sustainability of ships?
8. Give examples of critical marine structures where welding is indispensable.
9. How does welding support the construction of offshore oil platforms and rigs?
10. In what ways does welding facilitate repair and maintenance of marine vessels?
11. Why are lightweight alloys such as aluminum important in marine welding?
12. How is the integration of advanced materials shaping marine welding practices?
13. Describe how automation is transforming welding processes in modern shipyards.
14. What is the significance of digitalization and real-time monitoring in welding?
15. How does welding contribute to green shipbuilding and environmental goals?
16. What challenges do saltwater, humidity, and temperature pose for marine welds?
17. Why is corrosion resistance critical in material selection for marine applications?
18. Explain the complexities associated with underwater welding.
19. What difficulties arise when welding dissimilar metals in marine environments?
20. How do international standards and classification societies impact welding practices?
21. Name at least three classification societies that regulate marine welding standards.

22. What safety considerations are most critical in marine welding operations?
23. Explain the role of hydrogen in weld cracking and how it affects marine structures.
24. Why is non-destructive testing (NDT) important for marine weld quality assurance?
25. Compare ultrasonic testing and radiographic testing for marine welds.
26. How does automation improve worker safety in shipyard welding environments?
27. Describe recent advances in welding consumables for marine applications.
28. What innovations in NDT help inspection in harsh marine conditions?
29. How is robotics expected to influence future shipyard welding efficiency?
30. Summarize the overall importance of welding in modern marine engineering.

2 Materials Used in Marine Welding

2.1 INTRODUCTION

Materials play a decisive role in the performance, safety, and longevity of marine structures, and their selection directly influences the quality of welding in shipbuilding and offshore applications. Unlike conventional industries, marine environments expose materials to highly corrosive conditions, continuous loading, immersion in saltwater, humidity, and temperature fluctuations. As a result, weldable materials for ships, submarines, and offshore platforms must not only provide high strength and toughness but also maintain resistance to corrosion and fatigue over long service lifetimes. Steel, particularly high-strength low-alloy (HSLA) steels, remains the backbone of shipbuilding due to its favorable weldability and cost-effectiveness (Ghazanlou et al. 2024). At the same time, aluminum alloys are increasingly used in lightweight vessels for improved fuel efficiency and maneuverability. At the same time, the need for enhanced durability has driven the adoption of specialized corrosion-resistant alloys and advanced composite-metal hybrid structures. Understanding the metallurgical behavior, weldability, and performance of these materials under marine conditions is therefore critical for engineers and fabricators. This chapter explores the types of materials used in marine welding, their characteristics, and the factors guiding their selection for different maritime applications.

2.1.1 Role of Material Selection in Marine Welding

Material selection plays a pivotal role in marine welding because the performance, safety, and longevity of welded structures in harsh ocean environments depend heavily on the properties of the chosen materials. Marine structures such as ships, offshore platforms, and underwater pipelines are continuously exposed to seawater, high salinity, humidity, and fluctuating temperatures, all of which accelerate corrosion and structural degradation. Therefore, corrosion resistance becomes the foremost consideration. For instance, stainless steels like grade 316 are widely used in seawater piping systems and storage tanks because of their resistance to pitting and crevice corrosion. Similarly, nickel-based alloys such as Inconel are employed in subsea oil and gas equipment where prolonged exposure to high-pressure, chloride-rich environments would rapidly degrade conventional steels. In contrast, carbon steels are still extensively used in ship hulls and bulk carriers due to their low cost and availability, though they require protective coatings and cathodic protection to extend service life.

Mechanical strength and fatigue resistance are equally critical, as marine components endure cyclic stresses from waves, currents, and vibrations. HSLA steels are

DOI: 10.1201/9781003217350-2

frequently chosen for ship hulls, offshore wind turbine foundations, and naval vessels, where both strength and weldability are essential. The U.S. Navy, for example, uses HY-80 and HY-100 grades of HSLA steels in submarine hull construction because they combine high strength with resistance to brittle fracture. Weldability itself is a deciding factor in material selection, since not all alloys respond well to common marine welding techniques. Materials with controlled carbon content are preferred to reduce the risk of brittle microstructures forming in the heat-affected zone (HAZ), ensuring reliable joint performance in large welded assemblies (Mandal 2017).

Another important factor is low-temperature toughness, particularly for ships operating in polar regions or subsea structures at great depths. Icebreakers and Arctic LNG carriers, for example, use nickel alloy steels capable of maintaining toughness at temperatures well below freezing, preventing catastrophic brittle fractures. Offshore oil and gas pipelines in deep water are also made from specialized steels and corrosion-resistant alloys to withstand not only pressure but also extreme cold and corrosive conditions.

Compatibility with protective systems further influences material choice. Marine structures such as oil platforms and large container ships are commonly safeguarded by epoxy coatings and cathodic protection systems. Here, the choice of carbon steel must align with these protective methods to prevent accelerated galvanic corrosion. Stainless steels and aluminum, in contrast, often rely more on inherent corrosion resistance but still benefit from coatings in highly aggressive environments.

Cost and availability are practical concerns that shape decisions. While exotic alloys like titanium offer unmatched corrosion resistance and are used in specialized subsea applications, their high cost limits widespread adoption. Carbon steels remain the dominant choice for large-scale shipbuilding due to their affordability, despite requiring more frequent inspection and maintenance. Stainless steels find widespread use in marine fittings, pumps, and propeller shafts, while aluminum alloys are extensively applied in fast ferries, naval patrol boats, and luxury yachts, where weight reduction translates to higher speeds and improved fuel efficiency. For example, the high-speed trimaran warship USS *Independence* (LCS-2) incorporates aluminum alloys in its superstructure to reduce displacement and improve maneuverability.

Ultimately, the proper selection of materials directly impacts the service life and maintenance requirements of marine welded structures. An informed choice reduces the frequency of repairs, lowers operational costs, and ensures structural integrity in demanding environments. Conversely, poor material selection can lead to rapid corrosion, crack propagation, and expensive failures, as seen in historical cases where insufficiently corrosion-resistant steels led to premature pipeline failures in offshore oil fields. Thus, material selection in marine welding is a multidisciplinary decision that balances corrosion resistance, strength, weldability, toughness, protective compatibility, and cost-effectiveness, ensuring that marine structures remain reliable and efficient throughout their intended service life.

2.1.2 Influence of Marine Environment on Material Performance

The marine environment exerts a profound influence on the performance of engineering materials, especially those used in welded structures, ships, offshore platforms, and subsea pipelines. Unlike controlled industrial settings, the ocean represents one of the most aggressive natural environments due to the combined action of seawater, high salinity, dissolved oxygen, fluctuating temperatures, and biological activity. The most significant impact is corrosion, which occurs at a much faster rate in seawater than in air. Chloride ions present in saltwater penetrate protective oxide films on metals, leading to localized attack such as pitting and crevice corrosion. For example, stainless steel, though generally corrosion-resistant, can suffer chloride-induced stress corrosion cracking in marine applications if not properly alloyed or protected. Carbon steels, which form the bulk of ship hulls and offshore structures, are highly prone to rusting and thus rely heavily on protective coatings, cathodic protection, or alloying to extend their service life.

The marine environment also subjects materials to constant cyclic loading from waves, tides, and currents, which accelerates fatigue damage in welded joints. Offshore oil platforms and naval vessels are particularly vulnerable to fatigue cracking, where even minor material imperfections can grow into critical defects under repetitive stresses. Low temperatures in deep-sea or Arctic regions further aggravate material performance by reducing toughness and increasing the risk of brittle fracture. For instance, steels that perform well at room temperature may fail catastrophically when exposed to subzero seawater temperatures unless specially alloyed with nickel or designed with controlled microstructures.

Another environmental influence comes from biofouling, where marine organisms such as algae, barnacles, and mollusks attach to material surfaces. While often seen as a surface nuisance, biofouling can create crevices that trap seawater and accelerate localized corrosion, particularly in aluminum alloys and stainless steels. In addition, galvanic corrosion becomes a critical issue when dissimilar metals are joined in the presence of seawater, as the more active metal corrodes preferentially. For example, when steel is joined with copper alloys in a seawater piping system, the steel corrodes rapidly unless insulating materials or sacrificial anodes are used.

Temperature fluctuations, oxygen concentration differences, and exposure to ultraviolet radiation at the sea surface also impact materials. Oxygen-rich zones near the surface promote faster corrosion rates compared to oxygen-depleted deeper waters. Similarly, splash zones areas repeatedly wetted and dried by tides—are considered the most aggressive regions for corrosion, as they combine high oxygen availability with salt deposition. This makes material selection and protective strategies especially challenging for components like offshore platform legs and ship hulls.

Several real-world cases highlight these challenges. One notable example is the failure of the Aegir semi-submersible oil platform in the North Sea during the 1980s, where corrosion and fatigue cracking in welded joints of the legs led to extensive repair costs and downtime. Another case involves the USS Thresher submarine (1963), which was lost during sea trials due to suspected failure of piping systems made of inadequate alloys that could not withstand high-pressure seawater environments. In Arctic shipping, ice-class tankers operating in subzero waters have experienced brittle fractures in hull steels not designed for low-temperature toughness, prompting widespread adoption of nickel alloy steels for polar service. Similarly, the Alexander L. Kielland drilling platform disaster in 1980 off Norway underscored the dangers of fatigue cracks in welded joints, which propagated under harsh North Sea wave conditions and ultimately caused the structure to capsize, resulting in significant loss of life.

These incidents demonstrate that the marine environment reduces the mechanical, chemical, and structural reliability of materials if not properly addressed during design and selection. Materials must therefore be chosen not only for their intrinsic properties but also for their ability to withstand these complex environmental influences through protective coatings, alloy modifications, and regular maintenance. Failure to account for such influences has historically led to premature structural failures, costly downtime, and catastrophic accidents in the shipping and offshore industries.

2.2 STEEL ALLOYS FOR SHIPBUILDING

Steel remains the most widely used material in shipbuilding due to its strength, availability, cost-effectiveness, and suitability for large welded structures. Over decades of naval architecture and marine engineering, different grades of steel have been developed to balance the requirements of strength, weldability, corrosion resistance, and toughness in harsh marine environments (Adrian et al. 1977). The choice of steel alloy directly impacts the performance, durability, and safety of ships, making it essential to understand the characteristics and applications of various steel types.

2.2.1 Carbon Steel Grades and Applications

Carbon steels form the backbone of shipbuilding materials, especially in the construction of hulls, decks, and bulkheads. They are favored because of their low cost, good weldability, and adequate mechanical properties for general marine use. Low- and medium-carbon steels are commonly employed, with grades defined by standards such as ASTM, JIS, or IS. Typical grades like ASTM A36 or equivalent shipbuilding steels (e.g., AH36, DH36, EH36) are used for hull structures where strength and ductility are critical. While carbon steels offer simplicity and affordability, their major limitation lies in their poor corrosion resistance. To combat seawater-induced corrosion, shipbuilders apply protective coatings, galvanization, or cathodic protection systems. Despite these challenges, carbon steel remains the dominant choice for large vessel construction due to its favorable balance of strength and economy.

2.2.2 High-Strength Low-Alloy (HSLA) Steels

HSLA steels have gained importance in shipbuilding where higher strength-to-weight ratios are required. By adding small amounts of alloying elements such as manganese, nickel, chromium, or vanadium, HSLA steels achieve improved tensile strength, toughness, and resistance to fatigue without compromising weldability. These steels allow shipbuilders to reduce plate thickness in critical sections, resulting in weight savings and enhanced fuel efficiency (Chernyshov et al. 2016). For instance, grades such as AH32, DH36, and EH40 are widely employed in modern ship hulls, offshore platforms, and naval vessels. HSLA steels are especially important in naval architecture, where weight reduction translates into higher speeds, improved maneuverability, and greater payload capacity. In polar-class ships and offshore drilling platforms, HSLA steels with enhanced low-temperature toughness are preferred to prevent brittle fracture in freezing waters.

2.2.3 Welding Characteristics of Marine Steels

Welding plays a central role in ship construction, as ships are predominantly large welded assemblies. Carbon steels generally exhibit good weldability due to their low carbon content, but care must be taken to control the heat input to prevent distortion and residual stresses. HSLA steels, while also weldable, require stricter control of welding procedures because of their higher alloy content, which can lead to hard microstructures or hydrogen-induced cracking in the HAZ. Preheating, post-weld heat treatment, and careful selection of filler materials are often necessary to maintain joint integrity. Shipyards typically employ processes such as shielded metal arc welding (SMAW), gas metal arc welding (GMAW), and flux-cored arc welding (FCAW) for structural steel fabrication. The welding characteristics of marine steels are closely tied to their classification grades, as each grade is designed to withstand specific service conditions, from normal atmospheric exposure to highly corrosive seawater environments.

2.2.4 Standards and Classifications for Marine Steel

To ensure quality and safety in shipbuilding, marine steels are produced and classified according to internationally recognized standards. Classification societies such as the American Bureau of Shipping (ABS), Lloyd's Register (LR), Det Norske Veritas (DNV), Bureau Veritas (BV), and Indian Register of Shipping (IRS) define the requirements for chemical composition, mechanical properties, and welding performance of shipbuilding steels. Common categories include ordinary strength steels (grades A, B, D, and E) and higher strength steels (AH, DH, EH, and FH grades), which are further classified by yield strength and impact toughness. These standards ensure that materials can withstand demanding marine service conditions, from high stresses in hull structures to extreme low-temperature applications in Arctic waters. Compliance with classification society rules not only guarantees structural safety but is also mandatory for certification and seaworthiness of commercial and naval ships (Table 2.1).

2.3 ALUMINUM ALLOYS FOR SHIPBUILDING

Aluminum alloys are increasingly used in shipbuilding, particularly in high-speed vessels, naval patrol craft, passenger ferries, and luxury yachts, where weight reduction, corrosion resistance, and performance are crucial. Although steel remains dominant for large hull structures, aluminum has carved a significant niche due to its unique combination of lightness and durability in marine environments. Its application is often concentrated in superstructures, decks, and lightweight vessels, where reducing weight directly translates to improved fuel efficiency, greater speed, and enhanced payload capacity.

TABLE 2.1
Comparative Properties and Applications of Marine Steels

Steel Type	Typical Grades	Yield Strength (MPa)	Key Features	Applications in Shipbuilding
Carbon steel	A36, Grade A/B/D/E	~235–260	Low cost, good weldability, moderate strength, poor corrosion resistance	Hull plates, bulkheads, decks, general ship structure (with coatings/cathodic protection)
HSLA steel	AH32, DH36, EH36, EH40	315–460+	Higher strength-to-weight ratio, good toughness, controlled alloying for weldability	Hull girder sections, offshore platforms, naval vessels, polar-class ships
Ordinary strength steel	Grade A, B, D, E	~235	Basic strength, adequate ductility, good weldability	General-purpose ship construction, small vessels, secondary structures
Higher strength steel	AH/DH/EH/FH grades	315–390+	Enhanced toughness, resistance to brittle fracture, suitable for Arctic service	Large container ships, oil tankers, naval warships, offshore rigs

2.3.1 Common Marine-Grade Aluminum Alloys (5xxx, 6xxx Series)

The most widely used marine-grade aluminum alloys belong to the 5xxx (Al-Mg) and 6xxx (Al-Mg-Si) series. Alloys in the 5xxx series, such as 5052, 5083, and 5754, are known for excellent corrosion resistance in seawater, high strength, and good weldability. They are commonly used in hull plating, decks, and superstructures of small- to medium-sized vessels. The 6xxx series, including alloys like 6082 and 6061, offers higher strength and better machinability but requires additional care during welding because of their susceptibility to HAZ softening. These alloys are often used in structural members, fittings, and extruded profiles in shipbuilding. Both series are standardized under international classifications, ensuring their suitability for marine service.

2.3.2 Advantages of Aluminum in Ship Structures

The key advantage of aluminum in shipbuilding lies in its low density, approximately one-third that of steel, which significantly reduces overall vessel weight. This results in higher speeds, improved fuel efficiency, and increased cargo or passenger capacity, making aluminum alloys ideal for fast ferries, naval combat vessels, and high-performance yachts. Additionally, aluminum exhibits excellent corrosion resistance in seawater, reducing the reliance on protective coatings and lowering maintenance requirements over a ship's service life. Its ease of fabrication, good formability, and ability to be extruded into complex profiles make it suitable for diverse ship components. Furthermore, aluminum's non-magnetic properties are valuable for naval mine countermeasure vessels, where minimizing magnetic signatures is critical for safety.

2.3.3 Welding Methods for Aluminum in Marine Applications

Aluminum alloys are typically welded using gas tungsten arc welding (GTAW/TIG) and GMAW/MIG processes, both of which are well-suited for thin sections and provide high-quality joints. For large shipbuilding applications, MIG welding is preferred due to its higher deposition rates and productivity. Special filler alloys, usually from the 5xxx or 4xxx series, are selected to ensure compatibility and to minimize the risk of cracking. Welding aluminum requires strict control of joint preparation, surface cleanliness, and shielding gases (argon or helium) to prevent oxidation during the welding process. In modern shipyards, friction stir welding (FSW) is also gaining popularity for aluminum panels, as it produces defect-free welds with minimal distortion and superior mechanical properties compared to conventional fusion welding methods.

2.3.4 Challenges in Aluminum Welding (Distortion, Porosity)

Despite its advantages, welding aluminum in marine applications presents several challenges. One major issue is distortion, as aluminum has a high thermal conductivity and a low melting point, which leads to rapid heat flow and uneven thermal expansion. This often causes warping and dimensional inaccuracies in welded assemblies. Another common problem is porosity, resulting from hydrogen absorption during welding. Trapped gas bubbles weaken the weld and compromise structural performance if not carefully controlled. In addition, alloys from the 6xxx series are susceptible to HAZ softening, which reduces the mechanical strength of welded joints. Oxide film formation on aluminum surfaces also complicates welding, requiring thorough cleaning before fabrication. These challenges demand strict process control, skilled welders, and often post-weld inspections using non-destructive testing (NDT) techniques to ensure quality and reliability in marine service.

Comparative properties and applications of 5xxx and 6xxx aluminum alloys in shipbuilding are as shown in Table 2.2.

2.4 CORROSION-RESISTANT MATERIALS FOR MARINE APPLICATION

The marine environment is one of the most aggressive natural settings, characterized by high salinity, fluctuating temperatures, and the presence of dissolved oxygen and microorganisms. Corrosion not only weakens marine structures but also leads to costly repairs, increased downtime, and reduced service life of vessels and offshore installations. To counter these challenges, engineers employ specialized corrosion-resistant materials, including stainless steels, copper-nickel alloys, titanium, and advanced coatings or cladding methods (Wert and DiSabella 2006). Selecting the appropriate material depends on the nature of the exposure, structural requirements, and economic considerations.

TABLE 2.2
Comparative Properties and Applications of 5xxx and 6xxx Aluminum Alloys in Shipbuilding

Property/Feature	5xxx Series (Al-Mg Alloys)	6xxx Series (Al-Mg-Si Alloys)
Typical grades	5052, 5083, 5754	6061, 6082
Primary alloying element(s)	Magnesium (Mg)	Magnesium (Mg) and silicon (Si)
Strength	Medium to high strength	Medium to high strength (higher with heat treatment)
Corrosion resistance	Excellent, especially in seawater	Good, but slightly less than 5xxx
Weldability	Excellent, widely used for welded marine structures	Good, but prone to heat-affected zone (HAZ) softening
Formability	High (good for sheet and plate forming)	Moderate (better machinability, less ductile)
Fabrication methods	Rolling, sheet forming, extrusion (limited)	Extrusion, machining, and structural profiles
Common applications	Hull plating, decks, superstructures, tanks, bulkheads	Structural members, masts, fittings, extruded profiles
Special notes	Preferred for direct seawater exposure	Often used where higher mechanical strength and precision machining are needed

2.4.1 Stainless Steel Types for Marine Use (Austenitic, Duplex, Super Duplex)

Stainless steels are widely used in marine applications due to their chromium-enriched passive film, which provides inherent corrosion resistance:

Austenitic Stainless Steels (e.g., 304, 316L): Known for excellent corrosion resistance and formability. Grade 316L, with added molybdenum, is especially resistant to pitting in seawater. Common applications include handrails, tanks, and minor structural fittings.

Duplex Stainless Steels (e.g., 2205): Offer a balance of strength and corrosion resistance, with superior resistance to stress corrosion cracking compared to austenitic. They are often used in offshore oil platforms, seawater piping, and pressure vessels.

Super Duplex Stainless Steels (e.g., 2507): Provide exceptional resistance to pitting, crevice corrosion, and chloride stress cracking. They are used in highly demanding offshore environments, such as risers, subsea equipment, and desalination plants.

2.4.2 Copper-Nickel Alloys for Seawater Systems

Copper-nickel (Cu-Ni) alloys, typically containing 70%–90% copper and 10%–30% nickel, are renowned for their excellent resistance to seawater corrosion, biofouling, and macrofouling. They exhibit natural antimicrobial properties, which prevent the buildup of marine organisms in piping systems. The most common alloys are Cu-Ni 70/30 and Cu-Ni 90/10, both widely used in seawater cooling systems, condensers, heat exchangers, and piping for desalination plants. Their resistance to erosion-corrosion makes them a preferred choice in high-velocity seawater systems such as offshore firewater systems.

2.4.3 Titanium and Exotic Alloys in Offshore Structures

Titanium and its alloys, though costly, provide unmatched corrosion resistance in seawater. They form a stable oxide layer that prevents further degradation, even under prolonged exposure to chlorides. Titanium alloys are lightweight, strong, and resistant to biofouling, making them ideal for deep-sea exploration equipment, heat exchangers, and subsea structures. Exotic alloys, such as nickel-based alloys (e.g., Inconel, Monel, Hastelloy), are also used where both high strength and extreme corrosion resistance are required. Applications include subsea valves, fasteners, and high-performance offshore systems in oil and gas exploration.

2.4.4 Coatings and Cladding Techniques for Corrosion Resistance

In addition to selecting corrosion-resistant materials, protective coatings and cladding methods are extensively used to prolong the service life of marine structures.

Coatings: Epoxy-based paints, polyurethane systems, and advanced antifouling coatings protect steel surfaces from seawater exposure and biological growth. Thermal spray coatings (zinc or aluminum) are also widely applied.

Cladding: Involves metallurgically bonding corrosion-resistant materials (e.g., stainless steel or nickel alloy) onto carbon steel substrates, combining strength with corrosion protection.

Cathodic Protection: Often used in conjunction with coatings, employing sacrificial anodes (zinc, aluminum) or impressed current systems to mitigate corrosion in hulls, pipelines, and offshore platforms.

These methods enhance durability, reduce maintenance frequency, and ensure the safe operation of marine assets in aggressive environments.

Comparative properties and applications of corrosion-resistant materials in marine engineering are shown in Table 2.3.

2.5 MATERIAL COMPATIBILITY AND DISSIMILAR METAL WELDING

The demand for lightweight, high-strength, and corrosion-resistant vessels has encouraged the integration of multiple materials in modern shipbuilding and offshore engineering. However, welding dissimilar metals such as steel to aluminum or copper alloys to stainless steels poses significant challenges due to differences in physical, chemical, and metallurgical properties. Improper joining can lead to galvanic corrosion, cracking, or premature joint failure, which directly compromises

TABLE 2.3
Comparative Properties and Applications of Corrosion-Resistant Materials in Marine Engineering

Material Type	Key Properties	Advantages	Typical Marine Applications
Austenitic stainless steel (e.g., 316L)	Good corrosion resistance, good formability, moderate strength	Readily available, easy to fabricate, resistant to pitting (with Mo grades)	Handrails, tanks, minor fittings, general ship structures
Duplex stainless steel (e.g., 2205)	High strength, good resistance to stress corrosion cracking	Combines toughness with corrosion resistance, cost-effective vs. super alloys	Offshore oil platforms, seawater piping, pressure vessels
Super duplex stainless steel (e.g., 2507)	Very high corrosion resistance, high strength, resistant to chloride attack	Exceptional durability in aggressive seawater conditions	Subsea equipment, risers, desalination plants, offshore rigs
Copper-nickel alloys (Cu-Ni 70/30, 90/10)	Excellent seawater corrosion resistance, antimicrobial, erosion-resistant	Natural resistance to biofouling, low maintenance	Cooling systems, condensers, desalination piping, firewater lines
Titanium alloys	Excellent seawater and chloride resistance, lightweight, biofouling resistant	High strength-to-weight ratio, long service life, minimal maintenance	Subsea structures, heat exchangers, deep-sea exploration equipment
Exotic nickel-based alloys (e.g., Inconel, Monel, Hastelloy)	Extreme corrosion resistance, high strength at elevated temperatures	Suitable for highly demanding and corrosive environments	Subsea valves, fasteners, high-performance offshore systems

the safety and longevity of marine structures (Ganguly 2023). Therefore, understanding compatibility issues and adopting specialized techniques are essential in mixed-material construction.

2.5.1 Issues in Welding Steel to Aluminum and Other Metals

One of the most common dissimilar welding challenges in marine applications is joining steel to aluminum, particularly in high-speed vessels where aluminum superstructures are welded onto steel hulls. Direct fusion welding is not feasible because:

Different Melting Points: Aluminum melts at ~660°C, while steel melts at ~1,500°C, making simultaneous fusion difficult.

Intermetallic Compounds: Fusion often produces brittle Fe-Al intermetallics, which weakens the joint.

Thermal Expansion Mismatch: Aluminum expands nearly twice as much as steel, leading to residual stresses and cracking.

Other dissimilar combinations, such as copper-nickel alloys to stainless steel or titanium to steel, face similar issues with thermal mismatch, metallurgical incompatibility, and the risk of galvanic corrosion when exposed to seawater.

2.5.2 Techniques to Minimize Galvanic Corrosion

Galvanic corrosion can be minimized through proper material selection, design, and maintenance practices. The most effective approach is to use metals that are close to each other in the galvanic series, thereby reducing the potential difference that drives corrosion. Electrical insulation between dissimilar metals, using non-conductive gaskets, washers, or coatings, can prevent direct metal-to-metal contact and restrict current flow. Protective coatings such as paints, anodized layers, or polymer films are widely used; however, care must be taken to ensure the coating remains intact, as any damage can lead to localized corrosion.

Cathodic protection, either by using sacrificial anodes like zinc or magnesium or by employing impressed current systems, is another reliable technique to shift the potential of the protected metal to a more cathodic state. Environmental control, such as reducing the exposure of joints to moisture or salt-laden air and using sealing compounds, also helps in lowering corrosion rates. Design modifications play an important role in ensuring a larger anodic area relative to the cathodic one, avoiding direct contact of dissimilar metals, and improving drainage to prevent stagnant electrolytes are key considerations.

Additionally, the use of corrosion inhibitors and regular inspection is essential for long-term performance. In marine environments, where exposure to seawater accelerates galvanic action, combining these methods especially isolation, coatings, and cathodic protection provides effective and sustainable corrosion control.

2.6 SUMMARY

This chapter highlighted the critical role of material selection and welding practices in ensuring the performance, durability, and safety of marine structures operating in aggressive seawater environments. Steel alloys continue to dominate shipbuilding, with carbon steels used for general hull structures and HSLA steels applied in strength-critical zones, all governed by international standards. Aluminum alloys, particularly the 5xxx and 6xxx series, are increasingly employed in high-speed craft and superstructures due to their low density and corrosion resistance, though their welding requires careful control to overcome issues such as distortion, porosity, and HAZ softening. To resist seawater-induced degradation, corrosion-resistant materials such as stainless steels, copper-nickel alloys, titanium, and nickel-based super alloys are widely adopted, often in combination with protective coatings, cladding, and cathodic protection. The integration of dissimilar materials, such as steel-to-aluminum joints or copper-nickel-to-stainless steel connections, presents challenges, including galvanic corrosion and brittle intermetallic formation, which are mitigated using transition joints, insulating materials, and advanced welding or bonding techniques. Case studies from naval vessels, offshore oil platforms, and desalination systems illustrate the practical application of these materials and methods. Together, these strategies enable modern marine engineering to achieve an optimal balance of strength, corrosion resistance, weight efficiency, and long service life under demanding conditions.

2.7 REVIEW QUESTIONS

1. Why is material selection considered critical in marine welding applications?
2. Explain how seawater exposure influences the choice of materials for marine structures.
3. What factors must be considered when selecting base metals and filler materials for marine welding?
4. How does material selection impact maintenance costs and service life of ships?
5. Give two real-world examples where improper material selection led to marine structure failures.
6. Differentiate between ordinary carbon steels and HSLA steels in terms of strength and applications.
7. What are the advantages of using HSLA steels in shipbuilding compared to carbon steels?
8. Describe the welding characteristics of marine steels and the challenges they present.
9. Discuss the significance of standards and classifications (e.g., AH/DH/EH grades) for marine steels.
10. Using a diagram or schematic, explain where different grades of steel are typically applied in a ship's structure.
11. Compare the properties of 5xxx and 6xxx series aluminum alloys used in marine applications.
12. What are the major advantages of aluminum alloys in high-speed vessels and ferries?
13. Which welding methods are most suitable for aluminum in shipbuilding, and why?
14. Explain the role of friction stir welding (FSW) in modern marine aluminum structures.
15. Identify the main challenges faced during aluminum welding and suggest methods to overcome them.
16. Compare the marine applications of austenitic, duplex, and super duplex stainless steels.
17. Why are copper-nickel alloys preferred for seawater piping systems?
18. Discuss the role of titanium in offshore structures and deep-sea equipment.
19. What are the advantages and limitations of using exotic nickel-based alloys in marine systems?
20. Explain the differences between coatings, cladding, and cathodic protection in preventing corrosion.

21. What are the main challenges in welding steel to aluminum in marine construction?
22. How do intermetallic compounds affect the strength of dissimilar metal welds?
23. What is galvanic corrosion, and why is it a critical issue in marine applications?
24. Describe at least three techniques to minimize galvanic corrosion in mixed-metal systems.
25. Explain the function of bimetallic transition joints in shipbuilding.
26. What are the advantages of using dielectric isolation gaskets in seawater piping systems?
27. Give an example of how cathodic protection is applied in mixed-metal marine structures.
28. Discuss a case study of steel-aluminum integration in naval vessels and the solutions used.
29. Explain how copper-nickel piping is joined to stainless steel in desalination plants to avoid galvanic issues.
30. Why is material compatibility considered as important as material strength in marine engineering?

3 Welding Processes in the Marine Industry

3.1 INTRODUCTION

Welding is the backbone of modern marine fabrication, enabling the construction of ships, offshore structures, and subsea systems that can withstand the demanding conditions of ocean environments. Unlike mechanical fastening methods such as riveting or bolting, welding provides permanent joints with high strength, water tightness, and structural continuity, making it indispensable for ship hulls, bulkheads, pipelines, and critical load-bearing components. The marine industry employs a diverse range of welding processes, each tailored to the material, section thickness, and performance requirements of the structure being fabricated.

The choice of welding technique in shipyards is not only a matter of efficiency but also of safety and reliability. Factors such as production rate, cost-effectiveness, ease of automation, and compliance with classification society standards significantly influence process selection. Moreover, the aggressive marine environment characterized by high humidity, saltwater exposure, and fluctuating temperatures creates additional challenges that can compromise weld integrity (Anand and Khajuria 2015). These conditions demand specialized welding methods, corrosion-resistant materials, and strict quality assurance measures.

With advancements in automation, robotics, and non-destructive testing, welding in the marine industry continues to evolve, ensuring that vessels and offshore platforms meet the highest standards of strength, durability, and service life. This chapter examines the various welding processes used in marine applications, the criteria that guide their selection, and how the harsh oceanic environment affects welding techniques and outcomes.

3.1.1 Role of Welding Processes in Marine Fabrication

Welding plays a pivotal role in marine fabrication, as it provides the primary means of joining structural components to create robust, watertight, and durable ship structures. From hull plating to offshore platforms and subsea equipment, welding ensures structural continuity and integrity that bolted or riveted joints cannot provide. In modern shipbuilding, automated and semi-automated welding processes enhance productivity, reduce human error, and ensure compliance with stringent international standards (Anand and Khajuria 2015). The choice of welding technique directly influences the strength, reliability, and service life of vessels operating in harsh marine environments.

DOI: 10.1201/9781003217350-3

3.1.2 Factors Affecting Process Selection in Shipyards

The selection of welding processes in shipyards depends on a combination of technical, economic, and operational factors:

Material Type: Different base metals such as steels, aluminum alloys, and corrosion-resistant materials require distinct welding methods.
Thickness of Sections: Processes such as submerged arc welding (SAW) are better suited for thick sections, while gas tungsten arc welding (GTAW) or friction stir welding (FSW) is preferred for thinner alloys.
Production Rate: High-volume shipyards prioritize processes with faster deposition rates (e.g., flux-cored arc welding [FCAW], SAW).
Cost and Availability: Equipment, consumables, and labor skill levels heavily impact the feasibility of a welding process.
Quality Requirements: Certain sections, especially critical joints like pressure hulls in submarines, demand high-precision processes.

3.1.3 Influence of Marine Environment on Welding Techniques

The marine environment imposes unique challenges on welding operations. High humidity, salt-laden air, and temperature fluctuations accelerate corrosion and can lead to hydrogen-induced cracking, porosity, and weld metal embrittlement. Offshore and underwater welding conditions require specialized techniques such as wet welding, dry hyperbaric welding, or remotely operated robotic welding. Moreover, exposure to seawater demands the use of corrosion-resistant consumables and strict pre- and post-weld treatments. Protective coatings, cathodic protection, and advanced inspection techniques further ensure that welds can withstand long-term exposure to marine conditions without premature failure.

3.2 ARC WELDING PROCESSES

Arc welding is the most widely used joining method in the marine industry due to its versatility, cost-effectiveness, and ability to produce strong, durable welds across a variety of base metals. The process relies on an electric arc formed between the electrode and the workpiece, generating sufficient heat to melt and fuse materials. Marine structures—ranging from ship hulls and decks to offshore oil platforms—depend heavily on arc welding techniques to ensure water tightness, load-bearing strength, and long-term resistance to harsh oceanic conditions. Several arc welding variants are employed in shipyards, each suited to specific materials, section thicknesses, and production requirements.

3.2.1 Shielded Metal Arc Welding (SMAW)

3.2.1.1 Principle and Equipment

Shielded metal arc welding (SMAW), also known as manual metal arc (MMA) or stick welding, is one of the most widely practiced welding techniques in the marine

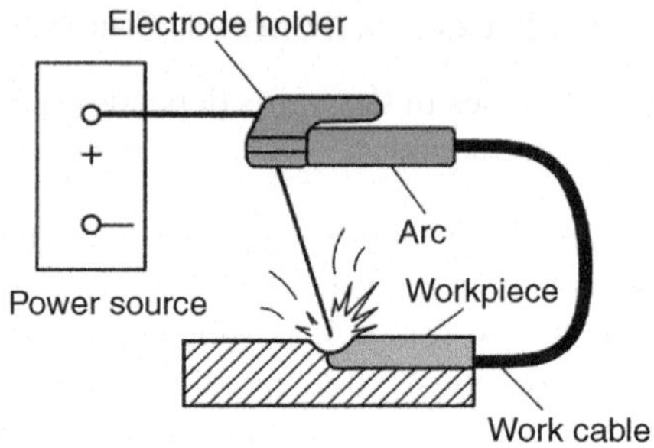

FIGURE 3.1 Principle and equipment for shielded metal arc welding (SMAW).

industry. The process operates on the principle of creating an electric arc between a consumable flux-coated electrode and the base metal. The heat generated by the arc melts both the electrode and the workpiece, forming a weld pool (Alkahla and Pervaiz 2017). The flux coating decomposes under heat, releasing gases that shield the molten metal from atmospheric contamination while also forming a protective slag layer that solidifies over the weld bead. The essential equipment for SMAW includes a power source (AC or DC), electrode holder, ground clamp, and flux-coated electrodes available in various compositions depending on the base material and service conditions.

Figure 3.1 illustrates the principle and equipment used in SMAW, a widely applied manual arc welding process. The setup consists of a power source, an electrode holder, a consumable coated electrode, a workpiece, and a work cable. The power source supplies either alternating or direct current to establish an electric arc between the tip of the electrode and the workpiece surface. This arc produces intense heat, which melts both the base metal and the electrode tip to form a molten weld pool. The electrode's flux coating decomposes under heat, generating shielding gases and a layer of slag that protects the molten metal from atmospheric contamination such as oxygen and nitrogen. As the welder moves the electrode along the joint, the molten metal solidifies to form a strong weld bead. The work cable completes the electrical circuit, ensuring continuous current flow throughout the welding operation.

3.2.1.2 Marine Applications

SMAW is extensively used in shipbuilding and repair due to its versatility and simplicity. It is particularly valuable in the following applications:

- **Hull Fabrication and Repairs**: Joining steel plates and structural members.
- **Onboard Maintenance**: Repairing cracks, corrosion-damaged areas, or broken fittings.
- **Confined and Difficult-to-Access Spaces**: Where portability and flexibility of equipment are critical.

- **Offshore Structures**: Maintenance welding in platforms and pipelines exposed to harsh conditions. Its adaptability makes SMAW especially suitable for field welding operations where environmental conditions cannot be fully controlled.

Advantages:

- Simple, portable, and inexpensive equipment.
- Suitable for a wide range of ferrous materials and thicknesses.
- Effective for outdoor and underwater applications when special electrodes are used.
- Ideal for ship repair and maintenance tasks.

Limitations:

- Slower deposition rate compared to automated methods like gas metal arc welding (GMAW) or SAW.
- Requires frequent electrode replacement, reducing productivity in large-scale fabrication.
- Dependent on welder's skill, leading to variability in weld quality.
- Produces slag, which must be removed after each pass.
- Less suitable for non-ferrous alloys like aluminum, which are increasingly used in modern vessels.

3.2.2 Gas Metal Arc Welding (GMAW)

3.2.2.1 Process Overview

GMAW, also known as metal inert gas (MIG) welding, is a semi-automatic or fully automatic arc welding process where a continuously fed consumable wire electrode is melted by an electric arc to join metals. An external shielding gas, typically inert (argon, helium) or active mixtures (argon-CO_2, argon-oxygen), protects the weld pool from atmospheric contamination. The process provides a stable arc, high deposition rates, and smooth weld bead appearance, making it a preferred method in industries requiring speed and quality. The equipment consists of a power source, wire feed unit, welding torch, shielding gas supply, and control system as shown in Figure 3.2. This figure illustrates the GMAW process, also known as MIG welding, which utilizes a continuous and consumable wire electrode fed through a welding gun. The setup consists of a power source, a wire feed unit, a welding gun, and the workpiece. The wire feed unit supplies the electrode wire at a controlled rate, while the power source provides the necessary current to establish an electric arc between the wire tip and the workpiece. As the arc melts both the electrode and the base metal, a molten weld pool is formed. Simultaneously, a shielding gas such as argon, carbon dioxide, or a gas mixture is emitted through the nozzle surrounding the arc, protecting the weld zone from atmospheric contamination. This process ensures high deposition rates and clean

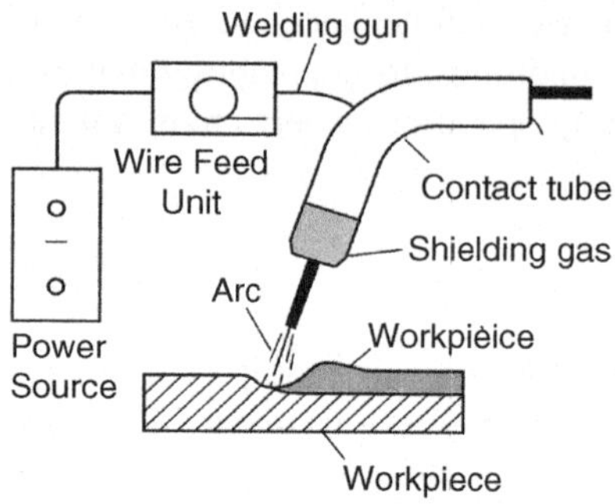

FIGURE 3.2 Schematic representation of gas metal arc welding (GMAW) process.

welds, and is widely used for welding ferrous and non-ferrous metals in both manual and automated applications.

3.2.2.2 Suitability for Shipbuilding

GMAW is highly suitable for modern shipyards due to its efficiency and compatibility with automation. Its advantages in shipbuilding include:

- **Large-Scale Hull Fabrication**: Ideal for long, continuous welds in flat or horizontal positions.
- **Aluminum Structures**: Particularly effective for welding 5xxx and 6xxx series alloys used in high-speed craft, ferries, and naval vessels.
- **Superstructures and Decks**: Where clean, visually appealing welds are required.
- **Integration with Robotics**: Supports automated welding lines in advanced shipyards, improving consistency and reducing reliance on manual labor.

  Although best suited for controlled environments (indoors or sheltered workshops), portable setups also enable field use with proper shielding measures.

3.2.2.3 Common Defects and Prevention

While GMAW provides high-quality welds, several defects can occur if parameters or conditions are not properly controlled:

- **Porosity**: Caused by poor shielding gas coverage, drafts, or surface contamination; prevented by ensuring proper gas flow, clean surfaces, and avoiding welding in windy conditions.
- **Lack of Fusion**: Due to low heat input, incorrect torch angle, or improper travel speed; mitigated by adjusting voltage, amperage, and electrode positioning.
- **Spatter**: Excessive spatter results from incorrect current settings or gas composition; minimized by using appropriate welding parameters and shielding gases.

- **Cracking**: May occur in aluminum alloys or high-strength steels due to rapid cooling; prevented by preheating where necessary and controlling heat input.
- **Burn-through**: Common when welding thin plates; prevented by lowering current and using proper wire feed speed.

Through careful parameter control, appropriate electrode and shielding gas selection, and skilled operation, these defects can be minimized, ensuring strong and reliable welds in marine environments.

3.2.3 Flux-Cored Arc Welding (FCAW)

FCAW is a versatile process that combines features of SMAW and GMAW. It uses a tubular wire electrode filled with flux, which provides shielding and slag formation during welding. Figure 3.3 illustrates the working principle of FCAW, showing the interaction between the electrode, shielding medium, and weld pool. At the core of the process is the tubular electrode, which contains a flux core that generates gases and slag during welding. Depending on whether the process is self-shielded or gas-shielded, an external shield gas may also be supplied to enhance the protection of the molten weld pool. As the arc melts the electrode and the base metal, the resulting molten metal fills the joint, forming the weld metal upon solidification. Simultaneously, a layer of solidified slag develops on top of the weld bead, protecting it from atmospheric contamination during cooling. This layered interaction between flux, shielding, molten pool, and slag ensures strong, clean welds suitable for marine fabrication and repair.

Figure 3.4 shows an FCAW machine, highlighting its main components and their functions in the welding process. The system consists of a power source, which provides the necessary electrical energy to sustain the arc, and a wire feed unit, which continuously supplies the tubular flux-cored electrode to the welding torch. The welding torch delivers both the electrode and, in gas-shielded setups, the shielding gas to the weld zone, ensuring stable arc formation and protection of the molten pool

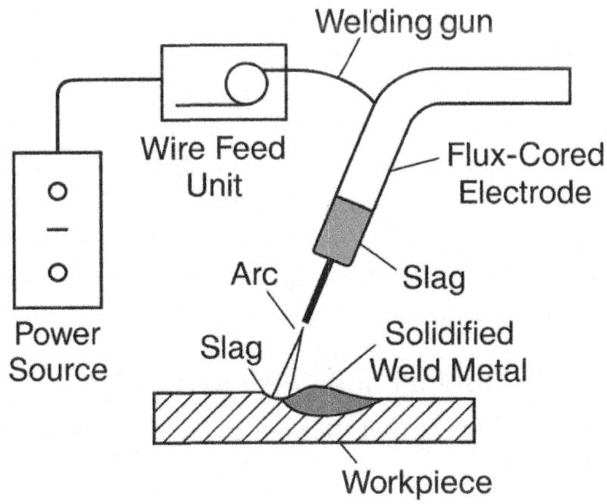

FIGURE 3.3 Schematic of flux-cored arc welding process.

FIGURE 3.4 Machine setup. (https://en.wikipedia.org/wiki/Flux-cored_arc_welding.)

(Meyer 1993). A ground clamp connects the workpiece to complete the electrical circuit, allowing current to flow effectively during welding. For gas-shielded FCAW (FCAW-G), a gas cylinder and regulator are included to control the flow of shielding gas, while the control panel allows the operator to adjust voltage, current, and wire feed speed according to the welding requirements. Together, these components make FCAW a highly adaptable process, capable of delivering strong and reliable welds in marine shipyards and offshore environments.

3.2.3.1 FCAW Is Available in Two Main Forms

Self-Shielded FCAW (FCAW-S): In this variant, the flux core itself generates the protective gases needed to shield the weld pool, eliminating the need for external shielding gas. It is highly portable and suitable for remote or windy environments where gas shielding may be disrupted. However, weld spatter and slag removal are more significant compared to gas-shielded variants.

Gas-Shielded FCAW (FCAW-G): Here, the flux core enhances weld performance, but an external shielding gas commonly CO_2 or argon-CO_2 mixtures is also supplied. This results in smoother welds with higher deposition rates and reduced spatter compared to self-shielded FCAW. FCAW-G is widely used in controlled shipyard environments where higher-quality welds and faster production are required.

3.2.3.2 Advantages in Outdoor and Offshore Work

FCAW offers several advantages that make it particularly suitable for outdoor and offshore marine applications:

- **Resistance to Environmental Disturbances**: Unlike GMAW, FCAW is less sensitive to wind and drafts, making it reliable for ship deck welding and offshore platform construction.

TABLE 3.1
Comparison between Self-Shielded FCAW (FCAW-S) and Gas-Shielded FCAW (FCAW-G)

Aspect	FCAW-S (Self-Shielded)	FCAW-G (Gas-Shielded)
Shielding method	Flux core generates protective gases; no external shielding gas required	Combination of flux core and external shielding gas (CO_2 or argon-CO_2)
Portability	Highly portable, suitable for remote or offshore sites	Less portable due to need for shielding gas cylinders
Weld quality	Adequate, but may have higher spatter and rougher appearance	Cleaner welds with smoother bead and lower spatter
Productivity	Good for fieldwork, slower than FCAW-G in controlled settings	Higher deposition rates, suitable for shipyard production lines
Environmental sensitivity	Performs well outdoors and in windy conditions	Shielding gas is vulnerable to wind/drafts; better suited for sheltered environments
Slag formation	Significant slag, requires thorough removal between passes	Moderate slag, easier cleanup compared to FCAW-S
Typical marine applications	Ship repair, offshore maintenance, emergency welding	Hull fabrication, deck sections, aluminum/steel superstructures in controlled shipyard environments

- **High Deposition Rates**: The process achieves faster welding speeds and deeper penetration, improving productivity in hull fabrication and heavy structural welding.
- **Versatility in Materials and Positions**: FCAW can be used effectively on a wide range of steels and performs well in vertical and overhead welding, which are common in ship repairs.
- **Robustness for Field Applications**: FCAW-S is often preferred for emergency or remote welding tasks on ships and rigs where shielding gas supply is impractical.

While FCAW does produce slag that must be removed between passes and requires skilled operation to avoid porosity, its adaptability and reliability under challenging conditions have made it a standard process in modern marine construction and repair.

Comparison between FCAW-S and FCAW-G is shown in Table 3.1.

3.2.4 Gas Tungsten Arc Welding (GTAW)

3.2.4.1 Precision Welding in Marine Components

GTAW, also known as tungsten inert gas (TIG) welding, is a process that uses a non-consumable tungsten electrode to produce an arc, with a separate filler metal added when required. The weld area is protected by an inert shielding gas, typically argon or helium, ensuring a clean and stable arc. GTAW is renowned for its ability

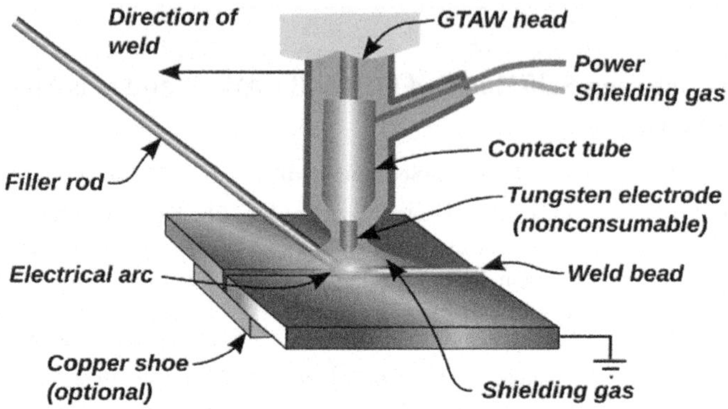

FIGURE 3.5 Schematic representation of gas tungsten arc welding (GTAW) process. (https://en.wikipedia.org/wiki/Gas_tungsten_arc_welding.)

to deliver high-quality, precise welds with excellent bead appearance and minimal spatter. In the marine industry, it is widely used for critical components where weld integrity is paramount, such as piping systems, valves, tanks, and pressure vessels. Its precision makes it particularly suitable for joining thin sections and achieving defect-free welds in high-stress or corrosion-sensitive environments (Short 2009). Figure 3.5 illustrates the GTAW (or TIG) process, highlighting its key components and working principle. In this setup, a non-consumable tungsten electrode generates an electric arc that melts the base metal, creating the weld pool. An inert shielding gas, usually argon or helium, flows through the nozzle to protect the molten weld from atmospheric contamination, preventing oxidation and porosity. When additional material is required, a separate filler rod is manually fed into the weld pool, ensuring strong and precise joints. The illustration also shows the formation of the molten pool, solidifying weld metal, and the protective shielding envelope around the arc zone. GTAW provides excellent control over heat input and weld quality, making it particularly suitable for thin sections, aluminum, and stainless steels in marine fabrication.

3.2.4.2 Use for Aluminum and Stainless Steels

GTAW is especially valuable in shipbuilding for **aluminum and stainless steel structures,** which are increasingly used due to their corrosion resistance and weight-saving benefits. For aluminum, GTAW enables strong, clean welds in the 5xxx and 6xxx marine-grade alloys used in high-speed ferries, naval vessels, and superstructures. For stainless steels including austenitic and duplex grades, GTAW ensures welds with excellent corrosion resistance, critical for piping systems and tanks exposed to seawater. The process also allows precise control of heat input, reducing the risk of warping, porosity, or sensitization in stainless steels. Although GTAW has slower deposition rates and requires higher operator skill compared to GMAW or FCAW, its unmatched precision and reliability make it indispensable in high-quality marine fabrication.

3.3 SUBMERGED ARC WELDING (SAW)

SAW is a widely used arc welding process, first patented in 1935. The process employs a continuously fed consumable electrode, which may be solid or tubular (metal-cored). During welding, both the molten weld pool and the arc zone are completely protected from atmospheric contamination by a thick blanket of granular fusible flux. This flux typically contains compounds such as lime, silica, manganese oxide, and calcium fluoride. When heated, the flux becomes partially molten and electrically conductive, creating a current path between the electrode and the workpiece. The flux not only stabilizes the arc but also forms a protective slag layer over the weld, which prevents spatter, suppresses fumes, and eliminates intense ultraviolet radiation common in processes like SMAW. As a result, SAW ensures high-quality welds with deep penetration and minimal operator exposure to harmful emissions (Singh et al. 2013).

Historically, the introduction of SAW revolutionized heavy fabrication industries by enabling high deposition rates and consistent weld quality, particularly for thick plates. In modern marine applications, SAW is extensively used in shipyards for welding longitudinal and transverse seams in hull structures, assembling large deck panels, and fabricating offshore oil and gas platforms. Its ability to produce deep, defect-free welds with high efficiency makes it a preferred choice for large-scale, critical marine construction, where both productivity and structural integrity are paramount.

Figure 3.6 illustrates the basic working principle of the SAW process. In this process, welding is carried out by establishing an electric arc between a continuously fed consumable electrode and the workpiece. The arc and the molten weld pool are completely covered by a layer of granular flux, which plays a crucial role in protecting the weld region.

As the electrode wire is fed toward the joint, the electric arc generates intense heat that melts both the electrode and the base metal, forming a molten weld pool. The granular flux surrounding the arc melts partially to form a slag layer, which covers the molten metal and isolates it from atmospheric gases such as oxygen and nitrogen. This protection prevents oxidation and contamination, thereby improving the metallurgical quality of the weld.

The flux also stabilizes the arc, enhances heat concentration, and reduces spatter and arc radiation losses. As the welding progresses, the molten weld metal solidifies behind the arc to form a uniform and smooth weld bead, while the solidified slag can be easily removed after cooling. Any unused flux can be recovered and reused, increasing the efficiency of the process.

SAW is characterized by high deposition rates, deep weld penetration, and excellent weld quality, making it especially suitable for welding thick plates and long, straight joints in heavy fabrication industries such as shipbuilding, pressure vessel manufacturing, pipelines, and structural steel construction.

3.3.1 Principle and Equipment

SAW is an arc welding process where the arc is formed between a continuously fed consumable electrode and the workpiece, but unlike open arc methods, the arc is

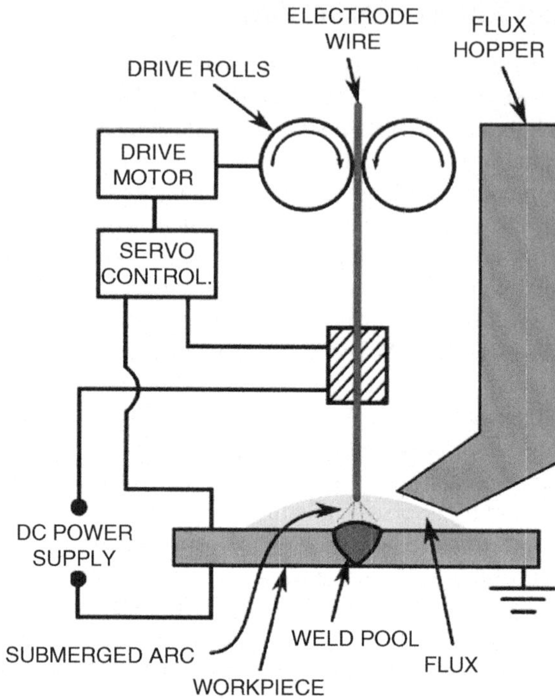

FIGURE 3.6 Schematic representation of submerged arc welding (SAW) process. (https://en.wikipedia.org/wiki/Submerged_arc_welding.)

completely submerged beneath a blanket of granular flux. This flux not only shields the molten weld pool from atmospheric contamination but also stabilizes the arc and contributes alloying elements to the weld metal. The key equipment includes a power source (usually DC or AC with high current capacity), a wire feeding system, a flux hopper for depositing granular flux, and a travel mechanism for automatic movement. Because the process is usually mechanized, SAW is ideal for producing long, straight, and consistent welds.

3.3.2 Advantages for Large Ship Panels and Offshore Platforms

SAW is particularly advantageous in shipbuilding and offshore construction because of its high deposition rate and deep penetration, which make it suitable for welding thick plates and large structural panels. In shipyards, it is widely used for joining longitudinal and transverse seams of hull panels, where productivity and efficiency are crucial. Offshore platforms also benefit from SAW's ability to produce defect-free welds in critical load-bearing joints, ensuring safety and reliability under harsh marine environments. Additionally, the flux blanket minimizes spatter and arc radiation, making the process safer and cleaner for operators.

Advantages and disadvantages of SAW are shown in Table 3.2.

TABLE 3.2
Advantages and Disadvantages of Submerged Arc Welding (SAW)

Advantages	Disadvantages
High deposition rate ensures faster welding of thick sections	Limited to flat and horizontal positions; unsuitable for vertical or overhead welding
Deep weld penetration provides strong joints for heavy structures like ship hulls and offshore platforms	Not suitable for thin materials due to excessive heat input and burn-through risk
Flux blanket prevents atmospheric contamination, resulting in high-quality, defect-free welds	Process restricted mainly to ferrous metals and certain nickel alloys; unsuitable for aluminum and most stainless steels
Minimal spatter, fumes, and arc radiation improve operator safety and cleanliness	Requires mechanized or automated setup, making it less flexible for small or complex welds
Slag cover and flux recycle reduce material wastage and improve efficiency	Flux contamination or improper handling can lead to porosity or inclusions in the weld
Suitable for long, continuous welds on large plates, improving productivity in shipyards	Equipment and flux handling systems add to initial setup and operational cost

3.3.3 Limitations and Quality Considerations

Despite its advantages, SAW has some limitations. It is not suitable for thin plates or out-of-position welding since the molten flux and weld pool rely on gravity. The process is limited to ferrous metals and certain nickel alloys, restricting its application in marine structures involving aluminum or stainless steel. Quality considerations include proper flux handling since contaminated or recycled flux can lead to porosity or inclusions and maintaining consistent parameters such as current, voltage, and travel speed. While SAW produces high-quality welds, its reliance on mechanized setups makes it less flexible compared to manual processes for complex geometries. Comparison of SAW with other arc welding processes is shown in Table 3.3.

3.4 RESISTANCE WELDING

Resistance welding can be defined as a solid-state welding technique in which a joint is formed by applying heat and pressure at the interface of two metal sheets. The heat is generated due to the electrical resistance at the contact surfaces when a high current passes through them, while pressure from the electrodes ensures fusion at the joint. Unlike fusion welding processes, resistance welding does not require filler material, flux, or shielding gas, making it cleaner and more efficient. The process is fundamentally a thermo-electric cycle, where the resistance at the faying surfaces creates localized heating, leading to the formation of a weld nugget under controlled force (Stavrov and Bersee 2005).

Despite its high efficiency, speed, and pollution-free nature, resistance welding has certain limitations. It is best suited for thin plates, typically less than 5 mm thick, as the heat generated is insufficient to weld thicker materials. The process also requires costly specialized equipment and electrodes, which can limit its use in

TABLE 3.3
Comparison of SAW with Other Arc Welding Processes

Aspect	Submerged Arc Welding (SAW)	Shielded Metal Arc Welding (SMAW)	Gas Metal Arc Welding/MIG (GMAW)	Flux-Cored Arc Welding (FCAW)	Gas Tungsten Arc Welding (GTAW)/TIG
Welding speed and deposition rate	Very high; suitable for thick plates and long welds	Low; manual and time-consuming	Moderate to high; faster than SMAW	High; especially with gas-shielded FCAW	Low; slow process, mainly for precision work
Weld quality	Excellent; deep penetration and minimal defects	Good, but highly dependent on operator skill	Good; stable arc and clean welds	Good; some risk of porosity in windy conditions	Excellent; precise and high-quality welds
Positions	Limited to flat and horizontal	All positions	All positions (with proper shielding)	All positions (especially FCAW-S)	All positions
Automation	Easily mechanized and automated	Primarily manual	Easily mechanized or semi-automatic	Semi-automatic or mechanized	Usually manual, but can be automated
Material suitability	Ferrous metals, some nickel alloys; not for Al or most SS	Wide range of steels	Steels, stainless steels, aluminum	Steels, stainless steels, nickel alloys	Almost all metals, including Al, SS, Ti
Shielding method	Granular flux blanket	Flux-coated electrode	Shielding gas (Ar, CO_2, mix)	Flux core (self- or gas-shielded)	Inert shielding gas (Ar, He)
Operator safety	Safer: arc and radiation covered by flux	More exposure to radiation, fumes, spatter	Moderate exposure	Moderate exposure	More exposure to radiation but clean welds
Applications in marine industry	Ship hull seams, deck panels, offshore platforms	On-site repairs, secondary structures	Structural fabrication, aluminum components	Offshore repair, outdoor fabrication	Precision components, piping, stainless and aluminum fittings
Limitations	Not flexible for complex shapes; limited to thick plates	Slow, labor-intensive, quality varies	Sensitive to wind (gas shielding loss)	Slag removal required, porosity risk	Slow, costly, requires high skill

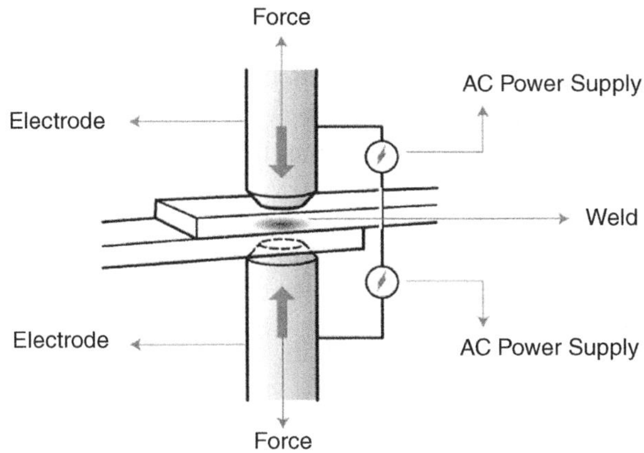

FIGURE 3.7 Schematic diagram of resistance welding process. (https://www.geeksforgeeks.org/electrical-engineering/electric-resistance-welding/.)

large-scale ship construction. Nevertheless, resistance welding finds valuable applications in marine fabrication for thin-sheet assemblies, such as ventilation ducts, storage tanks, and partition panels.

Figure 3.7 shows the principle of resistance welding, where two copper alloy electrodes apply both force and current to the metal sheets. The AC power supply generates the current needed to create heat at the interface, and the resulting pressure consolidates the weld nugget, completing the joint.

3.4.1 Working Principle of Resistance Welding

The process of resistance welding begins with the metal parts being placed in contact with each other and firmly held between two electrode tips. These electrodes serve a dual purpose: they apply the required pressure to maintain proper contact and alignment, while also conducting the welding current into the workpieces.

Once the setup is complete, an electric current is passed through the electrodes and into the workpieces. At the interface where the two metals meet, electrical resistance is encountered, which restricts the flow of current. This localized resistance results in the rapid generation of heat at the joint. The phenomenon is explained by Joule's heating effect, in which electrical energy is converted into thermal energy.

As the current flows, the concentrated heat causes the metal at the faying surfaces to soften and eventually melt, forming a localized weld nugget. At the same time, the pressure applied by the electrodes ensures proper fusion and consolidation of the joint. After cooling, a strong metallurgical bond is established between the two parts without the need for additional filler materials or shielding gases.

The heat generation in resistance welding is governed by the following equation:

$$H = I^2RT$$

where

H = Heat generated
I = Current flowing through the electrodes
R = Electrical resistance at the contact interface
t = Duration of current flow

This equation highlights that the amount of heat produced is directly proportional to the square of the current, the resistance at the contact point, and the time of current application.

3.4.2 Types of Resistance Welding

Different types of resistance welding are used in marine and industrial fabrication depending on the material thickness, joint design, and application requirements. The major types include the following.

3.4.3 Spot Welding

Spot welding is the simplest and most widely used form of resistance welding. In this process, the workpieces are pressed together by the force of copper alloy electrodes, which also conduct the welding current. When current passes through the sheets, electrical resistance at the faying surfaces generates localized heat. This resistance is particularly high at the contact points due to small air gaps and surface irregularities, causing rapid heating and melting of the interface. As the current flow stops, the electrode pressure is maintained briefly to allow the molten zone to solidify, forming a strong weld nugget. The nugget size typically depends on the electrode tip diameter, usually ranging between 4 and 7 mm. The electrodes are then repositioned to the next location, allowing successive spot welds to be made efficiently (Dickinson et al. 1980).

Figure 3.8 illustrates the labeled schematic diagram of the spot welding process. Two metal sheets are placed together and pressed between copper alloy electrodes, which apply mechanical pressure to hold the sheets in firm contact. When an electric current passes through the electrodes, the electrical resistance at the faying surfaces of the sheets generates intense localized heat. This heat causes the interface region to soften and partially melt, forming a molten weld nugget. After the current is switched off, the continued application of electrode pressure allows the molten nugget to solidify into a strong finished weld. The diagram clearly highlights the roles of the electrodes, metal sheets, pressure, and the final weld nugget, showing how spot welding produces small, discrete weld points without requiring filler material or shielding gas.

Advantages of Spot Welding:

- Provides high efficiency, repeatability, and consistency.
- Economical, with no need for filler metal or flux.
- Energy-efficient, using electrical power effectively.
- Very fast process, suitable for high-production applications.

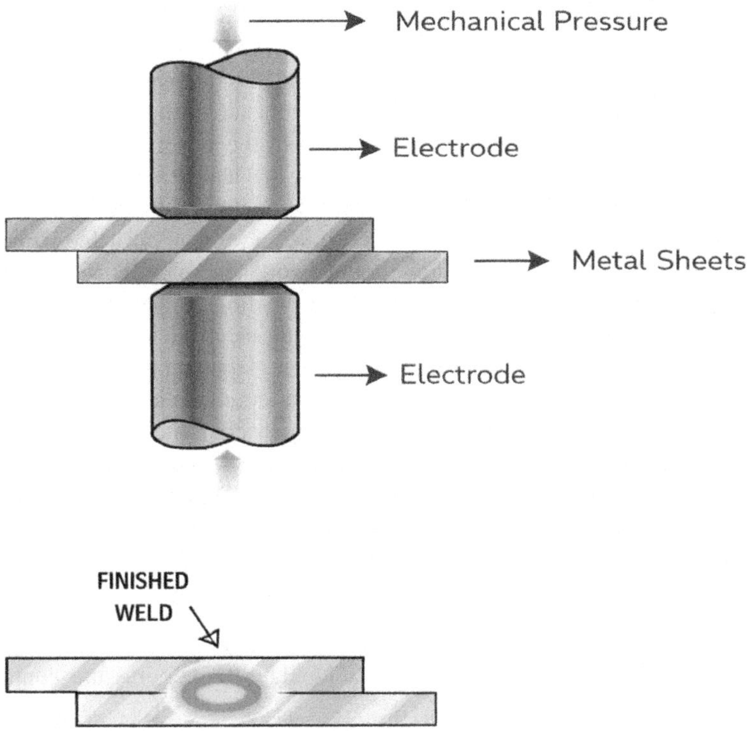

FIGURE 3.8 Schematic diagram of spot welding. (https://www.geeksforgeeks.org/electrical-engineering/electric-resistance-welding/.)

Disadvantages of Spot Welding:

- Requires a large working area and heavy equipment.
- Welding guns are bulky and can be difficult to handle, especially for aged or less-experienced welders.
- Limited to thin sheets; unsuitable for thicker materials.
- Electrode wear and maintenance are frequent issues.

Applications of Spot Welding:

- Widely used in industries such as automotive, aerospace, metal furniture, electrical equipment, and building construction.
- Suitable for high-volume production where speed and consistency are required.
- Commonly used to weld thin metal sheets together in subassemblies.
- Applied in the fabrication of sheet metal enclosures, boxes, transformer cores, and cases requiring high mechanical strength.

3.4.4 Seam Welding

Seam welding, often referred to as continuous spot welding, is a resistance welding technique in which roller-type electrodes are used instead of stationary electrodes. These rollers apply pressure and simultaneously conduct current across the workpieces, which are typically overlapping sheets. As the rollers move along the joint, they deliver a continuous or overlapping series of spot welds, creating a strong and leak-proof seam. The movement of the electrodes and the weld timing are carefully controlled to ensure proper overlap and to avoid overheating of the plates. In practice, seam welding can achieve high speeds, with airtight welds being produced at rates of up to 60 inches per minute (Tzeng 2000).

Figure 3.9 illustrates the principle of seam welding. In this process, two overlapping metal sheets are positioned between roller electrodes that apply both pressure and current. A transformer supplies the necessary voltage and current, which flow through the rollers into the sheets. As the rollers move across the surface, they generate a series of overlapping spot welds, which combine to form a continuous weld seam. The applied force ensures proper contact and fusion, while the controlled movement of the rollers maintains uniformity. This method produces airtight and watertight joints, making it highly suitable for applications such as tanks, pipes, and marine structures requiring leak-proof integrity.

Advantages of Seam Welding:

- Produces airtight and watertight joints, ideal for fluid or pressure-containing applications.
- High welding speed; easily automated for mass production.
- No need for filler or flux materials.
- Provides consistent, uniform weld quality.

Disadvantages of Seam Welding:

- Limited to straight or uniformly curved welds due to the roller electrode design.

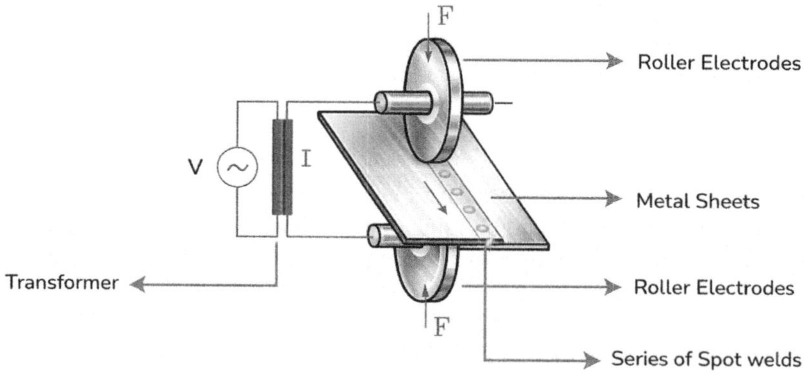

FIGURE 3.9 Principle of seam welding. (https://www.geeksforgeeks.org/electrical-engineering/electric-resistance-welding/.)

- Not suitable for metal sheets thicker than 3 mm.
- Requires precise control of current and timing to avoid burn-through or weak welds.

Applications of Seam Welding:

- Manufacturing of lap joints in sheet metals.
- Fabrication of sealed tanks such as gas tanks, oil containers, transformer tanks, and aircraft fuel tanks.
- Production of pipes and tubes with continuous leak-proof seams.
- Construction of ship fuel tanks, ballast tanks, and freshwater storage units, where water tightness is essential.
- Fabrication of marine HVAC ducts and exhaust systems.
- Assembly of pressure-tight chambers and hull sections in small vessels and submarines.
- Welding of marine pipelines used for onboard fuel and fluid transfer systems.

3.4.5 Projection Welding

Projection welding is a specialized form of resistance welding, similar to spot welding, but with the distinction that small projections or dimples are created on the workpieces at the desired weld points. During welding, the workpieces are held firmly between copper alloy electrodes, and a large amount of electric current is passed through the localized projections. These projections concentrate the current density at specific points, generating intense heat that melts the material at the dimples. Simultaneously, the applied electrode pressure flattens the projections, creating strong and consistent weld nuggets (Gould 1993). This process allows for precise localization of the weld and efficient heat distribution. Figure 3.10 illustrates the basic principle of projection welding. The workpieces to be joined are prepared with small projections on one of the metal sheets. These sheets are placed between flat copper alloy electrodes. When a large electric current (*I*) is passed through the

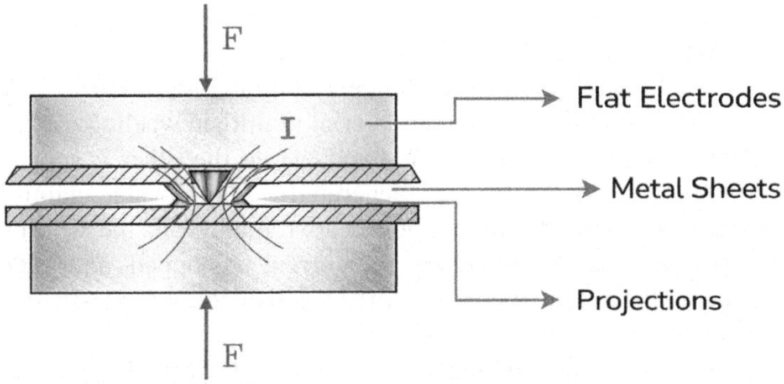

FIGURE 3.10 Principle of projection welding. (https://www.geeksforgeeks.org/electrical-engineering/electric-resistance-welding/.)

electrodes, the current concentrates at the projections due to their smaller contact area. This localized resistance heating melts the projections. At the same time, the electrode force (*F*) applied from both sides flattens the molten projections, resulting in a strong and localized weld nugget.

Advantages of Projection Welding:

- No limitation on the metal thickness; metals of varying thicknesses can be welded effectively.
- Multiple welds can be achieved simultaneously, improving productivity.
- Lower current and pressure requirements extend the life of electrodes.
- Provides good heat balance, reducing distortion.
- Produces welds with a neat and polished appearance.

Disadvantages of Projection Welding:

- The preparation of projections is complex and time-consuming.
- Requires highly skilled operators and careful setup.
- Not suitable for all types of workpieces or irregular geometries.
- Equipment for projection welding tends to be more expensive.

Applications of Projection Welding:

- Widely used for welding studs, nuts, and bolts onto plates.
- Extensively applied in shipbuilding, where large numbers of fasteners and fittings must be attached securely.
- Common in the automotive industry for joining components such as brackets and reinforcements.
- Used in the sheet metal industry for precise and efficient joining.
- Also employed in manufacturing appliances such as refrigerators, condensers, and radiators.

3.4.6 Flash Butt Welding

Flash butt welding (FBW) is a type of resistance welding widely used in the steel industry for joining tubes, rods, and other components. In this process, two workpieces are clamped firmly in electrode holders. A high pulsed current (up to about 100,000 amperes) is passed through the material to initiate welding.

In this setup, one electrode holder is fixed, while the other is movable. When current is supplied, sparks (flashes) are generated as the movable clamp is pressed against the fixed one. This flashing action heats the edges of the workpieces until they reach a plastic state. At this point, the current is stopped, and axial pressure is applied, causing the softened edges to fuse together through plastic deformation, forming a strong weld (Ichiyama and Kodama 2006).

Figure 3.11 shows the fundamental working principle of the FBW process, which is a resistance welding technique used to join two metal workpieces end to end. In the schematic, two workpieces labeled as the left workpiece and the right workpiece are positioned opposite each other with a small gap in between. When an electric

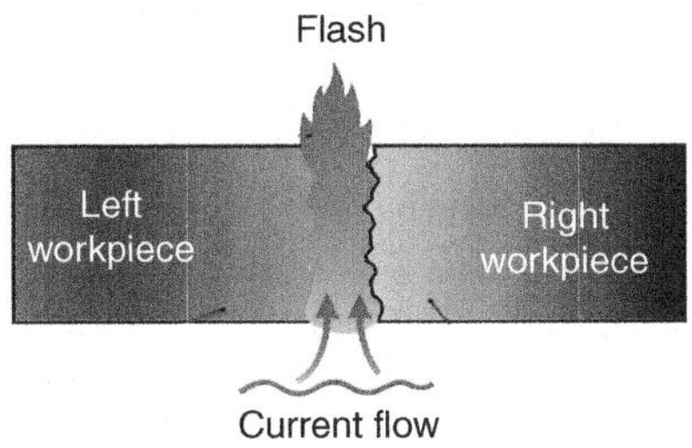

FIGURE 3.11 Working principle of flash butt welding. (https://www.geeksforgeeks.org/electrical-engineering/electric-resistance-welding/.)

current passes through the contact points, resistive heating occurs due to high electrical resistance at the joint interface. As the metal at the ends begins to melt and vaporize, a bright flash is produced, which removes surface oxides and contaminants.

The flashing stage continues until sufficient heat is generated throughout the interface. Then, the workpieces are rapidly brought together under axial pressure, causing forge welding and forming a solid joint. The color gradient from dark gray to bright orange indicates the temperature distribution, with the hottest zone at the flash region where the metal reaches its plastic or molten state. This process ensures a uniform, high-strength bond without filler material or flux.

Advantages of Flash Butt Welding:

- Faster welding process.
- No need for surface preparation.
- Lower power requirement compared to some other welding methods.
- Capable of joining dissimilar metals with different melting points.
- Cost-effective.
- Provides weld strength up to 100%.

Disadvantages of Flash Butt Welding:

- Higher risk of fire hazards due to flashing sparks.
- Loss of material during flashing and upsetting.
- Maintaining concentricity and straightness of workpieces can be challenging.

Applications of Flash Butt Welding:

- Welding of metal sheets, rods, bars, and fittings.
- Used in automotive and aircraft components.
- Found in household appliances, refrigerators, and agricultural tools.
- Applied in the marine industry for joining propeller shafts, anchor chains, and structural components of ships.

Comparison between resistance welding and electric arc welding and comparison between spot, seam, projection, and FBW are provided in Tables 3.4 and 3.5, respectively.

TABLE 3.4
Comparison Between Resistance Welding and Electric Arc Welding

Aspect	Resistance Welding (RW)	Electric Arc Welding (EAW)
Principle	Heat is generated by the **electrical resistance** of the workpieces to current flow	Heat is generated by an **electric arc** struck between electrode and workpiece
Heat source	Resistance of material + high current, low voltage	Electric arc with high temperature (up to ~6,000°C)
Current requirement	Very **high current** (up to 100,000 A), low voltage	**Moderate current** (50–600 A), higher voltage than resistance welding
Filler material	Usually not required (solid-state joining)	Often required (especially in SMAW, GMAW, etc.)
Surface preparation	Minimal preparation needed	Surface cleaning often required for quality welds
Type of weld	Localized welds (spots, seams, projections, flash)	Continuous weld beads or joints of various types
Automation	Easily automated, widely used in mass production	Automation possible but more complex (e.g., robotic arc welding)
Speed	Very fast process, suitable for thin sheets	Relatively slower than resistance welding
Strength of joint	High strength, comparable to base metal	Also strong, but depends on filler, technique, and heat input
Applications	Automotive bodies, aircraft parts, household appliances, marine shafts	Construction, pipelines, heavy machinery, pressure vessels, shipbuilding
Cost	Low cost per weld, but high initial machine cost	Lower equipment cost, but higher labor and filler costs

TABLE 3.5
Comparison Between Spot, Seam, Projection, and Flash Butt Welding

Aspect	Spot Welding	Seam Welding	Projection Welding	Flash Butt Welding
Principle	Two sheets are joined by localized heat and pressure at discrete spots	Similar to spot welding but uses roller electrodes to form a continuous weld seam	Localized projections/dimples concentrate current flow, creating welds at multiple points	Edges of two workpieces are brought together, flashed, and then forged under pressure
Electrodes	Two pointed or flat electrodes	Roller-type electrodes (copper wheels)	Flat electrodes pressing over projections	Heavy clamps/electrodes, one fixed and one movable

(Continued)

TABLE 3.5 *(Continued)*
Comparison Between Spot, Seam, Projection, and Flash Butt Welding

Aspect	Spot Welding	Seam Welding	Projection Welding	Flash Butt Welding
Current requirement	High current for short duration	Continuous current supply while rollers move	Current concentrated at projections	Very high pulsed current (up to 100,000 A)
Material loss	Negligible	Negligible	Small (at projections)	Some loss due to flashing and upsetting
Surface preparation	Not critical	Not critical	Not critical, but projections must be made beforehand	Not required
Weld strength	High at spots	Continuous weld, leak-proof, and strong	Strong welds at multiple points	Very high, up to 100% of base metal strength
Speed	Fast, suitable for mass production	Moderate, slower than spot but continuous	Fast for multiple welds in one cycle	Relatively slower due to flashing and upsetting
Applications	Automotive bodies, sheet metal assemblies	Fuel tanks, drums, radiators, aircraft tanks	Nuts, bolts, fasteners, wire mesh, cross members	Railway tracks, automotive axles, aircraft and marine shafts, rods, and pipes

3.5 FRICTION STIR WELDING (FSW)

3.5.1 Process Fundamentals

FSW is a solid-state joining process that uses a non-consumable rotating tool to join two facing workpieces without melting the material. Heat is generated by friction between the rotating tool and the workpieces, producing a softened region near the tool. As the tool traverses along the joint line, it mechanically stirs and forges the softened material under axial pressure, consolidating the weld. This mechanism is often compared to kneading clay or dough.

The welding tool consists of a shoulder and a pin (probe), where the probe penetrates the joint, while the shoulder provides frictional heat and containment. The probe is slightly shorter than the required weld depth, ensuring complete stirring without melting. As the tool moves forward, plasticized material is displaced from the leading edge to the trailing edge, consolidating into a defect-free weld (Mishra and Ma 2005).

Figure 3.12 illustrates the working principle of FSW. In this process, a non-consumable rotating tool with a shoulder and a probe is inserted into the joint line between two workpieces. As the tool rotates and moves forward, frictional heat is generated, which softens the material without melting it. The tool shoulder provides additional heat and forging action, while the probe stirs and intermixes the softened material. The joint has two distinct regions: the advancing side, where tool rotation

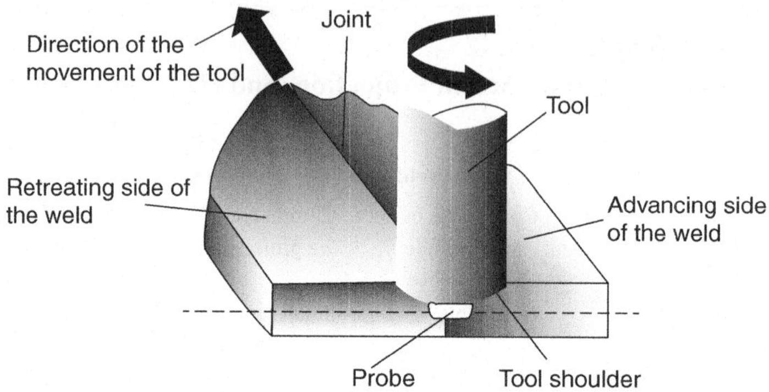

FIGURE 3.12 Schematic representation of friction stir welding process. (https://en.wikipedia.org/wiki/Friction_stir_welding.)

and travel directions coincide, and the retreating side, where they oppose each other. The softened material is displaced from the front of the tool to the rear, where it consolidates under pressure, forming a solid-state weld with a fine-grained microstructure and high strength.

In shipbuilding applications, this principle is particularly valuable for welding large aluminum structures such as hull panels, decks, bulkheads, and superstructures. The ability of FSW to produce long, defect-free welds with excellent strength and corrosion resistance makes it ideal for marine environments, where durability, lightweight construction, and resistance to seawater corrosion are critical.

The process produces severe plastic deformation and dynamic recrystallization, leading to a fine-grained weld microstructure with excellent mechanical properties.

Materials Joined: Aluminum alloys (primary), copper, magnesium, titanium, stainless steel, mild steel, and even polymers.

Dissimilar Metals: Successful joints have been demonstrated (e.g., aluminum-magnesium).

Key Applications: Shipbuilding, aerospace, railways, and automotive structures.

Historically, the concept of solid-state welding was patented in the Soviet Union by Yu. Klimenko in 1967, but it was not commercialized. The Welding Institute (TWI), UK, developed and patented FSW in 1991, leading to its widespread industrial adoption.

3.5.2 Applications in Aluminum Ship Structures

Aluminum alloys are extensively used in shipbuilding because of their light weight, high corrosion resistance, and fuel efficiency benefits. However, they are difficult to weld using fusion methods due to cracking and porosity. FSW has become a key enabling technology in the marine sector, with applications including:

- Hull panels (longitudinal and transverse joints).
- Deck structures and superstructures.

- Bulkheads and watertight panels, requiring leak-proof welds.
- Propeller housings, shafts, and casings, where precision and strength are critical.

  Shipyards across the world now employ FSW for producing large aluminum assemblies with improved durability and structural integrity.

3.5.3 Benefits for Strength and Corrosion Resistance

FSW offers multiple advantages over fusion welding in marine applications:

- **High mechanical strength** in the as-welded condition, often matching or exceeding base material strength.
- **Excellent corrosion resistance**, as welds are free from porosity, solidification cracking, or contamination.
- **No Consumables**: No filler material, shielding gas, or flux is needed.
- **Dimensional Stability**: Minimal distortion and residual stress, critical for large ship panels.
  - **Improved Safety**: No toxic fumes, spatter, or molten pool.
  - **Environmental Benefits**: Lower energy consumption and reduced CO_2 emissions.
  - **Good Weld Appearance**: Smooth surface, reducing the need for post-weld machining.

3.5.4 Challenges and Future Potential

Despite its advantages, FSW has some limitations:

- Exit hole left when the tool is withdrawn.
- Requirement of high axial force and rigid clamping, making the process less suitable for on-site or field welding.
- Limited flexibility for non-linear welds and varying thicknesses.
- Slower traverse rates than some fusion welding techniques (although fewer passes may be required).
- High initial equipment cost, particularly for large-scale marine panels.

3.5.4.1 Future Potential

- Development of advanced tool materials to extend tool life and enable welding of harder alloys.
- Wider adoption of robotic and automated FSW for complex marine and aerospace structures.
- Integration with hybrid processes (e.g., laser-assisted FSW) to increase speed and adaptability.
- Expansion to multi-material structures (e.g., aluminum steel for lightweight but strong ship superstructures).

3.6 PROCESS COMPARISON AND SELECTION GUIDELINES

3.6.1 Criteria for Choosing a Welding Process

The selection of an appropriate welding process in shipbuilding and offshore engineering depends on multiple factors such as the type of material, joint design, thickness of components, accessibility, and service conditions. Resistance welding methods like spot, seam, projection, and FBW are suitable for sheet metal, structural joints, and repetitive production, while solid-state processes like FSW are ideal for aluminum alloys that are sensitive to distortion and cracking. Arc-based techniques may be chosen for thick sections and high-strength steels where deeper penetration and flexibility are required. Considerations of safety, ease of automation, and compliance with classification society standards also play a key role.

3.6.2 Cost, Speed, and Quality Considerations

Cost-effectiveness is often evaluated in terms of material preparation, energy consumption, equipment setup, and labor requirements. For example, FBW is relatively inexpensive and fast for joining rods and rails but may lead to material loss during flashing. FSW, although requiring specialized tooling and machinery, ensures superior weld quality with reduced post-processing and minimal defects. Speed is another factor: resistance welding techniques allow rapid production with minimal preparation, while arc welding can be slower but offers greater flexibility. Quality aspects include mechanical strength, corrosion resistance, dimensional accuracy, and overall reliability of the joint in marine environments.

3.6.3 Case Studies from Shipbuilding and Offshore Projects

In modern shipbuilding, FSW has been extensively used for fabricating aluminum hulls and deck panels in naval and commercial vessels, ensuring lightweight structures with high corrosion resistance. FBW is applied in offshore platform construction for joining tubular steel members, where reliability and load-bearing capacity are critical. Spot and seam welding are widely employed in the fabrication of smaller ship components, storage tanks, and bulkheads. These case studies highlight that no single process is universally superior; instead, a combination of processes is selected based on the project's material requirements, economic considerations, and desired service life of the marine structure.

3.7 SUMMARY

The chapter highlights the wide range of welding techniques employed to meet the demanding requirements of shipbuilding and offshore engineering. Resistance welding methods such as spot, seam, projection, and FBW are discussed for their efficiency in joining sheets, rods, and structural components. FSW is emphasized as a modern solid-state process particularly suited for aluminum ship structures, offering high strength, minimal distortion, and superior corrosion resistance. Comparisons between resistance welding and electric arc welding underline the trade-offs in cost, speed, and weld quality, while selection guidelines provide practical insights into choosing the right process for marine applications. Case studies from shipbuilding and offshore projects demonstrate that the choice of welding method depends on material properties, structural design, and operational requirements, ensuring safe, durable, and cost-effective marine constructions.

3.8 REVIEW QUESTIONS

1. What are the primary requirements of welding processes in the marine industry?
2. Explain the principle of resistance welding.
3. Differentiate between resistance welding and electric arc welding.
4. What is spot welding, and where is it applied in shipbuilding?
5. Describe the working principle of seam welding.
6. Explain projection welding and its significance in marine fabrication.
7. How does flash butt welding differ from other resistance welding processes?
8. Discuss the advantages and limitations of resistance welding in marine structures.
9. What are the major applications of electric arc welding in shipbuilding?
10. Compare resistance welding processes in terms of cost, speed, and weld quality.
11. Explain the working principle of friction stir welding (FSW).
12. What role do the tool shoulder and probe play in FSW?
13. Why is FSW particularly suitable for aluminum alloys in ship structures?
14. Discuss the benefits of FSW in terms of strength and corrosion resistance.
15. What are the main disadvantages of friction stir welding?
16. Explain the application of FSW in joining dissimilar materials.
17. Provide examples of FSW applications in marine, aerospace, and rail industries.
18. Why does FSW leave an exit hole, and how is it managed in practice?
19. How does dynamic recrystallization occur during FSW?
20. Compare FSW with fusion welding processes.
21. What safety advantages does FSW have over arc welding?
22. How does clamping affect the success of FSW in shipbuilding?
23. What are the key selection criteria for choosing a welding process in marine applications?
24. Explain how cost considerations influence the choice of welding process.
25. What factors determine welding speed in large ship structures?
26. How does welding quality impact long-term performance in marine environments?
27. Provide a case study example of resistance welding used in shipbuilding.
28. Provide a case study example of FSW used in shipbuilding or offshore projects.
29. How does corrosion resistance influence welding process selection in marine industry?
30. Summarize the future potential of FSW in shipbuilding.

4 Metallurgical Aspects of Marine Welding

4.1 INTRODUCTION

Marine welding is a highly specialized field where metallurgical knowledge plays a vital role in ensuring structural integrity, durability, and safety of vessels and offshore structures. Unlike general welding practices, marine welding must account for the corrosive seawater environment, cyclic loading due to waves, and the demanding operational conditions ships and offshore platforms endure (Ibarra et al. 1988). A strong understanding of metallurgy helps in selecting suitable materials, controlling weld microstructure, and minimizing defects that could lead to premature failure.

4.1.1 Importance of Metallurgy in Marine Welding

Metallurgy is fundamental to marine welding, ensuring the structural integrity, longevity, and safety of marine vessels and offshore structures. The specific challenges posed by the marine environment, including constant exposure to seawater, corrosive ions, and dynamic loads, necessitate a deep understanding of how welding affects a material's properties. Without proper metallurgical control, welded joints can become the weakest point of a structure, leading to premature failure, catastrophic accidents, and costly repairs.

Key metallurgical considerations include:

- **Microstructural Changes**: The rapid heating and cooling cycles of welding can cause unfavorable changes in the microstructure of a metal, particularly in the heat-affected zone (HAZ), that reduce its mechanical strength and toughness.
- **Corrosion Resistance**: Welding can create galvanic differences between the weld metal (WM) and base metal (BM) that accelerate corrosion in a corrosive seawater environment.
- **Weld Defects**: Specialized techniques are required to prevent defects like porosity, inclusions, and hydrogen-induced cracking (HIC), which are exacerbated by the unique conditions of marine welding.
- **Mechanical Properties**: The weld and surrounding HAZ must meet or exceed the required mechanical properties, such as tensile strength, hardness, and fracture toughness, to withstand demanding service conditions.

DOI: 10.1201/9781003217350-4

4.1.2 Effect of Marine Environment on Welded Joints

The marine environment subjects welded joints to a variety of harsh conditions that directly impact their metallurgical behavior and performance.

- **Corrosive Medium**: Seawater is a highly aggressive, conductive electrolyte rich in chloride ions that accelerates corrosion. The difference in composition, microstructure, and residual stress between the WM, HAZ, and BM can create localized corrosion cells, a phenomenon known as galvanic corrosion.
- **Rapid Cooling**: In underwater welding, the surrounding water acts as a quenching medium, causing the weld and HAZ to cool extremely rapidly. This can result in the formation of brittle phases like martensite, which increase hardness but severely reduce ductility and fracture toughness.
- **Hydrogen Embrittlement**: Hydrogen is a significant threat in underwater welding. The high temperatures of the welding arc can dissociate water into its component hydrogen and oxygen. The hydrogen can then diffuse into the WM and HAZ, especially in high-strength steels. Trapped hydrogen can lead to cold cracking, a major concern that can occur hours or even days after welding.
- **Biofouling**: Microorganisms and other marine life attach to and colonize submerged surfaces, including welded joints. These biofilms can create differential aeration cells, accelerating corrosion in oxygen-deprived areas.
- **Loading and Stress**: Structures endure cyclic loading from waves and currents, which, combined with the residual stresses from welding, can contribute to stress corrosion cracking (SCC) and fatigue failure.

4.1.3 Overview of Common Marine Welding Materials

The selection of materials for marine applications is guided by their strength, corrosion resistance, and weldability in demanding environments.

- **Carbon and Low-Alloy Steels**: Due to their low cost and good mechanical properties, these are the most widely used materials for shipbuilding and offshore structures. In welding, they are prone to HIC, especially in underwater applications. Proper electrode selection and procedural controls are essential to minimize risk.
- **High-Strength Low-Alloy (HSLA) Steels**: Used for applications requiring a high strength-to-weight ratio, these steels are more susceptible to hydrogen cracking than their lower-strength counterparts. Welding them requires strict control of hydrogen levels and cooling rates to maintain good toughness.
- **Stainless Steels**: Valued for their exceptional corrosion resistance in marine environments, austenitic (e.g., 304, 316) and duplex stainless steels are commonly used. Welding stainless steel requires careful control of heat input

to prevent sensitization, a metallurgical change that can reduce corrosion resistance.

- **Aluminum Alloys**: Known for their light weight and corrosion resistance, aluminum alloys (especially the 5xxx and 6xxx series) are used for high-speed vessels and superstructures. Their high thermal conductivity and propensity for porosity make them challenging to weld. Techniques like friction stir welding (FSW) are sometimes used to manage heat input.
- **Titanium Alloys**: These alloys offer superior corrosion resistance and a high strength-to-weight ratio for specialized marine equipment and piping systems. However, their high reactivity requires inert gas shielding to prevent oxygen and nitrogen contamination during welding, which can lead to embrittlement.
- **Copper-Nickel Alloys**: Used in seawater piping and heat exchangers for their excellent resistance to biofouling and corrosion. Welding them requires careful control of heat input and dilution to avoid hot cracking and maintain properties.

4.2 HEAT-AFFECTED ZONE (HAZ) IN MARINE ALLOYS

4.2.1 Definition and Metallurgical Changes in HAZ

The HAZ is the area of BM surrounding a weld that has not melted but has been subjected to high temperatures during the welding process. This thermal cycle significantly alters the original microstructure and properties of the metal. Key metallurgical changes include (Adil and Bhole 1992):

- **Grain Growth**: Closest to the weld, where temperatures are highest, grains can coarsen. This larger grain structure reduces the material's strength and toughness.
- **Phase Transformations**: In some alloys, the thermal cycle can trigger phase changes. In steels, for instance, rapid cooling can create hard, brittle phases like martensite. In alloys strengthened by precipitation, the thermal cycle can dissolve or alter strengthening precipitates.
- **Residual Stresses**: Non-uniform heating and cooling during welding cause thermal expansion and contraction, which can lock in residual stresses in the HAZ. These stresses increase the risk of cracking and can reduce fatigue life.
- **Precipitation and Dissolution**: Certain alloys, such as stainless steels and some aluminum alloys, experience the precipitation or dissolution of specific phases that can degrade corrosion resistance and mechanical properties.

4.2.2 HAZ in Carbon and Low-Alloy Steels

In marine welding of carbon and low-alloy steels, the HAZ is prone to two primary issues, especially in high-strength variants:

- **Hydrogen-Induced Cracking (HIC)**: The introduction of moisture during underwater or wet welding processes increases the risk of HIC, also known as cold cracking. Atomic hydrogen is absorbed into the molten WM and diffuses into the HAZ upon cooling, where it concentrates in areas of high stress. The presence of hard, crack-sensitive microstructures, like martensite, exacerbates this risk, and cracking can be delayed for hours or days after welding.

- **Reduced Toughness**: The rapid cooling rates experienced in underwater environments can lead to the formation of brittle phases and the coarsening of grains in the HAZ. This significantly lowers the material's fracture toughness, a critical property for marine structures that must resist impact and fatigue loads. The coarse-grained HAZ (CGHAZ) is often the most susceptible to this embrittlement.

4.2.3 HAZ in Aluminum and Stainless Steels

- **Aluminum Alloys**: In high-strength aluminum alloys (e.g., 5xxx and 6xxx series), the HAZ can become the weakest part of the welded joint. Heat from welding causes the dissolution of strengthening precipitates or anneals the strain-hardened material, leading to a loss of strength. The higher the heat input, the softer and weaker the HAZ becomes.
- **Stainless Steels**: For austenitic and duplex stainless steels, HAZ degradation primarily manifests as a loss of corrosion resistance:
- **Sensitization**: Welding temperatures (typically between 400°C and 800°C) can cause chromium carbides to precipitate at the grain boundaries in the HAZ. This depletes the adjacent areas of chromium, making the material susceptible to intergranular corrosion. Low-carbon "L" grades, like 304L and 316L, and stabilized grades (321, 347) are designed to mitigate this issue.
- **Phase Imbalance**: In duplex stainless steels, a balanced austenite-ferrite microstructure is critical for optimal corrosion and mechanical properties. The rapid cooling associated with marine welding, especially underwater, can result in excessive ferrite content in the HAZ, which disrupts this balance and compromises both corrosion resistance and toughness.

4.2.4 Minimizing HAZ Degradation in Marine Fabrication

To mitigate the negative metallurgical effects in the HAZ of marine alloys, fabricators can use several strategies:

- **Control Heat Input**: Reducing the heat input minimizes the size of the HAZ and the degree of microstructural change. This can be achieved by using low-energy welding processes (like TIG or pulsed MIG), faster travel speeds, or smaller weld beads.
- **Filler Metal Selection**: Using specific filler metals can help overcome HAZ weaknesses. For high-strength aluminum alloys, selecting a filler with a higher solute content can improve as-welded strength. For stainless steels, using an appropriate filler metal (like ER2209 for duplex) helps restore the austenite-ferrite balance in the WM and HAZ.
- **Preheating and Post-Weld Heat Treatments**: In dry welding, preheating can slow the cooling rate to prevent the formation of hard, brittle

microstructures in steels and reduce residual stresses. Post-weld heat treatment (PWHT), when feasible, can relieve residual stresses, temper hardened microstructures, and restore properties, though it is often impractical for large marine structures. In steels, special "temper bead" techniques can achieve a similar effect by normalizing the HAZ of a previous pass.
- **Process Optimization**: Techniques like FSW for aluminum, or local cavity welding methods for underwater stainless steel repairs, minimize heat input and provide better control over the HAZ microstructure.
- **Material Selection**: When possible, selecting materials with inherent resistance to HAZ degradation, such as low-carbon stainless steels ("L" grades) or thermomechanically controlled processed (TMCP) steels, is the most straightforward solution.

4.3 WELDABILITY OF HIGH-STRENGTH STEELS

4.3.1 Types of High-Strength Steels in Marine Applications

High-strength steels are extensively used in marine construction for their superior strength-to-weight ratio, allowing for lighter, more efficient structures. However, their weldability is complex due to the precise microstructures that give them their strength (Billingham et al. 2003).

- **Thermomechanically Controlled Processed (TMCP) Steels**: These steels, including common grades like AH36, DH36, EH36, and FH36, are processed to achieve a fine, grain-refined microstructure with high toughness and good weldability. Their weld properties are sensitive to heat input, which must be controlled to prevent grain coarsening and softening in the HAZ.
- **Quenched and Tempered (Q&T) Steels**: These steels (e.g., S690QL, HY 130/150) are heat-treated to produce a hard, martensitic microstructure. While very strong, this microstructure is susceptible to HIC and HAZ softening during welding.
- **Ultra-High-Strength Steels (UHSS)**: A broad category, UHSS can include multi-phase steels (dual-phase, TRIP steels) and hardenable steels with complex chemistries. They pose significant challenges related to cracking and excessive HAZ softening due to the high sensitivity of their microstructures.

4.3.2 Factors Influencing Weldability (Preheat, Cooling Rate)

The unique properties of high-strength steels mean that welding parameters must be precisely controlled to avoid defects and degradation.

- **Preheat Temperature**: Applying heat to the BM before welding serves several critical functions:
  - **Slowing the Cooling Rate**: This prevents the formation of hard, brittle microstructures like martensite and promotes softer, more ductile phases like bainite.

  - **Hydrogen Removal**: Preheat allows more time for diffusible hydrogen to escape from the weld and HAZ, significantly reducing the risk of HIC.
- **Cooling Rate**: The cooling rate after welding is one of the most critical factors influencing the microstructure of the weld and HAZ.
  - **Underwater Welding**: In wet welding, the chilling effect of water causes extremely rapid cooling, leading to hard, brittle microstructures with poor toughness. In hyperbaric (dry) underwater welding, the controlled gaseous environment allows for much slower cooling and higher quality welds.
  - **Controlling Cooling**: For surface welding, controlling heat input, interpass temperature, and using appropriate preheating are essential for managing the cooling rate.

4.3.3 Controlling Microstructural Changes for Toughness

Maximizing the toughness of high-strength steel welds involves manipulating the microstructure to avoid brittle phases and stress concentrators.

- **Acicular Ferrite**: This fine, needle-like ferrite microstructure is highly desirable in steel WM because it provides both high strength and toughness by hindering crack propagation. Welding consumables are formulated to promote their formation.
- **Refining Grain Size**: The finer the grain structure, the better the toughness. TMCP steels are designed with a fine-grain structure, and welding procedures for these materials must be carefully controlled to avoid grain coarsening in the HAZ.
- **Preventing Hydrogen Embrittlement**: This is the most crucial aspect of welding high-strength steels, particularly in marine environments. Strategies include:
  - Using low-hydrogen welding consumables.
  - Employing dry welding techniques where possible.
  - Using preheat to allow hydrogen to diffuse away from the weld joint.
- **Avoiding Soft Zones**: In Q&T steels, the welding thermal cycle can cause tempering of the martensitic structure in the subcritical HAZ, creating a soft zone with reduced strength. Weld procedures must be optimized to limit this effect, which can be done by carefully controlling heat input.
- **Tempering**: Some welding procedures for Q&T steels use PWHTs or temper bead techniques to temper the hardened HAZ, reducing its hardness and improving toughness.

4.4 WELDABILITY OF CORROSION-RESISTANT ALLOYS

4.4.1 Stainless Steels (Austenitic, Duplex, Super Duplex)

Stainless steels are valued in marine environments for their inherent corrosion resistance, but maintaining this property during welding requires careful metallurgical control.

Austenitic Stainless Steels: The primary challenge is sensitization, which occurs when standard grades like 304 and 316 are heated into a temperature range (450°C–850°C), such as during welding. This causes chromium carbides to precipitate along grain boundaries, depleting the surrounding zones of chromium. These depleted areas become susceptible to intergranular corrosion. To mitigate this, low-carbon grades (e.g., 316L) minimize the carbon available for carbide formation, while stabilized grades (e.g., 321, 347) add elements like titanium or niobium to form more stable carbides.

Duplex and Super Duplex Stainless Steels: These alloys offer superior strength and chloride SCC resistance by balancing austenitic and ferritic phases. However, the heat from welding can disrupt this delicate balance. Rapid cooling in marine conditions can lead to an excessively high ferrite content in the HAZ, which compromises toughness and corrosion resistance. Conversely, slow cooling or excessive heat input can lead to the formation of brittle intermetallic phases, like sigma phase, which severely degrade mechanical properties and corrosion performance. Nitrogen-containing shielding and backing gases are used to promote austenite reformation during cooling and compensate for nitrogen loss.

4.4.2 Copper-Nickel and Titanium Alloys

These non-ferrous alloys present distinct challenges during welding due to their specific metallurgical properties.

Copper-Nickel (Cu-Ni) Alloys: Used for seawater piping and heat exchangers due to excellent resistance to corrosion and biofouling. The main challenge is controlling contamination from elements like sulfur, phosphorus, and lead, which can cause micro-fissuring and porosity. Rigorous pre-weld cleaning and dedicated, clean fabrication spaces are crucial. Because Cu-Ni alloys contain strong deoxidizers like manganese and titanium, they can be welded using a single-filler approach, with 70–30 Cu-Ni consumables typically recommended for both 90–10 and 70–30 base materials.

Titanium Alloys: Offer an exceptional strength-to-weight ratio and corrosion resistance for advanced marine applications. The primary difficulty is titanium's high reactivity with atmospheric gases (oxygen, nitrogen, hydrogen) at welding temperatures. This can lead to contamination and embrittlement of the weld. To prevent this, welding must be performed in a highly controlled, inert gas atmosphere using extensive shielding, purging, and sometimes vacuum chambers. Weld coloration is a simple indicator of potential contamination, with discolored welds requiring re-passivation or removal.

4.4.3 Special Considerations in Welding Non-Ferrous Marine Alloys

A variety of factors must be considered across non-ferrous alloys to ensure weld integrity in marine environments.

High Thermal Conductivity: Many non-ferrous alloys, including aluminum and copper alloys, require high heat input during fusion welding. This can lead to increased distortion and larger HAZs, which can reduce the strength of age-hardened alloys.

High Thermal Expansion: The high thermal expansion of non-ferrous alloys like aluminum, often double that of steel, can cause significant distortion and high residual stresses. Proper jigging, balanced welding sequences, and controlled heat input are used to manage this effect.

Filler Metal Selection: For welding dissimilar alloys, a compatible filler metal is crucial to accommodate variations in thermal properties and potential dilution effects. For example, when welding Cu-Ni to steel, a high-nickel filler can better tolerate iron dilution without cracking.

Solid-State Welding: Techniques like FSW are effective alternatives for non-ferrous alloys, as they join materials in the solid state. This avoids fusion-related defects like solidification cracking and minimizes HAZ degradation, particularly in heat-treated aluminum alloys.

Porosity Control: Many non-ferrous alloys are prone to porosity due to the high solubility of gases (like hydrogen) in the molten state. Mitigating this requires careful control of shielding gases, filler metal additions, and thorough cleaning of joint surfaces to remove contaminants and moisture.

4.5 HYDROGEN-INDUCED CRACKING (HIC)

4.5.1 Mechanism of Hydrogen Embrittlement in Marine Steels

HIC, also known as cold cracking or delayed cracking, occurs when three factors are simultaneously present: hydrogen, a susceptible microstructure, and residual tensile stress. Atomic hydrogen, being extremely small and mobile, diffuses into the steel's crystal lattice. It collects at imperfections or trap sites within the material, such as inclusions, voids, and grain boundaries, where it can recombine into molecular hydrogen. As molecular hydrogen is much larger than its atomic counterpart, this recombination creates immense pressure, building localized stress that can initiate cracking. High-strength steels and those with hard microstructures (like martensite) are more susceptible because they are less ductile and contain higher residual stress. In marine environments, HIC can also be exacerbated by galvanic corrosion, where the WM acts as an anode, promoting hydrogen infusion into the BM.

4.5.2 Sources of Hydrogen During Welding

In marine welding, hydrogen can be introduced from numerous sources, particularly in the wet and humid conditions of the environment.

Moisture: In underwater wet welding, the welding arc causes water molecules to dissociate into hydrogen and oxygen. This hydrogen can then be absorbed into the molten weld pool. The surrounding humidity in topside welding environments can also be a significant source.

Contaminants: Organic materials like oil, grease, paint, rust, and dirt on the BM or filler wire introduce hydrogen into the weld pool.

Consumables: Improperly stored or baked flux-coated electrodes and submerged arc welding (SAW) fluxes are a major source of hydrogen if they have absorbed moisture from the atmosphere.

4.5.3 Prevention and Control Techniques

The key to preventing HIC is to eliminate or control one or more of the three necessary factors (hydrogen, susceptible microstructure, and stress).

4.5.3.1 Hydrogen Control

Use low-hydrogen consumables, such as basic flux electrodes (properly baked and stored in sealed packages or ovens) or specific metal-cored and flux-cored wires.

Ensure proper joint preparation, including removing all contaminants like oil, grease, rust, and paint.

In wet welding, dry hyperbaric techniques, which isolate the welding area from the water, are the most effective way to eliminate hydrogen.

4.5.3.2 Microstructure Control

Preheat and Interpass Temperature: Preheating the BM and maintaining appropriate interpass temperatures slow the cooling rate of the weld and HAZ. This promotes the formation of softer, more ductile microstructures (like bainite instead of martensite) and allows more time for hydrogen to diffuse out of the weldment.

Post-Heating and PWHT: Applying heat immediatcly after welding (post-heating) or later (PWHT) can relieve residual stresses and further aid hydrogen diffusion.

4.5.3.3 Stress Control

Welding Sequence and Fit-Up: Carefully planned welding sequences can minimize the buildup of residual stresses. Proper fit-up is also important, as excessive root gaps or misalignment can increase stress concentration.

Temper Bead Technique: This specialized welding method uses subsequent weld passes to temper the underlying WM and HAZ, reducing hardness and associated stress.

4.6 STRESS CORROSION CRACKING (SCC) IN MARINE WELDS

4.6.1 Mechanism and Contributing Factors

SCC is a brittle fracture mechanism that occurs due to a specific combination of three factors: a susceptible material, a corrosive environment, and tensile stress. In marine welds, SCC can be an especially insidious problem because the weldment itself introduces metallurgical and geometric features that exacerbate the risk.

Susceptible Material: Some alloy systems are more prone to SCC than others in a specific environment. For marine welds, this includes certain stainless steels (especially austenitic grades), some high-strength aluminum alloys, and specific copper alloys.

Corrosive Environment: The marine environment, with its high concentration of chloride ions, is a potent promoter of SCC. Localized corrosion cells, such as those that form in crevices or at metallurgical interfaces, can also intensify the corrosive effects.

Tensile Stress: The stress can be either externally applied (e.g., operational loads, residual welding stresses) or internally generated. Welding is a primary source of residual tensile stress, and these stresses are often concentrated in the weld bead or the HAZ, creating ideal conditions for SCC initiation.

4.6.2 SCC in Stainless Steels and Aluminum Alloys

Stainless Steels: Austenitic stainless steels, particularly those with higher nickel content, are susceptible to chloride-induced SCC, especially at temperatures above 60°C. The residual tensile stresses from welding can act as the primary stress driver. Duplex and super duplex stainless steels, with their mixed austenite-ferrite microstructure, are much more resistant to SCC due to the presence of the ferrite phase. However, welding procedures must still be controlled to prevent sensitization or the formation of brittle phases that could create a susceptible metallurgical condition.

Aluminum Alloys: Certain high-strength, heat-treatable aluminum alloys (e.g., some 7xxx series) are susceptible to SCC in marine environments. The welding process can alter the microstructure in the HAZ, particularly in age-hardened alloys, leading to localized areas of reduced strength and increased susceptibility to cracking.

4.6.3 Mitigation Strategies for Marine Structures

To combat SCC in marine welded structures, a multi-pronged approach targeting one or more of the three key factors is necessary.

4.6.3.1 Material Selection

Choosing materials with inherent resistance to SCC is the most effective strategy. This includes using duplex or super duplex stainless steels instead of standard austenitic for many applications.

4.6.3.2 Stress Management

Post-Weld Heat Treatment (PWHT): Where feasible, PWHT can significantly reduce or eliminate the residual stresses from welding, thereby mitigating the risk of SCC.

Welding Procedure Control: Optimized welding sequences, minimizing weld passes, and proper fit-up can help reduce the magnitude of residual stresses.

Surface Peening: Shot or laser peening can induce a layer of compressive stress on the weld surface, which counteracts the harmful tensile stresses and can improve SCC resistance.

4.6.3.3 Environmental Control

Cathodic Protection: Applying cathodic protection can make the WM more resistant to corrosion, inhibiting the electrochemical reactions that drive SCC.

Protective Coatings: Applying specialized coatings can act as a physical barrier between the corrosive marine environment and the susceptible material.

Procedural Control: Techniques to limit corrosion, such as limiting crevices and avoiding overly acidic or alkaline solutions, also contribute to overall SCC resistance.

4.7 METALLURGICAL TESTING AND EVALUATION

4.7.1 Microstructural Analysis of Weld Zones

Metallurgical testing is a crucial step in assessing the quality and integrity of marine welds. Microstructural analysis uses microscopy to examine the various zones of a welded joint: the WM, the HAZ, and the unaffected BM.

Optical Microscopy (OM): This is a primary tool for initial evaluation, revealing the overall grain structure, any grain coarsening in the HAZ, and the presence of defects like porosity or inclusions. It can identify the different phases present in alloys like duplex stainless steels and confirm that they are within the specified limits.

Scanning Electron Microscopy (SEM): For more detailed analysis, SEM provides higher magnification and allows for examination of fine microstructures, fracture surfaces, and identification of precipitates. SEM is particularly useful for investigating the root cause of failures, such as fatigue or brittle fracture.

Energy-Dispersive X-ray Spectroscopy (EDS): Used in conjunction with SEM, EDS provides a chemical analysis of specific areas, confirming the composition of phases or precipitates. This can confirm, for example, the formation of chromium carbides in sensitized stainless steels.

X-Ray Diffraction (XRD): For multi-phase alloys like duplex stainless steels, XRD can quantitatively determine the volume fraction of different phases (austenite and ferrite). This ensures that the weldment meets the required phase balance, which is critical for toughness and corrosion resistance.

4.7.2 Hardness and Toughness Testing in Marine Welds

Assessing the mechanical properties of a weldment is vital for ensuring it can withstand the demanding conditions of the marine environment.

4.7.2.1 Hardness Testing

Vickers Hardness (HV): This method is commonly used to measure the hardness across different regions of the weldment. Hardness traverses across the WM, HAZ, and BM can reveal areas of excessive hardening (e.g., martensite in steel) or softening (e.g., over-tempering in high-strength steels). High hardness in the HAZ is often associated with brittle microstructures that are more susceptible to cracking.

4.7.2.2 Toughness Testing

Charpy V-Notch Impact Testing: This standard test measures the material's resistance to brittle fracture. It is particularly important for marine structures, which must withstand impact loads and operate in cold temperatures. Charpy specimens are often notched in specific locations (WM, HAZ, fusion line) to test the toughness of different zones.

Fracture Toughness Testing: For critical components, fracture toughness testing provides a quantitative measure of a material's resistance to crack propagation, giving a more precise assessment of its structural integrity.

4.7.3 Corrosion Testing in Simulated Seawater

To confirm the weldment's corrosion performance, specialized tests are conducted under controlled conditions.

Potentiodynamic Polarization: This electrochemical test measures the corrosion potential and corrosion current density of the weldment in simulated seawater. It can reveal differences in corrosion behavior between the WM, HAZ, and BM, which can indicate a risk of galvanic corrosion.

Electrochemical Impedance Spectroscopy (EIS): EIS can provide further details on the corrosion mechanisms at the weldment surface, including the resistance of passive films and the kinetics of corrosion reactions.

Simulated Immersion Tests: Immersion tests in standard saltwater solutions (e.g., ASTM B117) are used to evaluate the long-term corrosion performance and observe pitting, crevice corrosion, and other localized corrosion forms. For stainless steels, tests in ferric chloride solution can be used to assess pitting and crevice corrosion resistance.

Stress Corrosion Cracking (SCC) Tests: For materials susceptible to SCC, specialized tests like the slow strain rate test (SSRT) or tests involving pre-cracked specimens are conducted in simulated marine environments to evaluate their resistance to cracking under tensile stress.

4.8 SUMMARY

This chapter on *Metallurgical Aspects of Marine Welding* emphasizes the critical role of metallurgy in ensuring the strength, durability, and safety of welded joints in ships and offshore structures. It explains how the welding process alters the microstructure of the WM and HAZ, directly influencing mechanical properties and corrosion resistance. The chapter highlights the severe challenges posed by the marine environment, including saltwater corrosion, cyclic loading, and temperature variations, which accelerate fatigue and material degradation. A review of commonly used marine welding materials, such as steels, stainless steels, and aluminum alloys, is provided, outlining their suitability for different applications.

4.9 REVIEW QUESTIONS

1. Why is metallurgy important in marine welding applications?
2. How does the marine environment influence the performance of welded joints?
3. List common materials used in marine welding and their key properties.
4. Define the heat-affected zone (HAZ) in welded joints.
5. What metallurgical changes typically occur in the HAZ during welding?
6. How does the HAZ behave differently in carbon steels compared to low-alloy steels?
7. What challenges are associated with the HAZ in aluminum and stainless steels?
8. What techniques can be applied to minimize HAZ degradation in marine structures?
9. Name the main types of high-strength steels used in marine applications.
10. How does preheating affect the weldability of high-strength steels?
11. Why is cooling rate an important factor in marine welding metallurgy?
12. What microstructural changes reduce toughness in high-strength steels during welding?
13. What measures can improve weldability while maintaining strength and toughness?
14. Compare the weldability of austenitic, duplex, and super duplex stainless steels.
15. Why are copper-nickel alloys important in marine structures?
16. What welding challenges are associated with titanium alloys?
17. What special considerations are required when welding non-ferrous marine alloys?
18. Explain the mechanism of hydrogen embrittlement in marine steels.
19. What are the main sources of hydrogen during welding processes?
20. How can preheating help prevent hydrogen-induced cracking?
21. What are the most effective techniques for controlling hydrogen in welds?
22. What is the basic mechanism of stress corrosion cracking (SCC)?
23. Identify key environmental and metallurgical factors that contribute to SCC.
24. How does SCC manifest in stainless steel welds in marine environments?
25. What makes aluminum alloys susceptible to SCC?
26. List three strategies to mitigate SCC in marine structures.
27. Why is microstructural analysis important in evaluating marine welds?
28. How do hardness and toughness tests help assess weld quality?
29. What methods are used for corrosion testing of welds in simulated seawater?
30. How does metallurgical evaluation contribute to improving marine welding practices?

5 Robotic and Automated Welding in Shipbuilding

5.1 INTRODUCTION

Robotic and automated welding has emerged as a transformative technology in modern shipbuilding, addressing the growing demand for precision, efficiency, and cost-effectiveness in fabricating large and complex marine structures. Traditional manual welding, though flexible, is often limited by human fatigue, inconsistent quality, and slower production rates. In contrast, robotic and automated systems provide consistent weld quality, improved accuracy, and the ability to operate continuously with minimal errors. These technologies are particularly valuable in shipyards, where large volumes of repetitive welds are required on steel and aluminum structures under strict safety and performance standards. Automation also enhances worker safety by reducing human exposure to hazardous fumes, heat, and confined spaces (Lee et al. 2011). As shipbuilding evolves toward digitalization and Industry 4.0 practices, robotic welding is increasingly integrated with advanced sensors, real-time monitoring, and adaptive control systems, making it a key enabler of smart shipyard operations and global competitiveness.

5.1.1 Evolution of Welding Automation in Shipbuilding

The evolution of welding automation in shipbuilding reflects the broader advancement of welding technology in heavy industries. In the early 20th century, shipbuilding primarily relied on, which were labor-intensive and time-consuming. By the mid-20th century, shielded metal arc welding (SMAW) and submerged arc welding (SAW) became standard practices, offering stronger joints and enabling the construction of larger steel ships. Semi-automated welding machines soon followed, particularly in the 1960s and 1970s, to improve productivity in repetitive tasks such as panel welding (Lee et al. 2011).

The introduction of computer numerical control (CNC) systems in the late 20th century marked a turning point, enabling precise mechanization and integration of welding with shipyard production lines. In the 1980s and 1990s, major shipbuilding nations like Japan and South Korea pioneered the adoption of robotic welding arms for tasks such as block assembly and hull fabrication, significantly reducing production time. For example, Hyundai Heavy Industries (HHI) in South Korea implemented robotic panel welding lines that increased throughput and minimized defects in large-scale ship sections.

Today, shipyards use advanced robotic welding systems integrated with real-time sensors, artificial intelligence (AI), and adaptive control technologies. These systems not only ensure consistent weld quality across large and complex structures but also

DOI: 10.1201/9781003217350-5

allow automation in challenging environments such as confined spaces and curved hull sections. The evolution demonstrates a shift from purely manual methods to digitally driven, intelligent robotic systems, making welding automation a cornerstone of modern "smart shipyards."

5.1.2 Benefits and Drivers for Automation in Marine Fabrication

Automation in marine welding offers significant advantages that address both technical and economic challenges in shipbuilding. One of the primary benefits is consistent weld quality, as robotic systems minimize human error, ensuring uniform strength and reducing defects such as porosity and incomplete fusion. Automation also leads to higher productivity, since robotic systems can operate continuously with fewer interruptions, achieving faster production cycles compared to manual welding. Another benefit is improved safety, as workers are less exposed to hazards like toxic fumes, ultraviolet radiation, heat, and confined spaces. Additionally, automated systems reduce rework and material wastage, which translates to lower overall costs.

The drivers pushing shipyards toward welding automation include the growing complexity of ship structures, which demand precise and reliable joints; stricter international quality and safety standards, requiring defect-free welds; and the shortage of skilled welders in many industrial regions. Furthermore, global competition and pressure for shorter delivery schedules compel shipyards to adopt technologies that increase efficiency. Rising labor costs in advanced shipbuilding nations also serve as a strong economic driver. Together, these factors highlight automation not just as an option but as a necessity for modern shipyards to achieve higher quality, productivity, and competitiveness.

5.1.3 Overview of Global Trends in Shipyard Robotics

Globally, shipbuilding nations are rapidly embracing robotics and automation to meet increasing demands for cost efficiency, safety, and high-quality construction. South Korea, Japan, and China lead the way, with large shipyards such as HHI, Mitsubishi Heavy Industries, and COSCO Shipyard employing thousands of industrial robots for panel welding, hull assembly, and block construction. These countries integrate welding automation with digital shipyard systems, where robotics, computer-aided design (CAD), and computer-aided manufacturing (CAM) are seamlessly linked for precision and reduced cycle times.

In Europe, shipyards in countries like Germany, Norway, and Finland are focusing on automation for specialized vessels, such as cruise ships and offshore platforms. Their emphasis lies on flexible robotic welding systems capable of handling complex geometries and customized designs. Meanwhile, U.S. shipyards are investing in robotic welding for naval and defense applications, with a focus on reliability and quality assurance.

Recent global trends include the adoption of collaborative robots (cobots) that work alongside human welders, adaptive welding systems with AI-based real-time parameter adjustments, and the use of augmented reality (AR) and digital twins for monitoring and simulation. The growing push toward Industry 4.0 and smart shipyards is transforming welding robotics into interconnected systems with predictive maintenance, advanced sensors, and data-driven optimization. These trends indicate a clear trajectory toward increased automation, ensuring competitiveness and sustainability in the global shipbuilding industry.

5.2 ROLE OF AUTOMATION IN SHIPYARD WELDING

Automation plays a pivotal role in modern shipyard welding, where efficiency, precision, and safety are critical to meeting the demands of large-scale marine construction. Shipbuilding involves extensive welding operations across massive steel and aluminum structures, requiring thousands of meters of joints that must be both durable and defect-free. Manual welding alone often struggles to deliver consistent quality and speed under such conditions, especially in repetitive or difficult-to-access areas.

Automated welding systems address these challenges by standardizing weld quality, reducing variability caused by human factors, and enabling continuous operation. They also allow for the handling of complex geometries in ship structures, such as curved hulls, stiffened panels, and confined compartments, where accuracy is essential. Furthermore, automation improves workplace safety by minimizing human exposure to hazardous fumes, high temperatures, and restricted spaces.

In addition to technical advantages, automation supports cost-effectiveness in shipbuilding. By reducing rework, minimizing defects, and optimizing material use, it lowers overall production costs while shortening delivery schedules. With integration into digital shipyard ecosystems, robotic welding systems also contribute to real-time monitoring and quality assurance, supporting predictive maintenance and long-term reliability. Ultimately, automation is not just a supporting tool but a strategic enabler, driving shipyards toward higher productivity, global competitiveness, and the realization of smart shipyard concepts.

5.2.1 Process Standardization and Repeatability

One of the most significant contributions of automation in shipyard welding is the achievement of process standardization and repeatability. Manual welding, though highly skilled, is often influenced by human factors such as fatigue, variations in technique, and environmental conditions. This leads to inconsistencies in weld quality, especially in large-scale shipbuilding where thousands of welds are performed.

Robotic and automated welding systems eliminate these uncertainties by operating within preprogrammed parameters, ensuring that each weld, whether on flat panels, curved hull sections, or stiffened structures, is produced with uniform precision. Standardization not only enhances the structural integrity of marine vessels but also ensures compliance with strict classification society requirements and international quality standards.

Repeatability further brings efficiency to fabrication. Automated systems can reproduce complex welding patterns with minimal error across multiple ships of the same design, reducing the time needed for inspection, repair, and rework. This consistency also supports better predictability in project timelines and costs, which is vital in highly competitive global shipbuilding markets.

5.2.2 Enhancing Weld Quality and Reducing Human Error

Weld quality is a decisive factor in the performance and safety of marine vessels, where joints are constantly exposed to harsh loads, vibrations, and corrosive seawater. Traditional manual welding, while versatile, is prone to human error arising from fatigue, skill variation, or inconsistent environmental conditions within the shipyard. Such deviations can lead to defects like porosity, lack of fusion, or uneven penetration, which compromise the long-term durability of structures.

Automation directly addresses these challenges by employing precisely controlled robotic systems and advanced sensors that regulate current, voltage, travel speed, and torch angle with high accuracy. This ensures consistent penetration, uniform bead geometry, and defect-free welds, even in complex ship structures such as double hulls or bulkheads. In addition, automated welding equipment can continuously monitor weld quality in real time, using machine vision and adaptive control to detect and correct irregularities instantly something that is difficult to achieve manually.

By minimizing human intervention, automation reduces the likelihood of errors, rework, and quality non-conformance. The result is stronger, more reliable welds, improved resistance to marine environmental stresses, and compliance with stringent classification society standards. This higher degree of precision not only boosts ship safety and lifespan but also enhances shipyard competitiveness in global markets.

5.2.3 Productivity Gains in Large-Scale Ship Fabrication

Shipbuilding involves joining thousands of large and complex components, such as hull plates, stiffeners, and bulkheads, where delays in welding directly impact overall project timelines. Traditional manual welding, while flexible, is labor-intensive and time-consuming, especially for repetitive and long weld seams found in large vessels.

Automation significantly enhances productivity in large-scale fabrication by enabling continuous, high-speed welding with minimal downtime. Robotic systems can operate for extended periods without fatigue, maintaining consistent travel speeds and deposition rates. This allows multiple sections of a ship to be welded simultaneously, often in parallel workstations, thereby shortening production cycles. Advanced systems such as panel line welding robots and automated gantry welders are already widely used in modern shipyards to process large hull panels efficiently.

Moreover, automation reduces the need for extensive post-weld inspection and rework by ensuring high accuracy from the outset. The reduction of errors translates into faster throughput and cost savings. With the integration of digital manufacturing tools, robotic welding also enables better planning, sequencing, and coordination of welding tasks, further improving workflow efficiency.

In essence, automation transforms ship fabrication from a labor-driven process into a streamlined, high-output operation, allowing shipyards to meet increasing global demands for commercial, naval, and offshore vessels while maintaining competitive delivery schedules.

5.2.4 Examples of Automated Welding Systems in Leading Shipyards

Leading shipyards around the world have widely adopted automated welding systems to improve efficiency, accuracy, and competitiveness. In South Korea, major shipbuilders like HHI, Samsung Heavy Industries (SHI), and Daewoo Shipbuilding & Marine Engineering (DSME) have developed highly automated panel line systems, where robotic welding gantries handle large flat sections of ship hulls with minimal manual intervention. These systems combine SAW and robotic gas metal arc welding (GMAW) to ensure high-quality weld seams in large panels.

In Japan, shipyards such as Mitsubishi Heavy Industries employ robotic welding manipulators and track-guided welding robots to fabricate curved structures and double-hull sections. Their robotic solutions are designed to handle complex geometries, particularly in liquefied natural gas carriers and tankers, where precision is critical.

European shipyards, such as Meyer Werft (Germany) and Fincantieri (Italy), have also embraced automation through CNC-controlled welding stations and adaptive robotic arms. These systems integrate real-time sensors to adjust welding parameters dynamically, ensuring consistent weld quality in cruise ships and naval vessels.

In China, large-scale shipbuilders like China State Shipbuilding Corporation (CSSC) are investing heavily in intelligent shipyards, incorporating robotic welding, automated material handling, and digital monitoring systems. These developments are part of a broader push toward "smart manufacturing" and Industry 4.0 practices.

Together, these examples demonstrate how global leaders in shipbuilding use automation not just as a tool for efficiency, but as a strategic enabler of advanced marine fabrication, helping to meet the demands of larger, more complex, and environmentally compliant vessels.

5.3 TYPES OF ROBOTIC WELDING SYSTEMS IN SHIPBUILDING

5.3.1 Fixed Welding Robots

Fixed welding robots are stationary robotic systems installed in specific locations within shipyards, typically in panel lines, block assembly areas, or welding stations. Unlike mobile or track-guided systems, these robots remain anchored at a fixed point and work within a defined workspace. They are commonly used for repetitive, high-volume welds on flat panels, stiffeners, and bulkheads, where the geometry is standardized and predictable.

In shipyards, fixed welding robots are integrated into automated panel lines, where they perform continuous SAW or GMAW on large steel plates and longitudinal stiffeners. Their precision and repeatability significantly reduce welding defects, improve joint quality, and minimize the need for rework.

Examples include the robotic gantries used in South Korean and Japanese shipyards, where multiple fixed robots are synchronized to weld large hull sections simultaneously. These systems ensure consistent weld penetration, reduced distortion, and high deposition rates, which are essential for large-scale ship fabrication.

FIGURE 5.1 Fixed welding robots in shipbuilding. (https://www.inrotech.com/inrotech-solve-well-known-challenges-in-shipbuilding/.)

Although limited in flexibility compared to mobile or multi-axis robotic systems, fixed welding robots remain a cornerstone of high-throughput, standardized fabrication processes in modern shipbuilding, contributing to both productivity and cost efficiency.

Figure 5.1 illustrates a fixed welding robot system installed in a shipyard environment for the automated welding of structural components. The robot is mounted on a fixed rail system that allows precise, controlled movement along the weld line. A robotic arm is positioned to perform the welding process between stiffened steel sections of the ship's hull structure.

The robot combines mechanized guidance with an automated welding torch, ensuring uniform heat input and consistent weld bead formation. This setup eliminates variability caused by manual welding, enabling high-quality joints in repetitive, large-scale fabrication tasks such as panel assembly, bulkhead welding, and stiffener attachment.

The arc indicated in the figure highlights the active welding process, illustrating how the robot performs continuous welds along long seams. By maintaining accurate torch angles, travel speed, and penetration depth, the system enhances repeatability, structural integrity, and productivity in shipbuilding operations.

5.3.2 Mobile Welding Robots

Mobile welding robots are designed with the flexibility to move across large shipyard workspaces, making them well suited for the complex and non-linear geometries commonly found in shipbuilding. Unlike fixed robots that operate in designated stations, mobile systems can travel along rails, tracks, or autonomous navigation platforms to reach different sections of a vessel.

These robots are typically equipped with multi-axis robotic arms, sensors, and vision systems that allow them to adapt to irregular weld paths, curved surfaces, and confined spaces inside ship hulls. In advanced shipyards, mobile welding robots are integrated with laser-guided navigation and real-time seam tracking technologies, enabling them to deliver precise welds even in challenging positions.

TABLE 5.1
Difference between Fixed Welding Robots and Mobile Welding Robots

Aspect	Fixed Welding Robots	Mobile Welding Robots
Mobility	Stationary; installed in specific welding stations or panel lines	Move along rails, tracks, or autonomously across shipyard floors and hull sections
Application area	Best suited for **standardized, repetitive welds** such as flat panels, stiffeners, and bulkheads	Suitable for **complex, non-linear welds** on curved hulls, large blocks, and confined spaces
Flexibility	Limited flexibility; restricted to predefined work zones	Highly flexible; capable of adapting to different geometries and welding positions
Precision and repeatability	Very high precision and repeatability in standardized tasks	High precision with advanced seam tracking and sensor integration, but performance can vary with surface complexity
Setup and installation	Requires **dedicated infrastructure** like panel lines and gantry systems	Requires **navigation systems, sensors, and mobility platforms**, but less fixed infrastructure
Productivity	Extremely efficient for high-volume, repetitive tasks with minimal variation	More versatile in handling different tasks, reducing material handling and repositioning
Examples in shipyards	Automated panel line robots in South Korea and Japan for submerged arc welding	KUKA or ABB mobile welding robots with vision systems used in European and Chinese shipyards
Limitations	Not suitable for irregular geometries or in confined ship sections	Slower in high-volume repetitive welding compared to fixed systems; higher cost and complexity

Mobile robots are particularly effective in welding large hull blocks, internal stiffeners, and curved panels, where flexibility and adaptability are essential. They also reduce the need for excessive repositioning of heavy steel structures, which lowers material handling costs and speeds up fabrication.

By combining mobility with automation, these robots enhance welding efficiency, reduce human intervention in hazardous areas, and improve overall productivity in shipyard operations. Their growing use reflects the trend toward smart, adaptive manufacturing systems in modern marine fabrication.

Difference between fixed welding robots and mobile welding robots is shown in Table 5.1.

5.3.3 Gantry and Track-Mounted Systems for Hull Construction

Gantry and track-mounted robotic welding systems are widely adopted in shipbuilding for the fabrication of large hull structures and block assemblies. These systems are designed to move along rails or tracks, providing precise, automated welding on long seams and stiffeners in flat panels and curved hull sections. Mounted on overhead gantries or floor-level tracks, they offer extensive coverage of large workpieces without the need for frequent repositioning of heavy steel components.

In hull construction, these systems typically employ SAW or GMAW processes, ensuring deep penetration, high deposition rates, and minimal distortion. Equipped with multi-axis robotic arms, seam tracking sensors, and automated torch adjustment, they can follow complex weld paths with high accuracy.

A major advantage of gantry and track-mounted systems is their ability to handle continuous, long welds across large panels, which improves productivity and reduces cycle time in block fabrication. They also enhance safety by reducing manual labor in hazardous environments such as confined hull sections.

Leading shipyards in South Korea, Japan, and Europe have integrated these systems into automated panel lines and block assembly workshops, enabling them to achieve world-class efficiency, consistent weld quality, and cost competitiveness in large-scale shipbuilding projects.

Figure 5.2 shows a large gantry-mounted system used in shipbuilding for lifting, positioning, and assisting in the welding and assembly of massive hull blocks. The gantry structure spans across the shipyard, supported by tall vertical legs and running on tracks laid along the ground. This allows the entire system to move along the length of the construction site, covering large work areas without requiring relocation of heavy ship sections.

In hull construction, gantry and track-mounted systems often integrate automated welding units or robotic arms that can travel along the gantry to perform long, continuous welds on flat panels, stiffeners, and block joints. The high load-carrying capacity also enables precise handling of steel sections, reducing the need for manual alignment.

By combining structural support, mobility, and automation, such systems enhance efficiency, safety, and weld accuracy in large-scale ship fabrication. They are particularly valuable in modern shipyards where block assembly and panel welding require both high precision and heavy lifting capability.

FIGURE 5.2 Gantry crane for shipyard. (https://www.weihuagantrycrane.com/product/Gantry-Crane-for-Shipyard-Shipbuilding.html.)

5.3.4 Collaborative Robots (Cobots) in Ship Sections

Collaborative robots, or cobots, are transforming modern shipbuilding by working directly alongside human operators in ship sections, where they enhance both productivity and safety. Unlike traditional industrial robots that require isolation within safety cages, cobots are designed with advanced sensors, force-limiting technologies, and intelligent programming, allowing them to operate in close proximity to workers without compromising safety. Their integration into shipyards addresses the growing demand for efficient, flexible, and worker-friendly automation in environments where tasks often involve heavy materials, confined spaces, and repetitive operations.

Cobots bring a wide range of benefits to ship section fabrication. They significantly improve efficiency by carrying out repetitive jobs such as machine loading and unloading, which streamlines workflow and minimizes downtime. Safety is also enhanced, as cobots are equipped with sensors capable of detecting human presence and instantly adjusting or halting their motion to avoid accidents. In addition, they automate physically strenuous tasks such as lifting or transporting heavy components, thereby reducing worker fatigue and the risk of injury. Instead of replacing skilled workers, cobots serve as assistants by handling monotonous, repetitive, or time-consuming jobs, which allows shipbuilders to focus on higher-value tasks like complex welding, inspection, and quality control. Their ease of programming and redeployment makes them especially valuable in shipyards, where tasks and layouts frequently change during different stages of ship section construction.

The specific applications of cobots in ship sections are broad and practical. They can transport components across assembly points, ensuring smoother logistics and better workflow coordination. Cobots also play a role in machine tending, efficiently loading and unloading parts from equipment to maximize throughput. During assembly, they can assist with operations such as fastening, screwing, and positioning components, thereby reducing human strain in repetitive tasks. Furthermore, when equipped with vision systems, cobots contribute to inspection and quality assurance by detecting defects in welds and components at early stages, improving overall reliability.

Key characteristics make cobots uniquely suited for shipbuilding. Their high-sensitivity sensors give them the ability to "feel" their surroundings and safely interact with humans. Their lightweight and compact design, combined with intuitive interfaces, allows for quick deployment and integration into existing workflows without extensive reconfiguration. Above all, cobots are built for collaboration, enhancing teamwork and efficiency rather than isolating workers from the process.

Figure 5.3 illustrates a collaborative robot (cobot) actively engaged in a welding operation within a shipbuilding setup. The cobot, equipped with articulated robotic arms and precision sensors, is performing automated welding on a metallic component. Unlike traditional industrial robots that operate in isolated environments, this cobot is designed to work safely alongside human workers.

The figure highlights how cobots are applied in ship section fabrication, where they handle repetitive and high-precision tasks such as welding, assembly assistance, and inspection. The compact and flexible design of the cobot allows it to adapt to

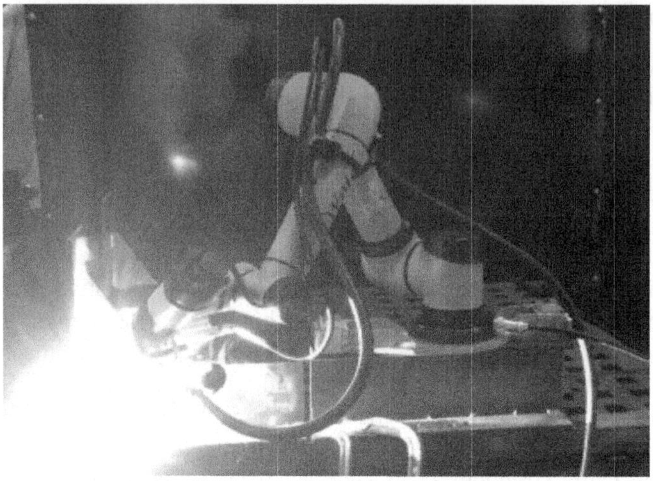

FIGURE 5.3 COBOT engaged in welding. (https://vimeo.com/528392166/d6e4eb581d.)

confined or complex shipyard spaces, while its advanced programming ensures accuracy and consistency in weld quality.

5.3.5 Underwater and Remote-Controlled Welding Robots

Underwater and remote-controlled welding robots are advanced systems designed to perform welding tasks in submerged and hazardous marine environments where human divers face risks such as high pressure, low visibility, and limited working time. These robots are typically equipped with cameras, sensors, and manipulator arms that allow precise control of welding torches, either autonomously or through remote operation from a control station.

In underwater ship repair and offshore structures, these robots are crucial for welding hull cracks, pipelines, and oil rigs without the need for costly dry docking. They ensure continuous operations and minimize downtime for vessels and marine platforms. Remote-controlled robots are operated by skilled technicians on the surface, who guide the welding process via real-time video and sensor feedback. Autonomous versions, enhanced with AI, can perform repetitive welds or inspections with minimal human intervention.

Key advantages include improved safety by eliminating diver exposure, higher efficiency in challenging underwater environments, and consistent weld quality under extreme conditions. Additionally, their deployment reduces repair costs and increases the operational lifespan of ships and offshore infrastructure (Miwa et al. 2002).

Figure 5.4 illustrates an underwater welding operation, showcasing how welding is performed in submerged marine environments such as ship hull repairs, pipelines, or offshore structures. The figure highlights a diver equipped with a specialized helmet, protective gear, and an underwater welding torch, generating an intense arc that produces bright sparks even under water.

FIGURE 5.4 Underwater welding. (https://toolbox.igus.com/143/underwater-welding-robotics.)

This process represents the principle behind underwater and remote-controlled welding robots, which replicate such tasks using robotic arms and remotely operated systems. While divers demonstrate the manual approach, advanced robotic systems are increasingly being deployed to perform similar welding jobs with greater safety, consistency, and efficiency. These robots are equipped with cameras, manipulators, and sensors, allowing operators on the surface to control them in real time, thus avoiding direct human exposure to hazardous underwater conditions.

5.4 AI AND MACHINE LEARNING IN WELDING AUTOMATION

AI and machine learning (ML) are revolutionizing welding automation by enabling real-time monitoring, analysis, and adaptive control of the welding process. Through the integration of sensors, cameras, and vision systems, AI systems continuously collect data on welding parameters such as arc length, temperature, and travel speed, while ML algorithms process this information to make immediate adjustments that maintain optimal conditions. This ensures consistent weld quality and minimizes the risk of defects. In addition, machine vision powered by ML plays a critical role in quality assurance by identifying imperfections such as cracks, undercuts, or misalignments at an early stage, thereby preventing faulty components from progressing in the production line.

AI and ML also contribute to process optimization by analyzing both historical and real-time data to predict weld bead geometry and refine welding parameters. This approach increases first-pass yield rates, reduces rework, and minimizes material waste. Intelligent robot learning further enhances welding automation, as AI can mimic the movements and techniques of skilled human welders, transferring their expertise into robotic systems that consistently replicate high-quality performance. Moreover, adaptive welding capabilities allow AI-driven robots to handle material variations, joint misalignments, and inconsistent conditions that traditional fixed-program systems cannot easily manage (Moinuddin et al. 2024).

The benefits of AI and ML in welding are significant. They ensure enhanced weld quality by reducing human error and optimizing parameters, while also increasing efficiency through faster production cycles and higher success rates on the first pass. Cost savings are realized through fewer defects, reduced rework, and minimized inspection requirements. Safety is also improved, as AI-powered robots can operate in hazardous or inaccessible environments, reducing risks to human workers. Importantly, these technologies complement rather than replace skilled welders, taking over repetitive or dangerous tasks and allowing human operators to focus on complex, value-added operations. By bridging the skill gap and facilitating knowledge transfer, AI and ML strengthen the overall welding ecosystem in shipbuilding, aerospace, and other demanding industries.

5.4.1 Role of AI in Real-Time Weld Parameter Adjustment

AI plays a crucial role in real-time weld parameter adjustment by enabling welding systems to continuously monitor, analyze, and optimize process conditions during operation. Traditional welding often suffers from variations in heat input, arc length, travel speed, and material inconsistencies, which can result in defects or quality variations. AI overcomes these challenges by using sensor data and machine vision to track parameters such as temperature, voltage, current, and weld pool dynamics. The collected information is processed instantly, allowing the system to adjust inputs such as torch speed, wire feed rate, or energy levels to maintain optimal welding conditions.

By making these dynamic adjustments, AI ensures greater consistency in weld quality, minimizes defects, and reduces the need for rework. For example, if joint misalignment or material thickness variation is detected, AI algorithms can automatically fine-tune the heat input to avoid undercutting or incomplete penetration. In high-volume shipbuilding environments, this capability is particularly valuable, as it allows welding robots to adapt to large and complex structures without manual intervention. Moreover, the integration of AI-driven adaptive control enhances productivity by maintaining continuous operation, reducing downtime, and ensuring a higher first-pass success rate. Ultimately, AI-driven real-time parameter adjustment not only improves quality and efficiency but also enhances safety by reducing human involvement in hazardous welding environments.

5.4.2 Machine Vision for Seam Tracking and Defect Detection

Machine vision has become a transformative tool in welding automation, particularly for seam tracking and defect detection. In traditional welding, maintaining precise alignment along the seam and ensuring defect-free joints are challenging due to factors such as material variations, thermal distortions, and operator dependency. Machine vision, powered by high-resolution cameras and advanced algorithms, addresses these challenges by continuously scanning the weld joint and monitoring the process in real time. This enables the system to accurately identify seam paths, even in complex geometries, and guide the welding torch with high precision, reducing the risk of misalignment.

Beyond seam tracking, machine vision is vital for early defect detection. Using image processing and AI-based classification, the system can identify surface and subsurface anomalies such as porosity, cracks, undercuts, and incomplete fusion. Detecting such flaws during welding allows for immediate corrective measures, preventing defective components from progressing to later stages of production. In industries like shipbuilding and aerospace, where weld integrity is critical, this capability ensures structural safety and compliance with stringent standards. By integrating seam tracking with defect detection, machine vision not only enhances weld accuracy and consistency but also reduces inspection time, material wastage, and rework, ultimately improving overall productivity and reliability in automated welding systems.

5.4.3 Predictive Maintenance of Welding Equipment

Predictive maintenance, powered by AI and ML, is increasingly applied to welding equipment to enhance reliability and reduce unplanned downtime. Unlike traditional preventive maintenance, which relies on fixed schedules, predictive maintenance continuously monitors the health of equipment through sensors that track parameters such as current flow, temperature, vibration, and wear on consumables. Data collected from welding machines and robotic systems are analyzed in real time, enabling AI algorithms to identify patterns and predict potential failures before they occur. For example, changes in arc stability, wire feed irregularities, or unusual motor vibrations can signal early signs of torch wear, cable damage, or mechanical misalignment.

By addressing these issues proactively, predictive maintenance ensures consistent weld quality, extends equipment lifespan, and reduces costly interruptions during shipbuilding or large-scale fabrication projects. It also optimizes spare part management by predicting exactly when components will need replacement, minimize inventory costs, and avoid unnecessary downtime. In marine and offshore applications, where productivity and safety are paramount, predictive maintenance helps maintain continuous operations even in harsh environments. Overall, the integration of predictive maintenance in welding automation not only improves equipment performance but also contributes to higher efficiency, lower costs, and greater reliability across the entire production cycle.

5.4.4 Intelligent Quality Control Systems in Shipyards

Intelligent quality control systems powered by AI and ML are transforming inspection and defect detection in modern shipyards. Traditional quality checks often rely on manual inspection, which can be time-consuming, labor-intensive, and prone to human error, especially in large-scale ship sections with complex welds. By contrast, AI-driven systems integrate high-resolution cameras, laser scanners, and ultrasonic sensors to capture real-time data during and after welding. ML algorithms analyze these data to detect defects such as porosity, cracks, misalignment, and incomplete penetration with a higher degree of accuracy and consistency than manual methods.

These systems not only identify defects but also classify their severity, predict their potential impact on structural integrity, and provide actionable recommendations for corrective measures. In some advanced shipyards, intelligent quality control is linked directly to welding robots, enabling immediate adjustments to prevent further errors. This closed-loop feedback system ensures that quality is built into the welding process rather than verified only after completion.

The adoption of such intelligent systems enhances productivity by reducing rework, ensures compliance with stringent marine safety standards, and supports traceability by maintaining detailed digital records of every weld. For shipyards aiming for higher efficiency, improved safety, and adherence to international classification society requirements, intelligent quality control systems provide a powerful tool to guarantee consistent, high-quality fabrication in demanding marine environments.

5.5 INTEGRATION OF AUTOMATION WITH DIGITAL SHIPBUILDING

The integration of automation with digital shipbuilding represents a major leap in how modern vessels are designed, fabricated, and maintained. Digital shipbuilding relies on advanced technologies such as 3D modeling, digital twins, and simulation tools to create a comprehensive virtual representation of the ship before physical construction begins. When combined with automated welding and fabrication systems, this approach enables seamless transition from design to production, reducing errors and ensuring that weld paths, sequences, and material usage are optimized.

Automation systems, guided by digital ship models, can execute welding tasks with precision, ensuring adherence to design specifications while minimizing waste and rework. Digital twins also allow real-time monitoring of welding performance by comparing actual weld data with the digital model, which helps in predicting potential defects and optimizing future processes. Furthermore, integration with Internet of Things (IoT)-enabled sensors and AI-driven analytics enhances coordination across design, production, and quality control phases, creating a closed-loop system that continuously improves efficiency.

In shipyards, this synergy between automation and digital shipbuilding accelerates project timelines, reduces costs, and supports greater customization for complex ship designs. It also ensures compliance with international standards by maintaining traceable, digital records of each weld and fabrication activity. Ultimately, this integration not only improves productivity and quality but also fosters a smarter, more sustainable, and future-ready shipbuilding ecosystem.

5.5.1 CAD/CAM-to-Robot Interface

The integration of **CAD/CAM** systems with robotic welding platforms forms a crucial link in digital shipbuilding and automated fabrication. In this process, the digital ship model created in CAD software is directly translated into executable instructions for robotic systems, eliminating manual programming and reducing the chances of error. CAM software interprets design data, optimizes weld paths, and generates the necessary tool trajectories that robots can follow with precision.

This interface ensures that weld sequences, joint geometries, and tolerances are accurately transferred from the virtual model to the physical work environment. Advanced systems further allow real-time feedback from the robot to the CAD/CAM environment, creating a closed-loop workflow where adjustments can be made automatically if deviations are detected. For example, sensor-based data collected during welding can be fed back into the digital model, allowing adaptive corrections and improved future planning.

In shipbuilding, the CAD/CAM-to-robot interface significantly shortens the design-to-production cycle, enhances repeatability, and improves productivity in handling complex hull geometries and large structural components. It also supports customization by enabling quick reprogramming for design variations without disrupting the overall production flow. As shipyards adopt Industry 4.0 principles, this interface becomes a key enabler of smart manufacturing, bridging digital design and physical fabrication seamlessly.

5.5.2 Digital Twin Applications for Welding Planning

In modern digital shipbuilding, digital twins play a pivotal role in automating welding planning and execution. By creating a virtual replica of the ship structure, welding processes can be simulated and optimized before physical implementation, reducing errors and minimizing material waste. Digital twins enable precise weld sequence planning, predicting heat-induced distortions, residual stresses, and metallurgical changes across complex assemblies. Integration with robotic welding systems allows automated execution of pre-planned weld paths with optimized parameters such as voltage, current, travel speed, and torch angle, ensuring consistent weld quality across large-scale ship structures. Real-time sensor feedback from the physical welding process continuously updates the digital twin, enabling predictive quality control and proactive defect detection. This seamless integration of simulation, automation, and monitoring streamlines production scheduling, enhances structural integrity, and supports data-driven decision-making for shipyard operations. Additionally, digital twins provide a platform for operator training and process validation, accelerating the adoption of advanced automated welding technologies in shipbuilding while reducing production time and costs.

Figure 5.5 illustrates the digital twin workflow used in automated shipbuilding welding. The process begins with the creation of a digital twin, a virtual replica of the ship structure that integrates design data, material properties, and fabrication parameters. This digital model enables engineers to simulate and optimize the welding process before physical execution.

The next stage is weld sequence planning, where the digital twin assists in determining the optimal welding path, sequence, and parameters to minimize distortion, residual stresses, and rework. This plan is then communicated to the robotic welding system, which executes the operations autonomously based on the predefined virtual model.

Throughout the process, real-time monitoring is carried out using sensors and data acquisition systems to track parameters such as arc temperature, weld bead

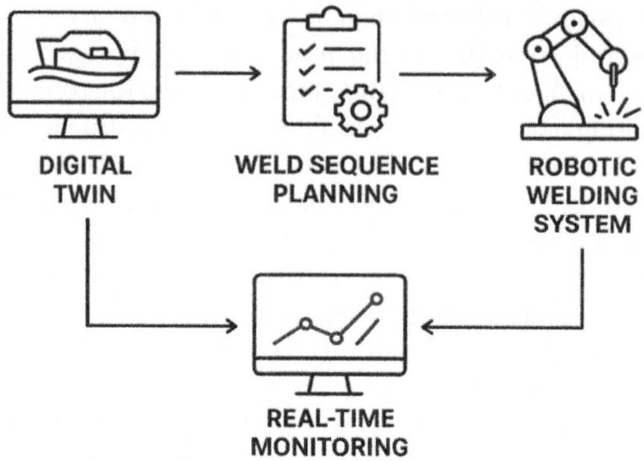

FIGURE 5.5 Digital twin applications for welding planning.

geometry, and robotic motion. The collected data are continuously fed back into the digital twin, enabling adaptive control, predictive maintenance, and continuous process optimization. This closed-loop integration ensures higher weld quality, improved efficiency, and reduced production costs in modern digital shipbuilding environments.

5.5.3 Data Analytics for Continuous Improvement

Data analytics plays a pivotal role in enabling continuous improvement and automation across the shipbuilding production life cycle. By collecting real-time data from sensors, welding robots, CNC machines, and other automated equipment, shipyards can monitor operational performance, detect deviations, and identify inefficiencies immediately. Advanced analytics tools, including predictive modeling and ML algorithms, process these data to forecast potential equipment failures, optimize resource allocation, and reduce rework, thereby improving overall productivity. Furthermore, the integration of digital twins with data analytics allows simulation of complex workflows, enabling scenario testing and decision-making before implementing changes on the shop floor. This closed-loop approach ensures that insights derived from historical and real-time data continuously inform process adjustments, leading to more efficient, accurate, and cost-effective shipbuilding operations while supporting the automation of quality control, maintenance planning, and production scheduling.

Figure 5.6 demonstrates a data-driven continuous improvement automation framework, highlighting the integration of data analytics and digital technologies in manufacturing and shipbuilding processes. The workflow begins with data collection, where information is gathered from sensors, machines, and production systems. These datasets include parameters such as temperature, vibration, weld quality, and operational efficiency.

The collected data are then processed through data analytics, where advanced computational methods such as statistical modeling, AI, and ML are applied to identify trends, patterns, and potential process inefficiencies. The insights obtained

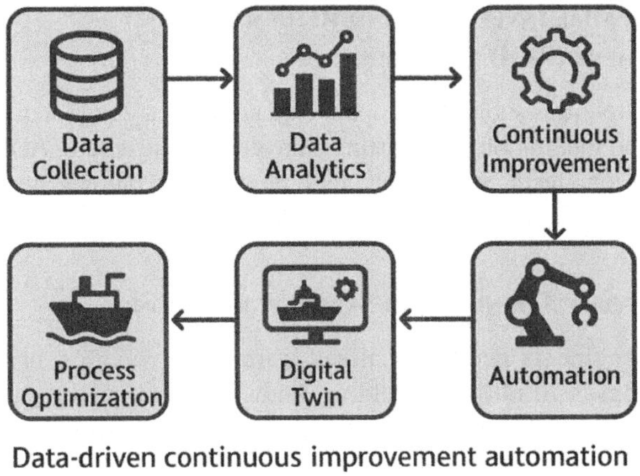

FIGURE 5.6 Data analytics for continuous improvement.

support continuous improvement, enabling manufacturers to refine processes, reduce defects, and enhance productivity based on real-time evidence.

Next, automation leverages these insights to autonomously adjust process parameters, improving precision and repeatability in production tasks. The digital twin component serves as a virtual simulation platform, replicating real-world operations and validating optimization strategies before implementation. Finally, process optimization ensures that the entire system operates at peak performance, feeding improved data back into the cycle.

5.6 CHALLENGES IN IMPLEMENTING ROBOTIC WELDING

Challenges in implementing robotic welding primarily arise from technical, operational, and organizational factors that influence its successful adoption in shipbuilding and other industries. One of the foremost challenges is the high initial investment cost associated with robotic welding systems, which includes the purchase of robots, programming software, sensors, and integration with existing infrastructure. Additionally, complexity in programming and setup poses difficulties, especially for large and varied welding tasks such as those in shipbuilding where joint geometries are not standardized. Ensuring consistent weld quality can also be challenging due to variations in material thickness, surface conditions, and part positioning. Another significant hurdle is the requirement for skilled workforce operators and engineers who must be trained not only in welding fundamentals but also in robotics, automation, and data analytics. Moreover, integration with digital systems such as CAD/CAM, digital twins, and production management platforms often demands additional customization. Operational challenges include limited flexibility in adapting to small batch or highly customized production and difficulties in real-time defect detection and correction. Finally, maintenance and downtime of robotic systems can disrupt workflows if not supported by predictive maintenance and robust technical support, making implementation a multifaceted challenge.

5.6.1 High Initial Investment and Return on Investment (ROI) Considerations

Robotic welding requires substantial capital expenditure for equipment, automation software, sensors, and integration into existing shipyard infrastructure. Achieving a positive ROI depends on long-term cost savings, high production volumes, and efficient system utilization, which can be uncertain in shipbuilding projects with varying demands.

5.6.2 Workforce Training and Skill Upgradation

Operators and engineers need specialized training in robotics, programming, and automation alongside traditional welding knowledge. Upskilling the workforce is time-intensive and costly, yet essential to ensure smooth operation, troubleshooting, and adaptation of robotic systems.

5.6.3 Adaptation to Complex Marine Structures

Ships feature large, irregular, and non-standardized geometries, making it difficult for robots to access and weld intricate joints. Advanced sensors, flexible robotic arms, and adaptive programming are required to manage these complex assemblies, which increases both cost and technical demands.

5.6.4 Integration with Digital Shipbuilding Systems

Robotic welding must be seamlessly integrated with CAD/CAM, digital twins, and production management platforms. Ensuring compatibility across software and hardware ecosystems requires significant customization and technical expertise, creating implementation hurdles.

5.6.5 Real-Time Quality Assurance

Although robots improve consistency, welding quality can still be affected by misalignment, surface impurities, or environmental factors. Real-time defect detection and correction systems, often relying on AI and machine vision, add complexity and cost but are essential for maintaining standards in marine applications.

5.6.6 Maintenance and Reliability Issues

Robotic systems demand regular preventive maintenance and robust technical support to avoid downtime. In shipyards where production schedules are tight, any breakdown or malfunction of welding robots can disrupt operations and delay project delivery.

5.6.7 Flexibility for Low-Volume or Customized Production

Shipbuilding projects often involve smaller production runs or highly customized components. Unlike automotive manufacturing, where robotic welding thrives on repetition, shipyards face challenges in adapting robots efficiently to varied tasks without incurring additional setup and programming costs.

5.6.8 Safety and Workspace Constraints

Integrating robots into crowded shipyard environments poses challenges related to worker safety, workspace optimization, and coordination with manual welders. Establishing safe operating zones and collaborative systems requires additional planning and investment.

5.7 FUTURE TRENDS IN MARINE WELDING AUTOMATION

The future of marine welding automation is being shaped by the convergence of advanced robotics, AI, and digital shipbuilding technologies. One key trend is the integration of AI and ML into welding systems, enabling adaptive parameter control, predictive maintenance, and intelligent defect detection in real time. Another important direction is the use of digital twins, which will allow shipyards to simulate welding processes virtually, optimize parameters before execution, and continuously update the model with real-world data for improved accuracy.

Collaborative robots (cobots) are also emerging as a practical solution, working alongside human welders in complex shipyard environments where full automation is challenging. These systems improve safety, efficiency, and flexibility in welding operations. In addition, advances in sensor fusion and machine vision will enable robots to handle highly irregular marine structures by recognizing joint geometries, adjusting weld paths automatically, and ensuring consistent quality.

The adoption of Industry 4.0 and smart manufacturing principles will further connect welding robots to IoT platforms, enabling real-time monitoring, cloud-based analytics, and automated decision-making across distributed shipyard facilities. Lightweight and modular robotic platforms are expected to be developed for mobility and adaptability in confined ship sections. Finally, sustainable welding practices such as energy-efficient processes and reduced material wastage will gain importance as shipyards align with global environmental regulations.

Together, these trends point toward a highly intelligent, flexible, and data-driven future for marine welding automation, where human expertise is enhanced by digital technologies to achieve higher productivity, precision, and sustainability.

5.8 SUMMARY

Robotic and automated welding is reshaping the future of shipbuilding by enhancing precision, productivity, and safety in an industry traditionally dominated by manual labor. The chapter highlighted how automation integrates with digital shipbuilding concepts, leveraging robotics, data analytics, and digital twins to achieve continuous improvement in welding quality and efficiency. Despite its advantages, implementation faces challenges such as high initial investment, workforce training requirements, adaptation to complex marine structures, and the need for robust integration with digital platforms.

Emerging trends indicate a strong shift toward intelligent systems, including autonomous welding, AI-driven process optimization, and collaborative human–robot models. These are supported by advances in sensor fusion, machine vision, and Industry 4.0 principles that enable predictive maintenance, real-time defect detection, and flexible operation across diverse shipyard environments. Sustainable practices and modular robotic platforms are also expected to align shipbuilding with environmental and operational goals.

Overall, robotic and automated welding is not a replacement for human expertise but a powerful complement that redefines roles in modern shipyards. By combining skilled labor with intelligent automation, the shipbuilding industry is moving toward a more efficient, adaptive, and future-ready manufacturing ecosystem.

5.9 REVIEW QUESTIONS

1. How has welding automation evolved in shipbuilding from early mechanization to modern robotics?
2. What are the primary factors driving the adoption of welding automation in marine fabrication?
3. Discuss key global trends in shipyard robotics and their influence on the competitiveness of shipbuilding nations.
4. How does process standardization improve welding consistency in automated shipyards?
5. In what ways does automation reduce human error in welding operations?
6. Explain how robotic welding contributes to productivity gains in large-scale ship fabrication.
7. Provide examples of advanced automated welding systems implemented in leading global shipyards.
8. Compare the advantages and limitations of fixed vs. mobile welding robots in shipbuilding.
9. How are gantry and track-mounted systems utilized in hull construction?
10. What role do collaborative robots (cobots) play in ship section fabrication?
11. Describe the applications and challenges of underwater and remote-controlled robotic welding.
12. How does AI support real-time adjustment of weld parameters?
13. Explain the role of machine vision in seam tracking and defect detection.
14. How is predictive maintenance applied to welding equipment in automated shipyards?
15. Discuss the use of intelligent quality control systems in ensuring consistent weld performance.
16. How do CAD/CAM systems interface with robotic welding in shipyards?
17. Explain how digital twin technology is applied for welding planning and optimization.
18. What is the significance of data analytics in driving continuous improvement in automated welding processes?
19. Why is high initial investment a major barrier to adopting robotic welding in shipyards?

20. How can workforce training and skill upgradation help overcome implementation challenges?
21. Why is it difficult to adapt robotic welding systems to complex marine structures?
22. What additional technical or operational challenges limit the full-scale adoption of robotic welding in shipbuilding?
23. What are autonomous welding systems, and how are they expected to transform ship fabrication?
24. How can AI-driven process optimization improve efficiency and quality in marine welding?
25. Discuss the benefits and limitations of hybrid human–robot collaboration models in shipyards.
26. Which future technologies (e.g., sensor fusion, IoT, sustainability practices) are likely to further enhance marine welding automation?
27. How can collaborative robots improve safety and efficiency in confined shipyard environments?
28. What role will sustainability and energy-efficient welding practices play in the future of shipbuilding?
29. How might modular and mobile robotic welding platforms enhance flexibility in shipyards?
30. In what ways can continuous improvement frameworks integrate with automation to create smarter, adaptive shipyards?

6 Hybrid Welding Technologies for Marine Applications

6.1 INTRODUCTION

Hybrid welding technologies have emerged as a transformative solution in marine fabrication, addressing the limitations of conventional welding methods while meeting the growing demands for efficiency, precision, and structural integrity in shipbuilding. By combining the advantages of multiple welding processes, such as gas metal arc welding (GMAW), gas tungsten arc welding (GTAW), laser beam welding (LBW), and plasma arc welding (PAW), hybrid systems achieve deeper penetration, reduced distortion, and improved mechanical properties in welded joints. This makes them particularly well suited for large and complex marine structures, where weld quality directly impacts vessel performance and safety (Walz et al. 2001).

In shipbuilding, hybrid welding enhances productivity by enabling higher travel speeds, minimizing rework, and allowing for automation in both block assembly and hull construction. The ability to weld thicker sections in a single pass, while maintaining excellent metallurgical properties, reduces fabrication time and cost. Additionally, hybrid approaches provide flexibility to adapt to diverse materials such as high-strength steels and aluminum alloys, which are increasingly used in modern ships for weight reduction and fuel efficiency.

As the industry moves toward digital shipbuilding and smart manufacturing, hybrid welding technologies are also being integrated with robotic systems, sensors, and real-time monitoring tools to ensure consistent quality and process optimization. These advancements make hybrid welding not just a complementary technique but a critical enabler for the next generation of marine fabrication.

6.1.1 Concept and Importance of Hybrid Welding in the Marine Industry

The concept of hybrid welding lies in the strategic combination of two or more welding processes within a single operation to leverage their complementary advantages and overcome individual limitations. For example, hybrid techniques such as laser-GMAW or plasma-GMAW integrate the deep penetration capability of high-energy beam processes with the filler metal adaptability and gap-bridging ability of arc welding. This synergy allows the creation of high-quality welds with superior mechanical strength, minimal distortion, and greater process stability compared to using either process alone (Walz et al. 2001).

DOI: 10.1201/9781003217350-6

In the marine industry, where ship structures demand long, continuous, and defect-free welds across thick and complex materials, hybrid welding plays a critical role. It significantly reduces fabrication time by enabling higher welding speeds and fewer passes, while ensuring reliable penetration in thick plates common in hull and deck construction. The process also enhances productivity and cost efficiency, helping shipyards meet tight schedules and reduce rework. Importantly, hybrid welding improves the ability to work with advanced materials such as high-strength steels and lightweight aluminum alloys, which are vital for building fuel-efficient, durable, and environmentally compliant vessels.

Moreover, the importance of hybrid welding extends beyond performance, and it aligns with the industry's transition toward digital shipbuilding and automation. When integrated with robotic systems, sensors, and artificial intelligence (AI)-driven monitoring, hybrid welding ensures consistency, adaptability to varying joint geometries, and real-time quality assurance. As shipbuilding continues to demand greater efficiency, sustainability, and safety, hybrid welding technologies are becoming an indispensable pillar of modern marine fabrication.

6.1.2 Advantages of Hybrid Welding over Conventional Processes

Hybrid welding offers a range of benefits that make it superior to conventional single-process welding methods, particularly in the marine industry where large, complex, and high-strength structures are fabricated. The key advantages include the following.

6.1.2.1 Deeper Penetration with Fewer Passes

High-energy beam processes (e.g., laser or plasma) combined with arc welding allow deep weld penetration, enabling thick plates to be welded in fewer passes compared to traditional arc welding.

6.1.2.2 Higher Welding Speeds

Hybrid processes significantly increase travel speed while maintaining weld quality, reducing fabrication time in large-scale shipbuilding projects.

6.1.2.3 Improved Gap-Bridging Capability

While laser or plasma welding alone struggles with gaps and misalignments, the addition of arc welding provides filler material, improving tolerance to joint preparation variations common in shipyard environments.

6.1.2.4 Reduced Distortion and Residual Stresses

Concentrated energy input from the hybrid process minimizes heat-affected zones (HAZs), leading to less distortion and reduced rework, which is critical for large hull and deck assemblies.

6.1.2.5 Enhanced Weld Quality and Mechanical Properties

Hybrid welding produces joints with superior strength, toughness, and metallurgical characteristics compared to conventional processes, ensuring reliability in marine applications.

6.1.2.6 Flexibility for Different Materials

Hybrid systems adapt well to advanced shipbuilding materials such as high-strength steels and aluminum alloys, supporting lightweight and fuel-efficient vessel construction.

6.1.2.7 Cost and Time Efficiency

By reducing the number of passes, distortion correction, and rework, hybrid welding lowers overall production costs and improves throughput in shipyards.

6.1.2.8 Compatibility with Automation and Digital Shipbuilding

Hybrid processes can be easily integrated with robotic systems, sensors, and digital twins, enabling real-time monitoring, adaptive control, and consistent quality in automated shipyard environments.

6.1.3 Scope of Hybrid Techniques in Shipbuilding and Offshore Structures

Hybrid welding techniques hold significant potential in advancing both shipbuilding and offshore engineering by addressing the unique fabrication challenges of large, complex, and safety-critical structures. In shipbuilding, hybrid processes are particularly valuable for hull and deck fabrication, where thick plates require high-quality, long welds with minimal distortion. By enabling deep penetration and higher welding speeds, hybrid systems reduce the number of passes needed, ensuring efficiency while maintaining structural integrity. They are also well suited for block assembly and panel welding, where productivity gains directly translate into shorter construction cycles and cost savings for shipyards.

In the construction of offshore structures such as oil platforms, floating production storage units (FPSOs), and wind turbine foundations, hybrid welding ensures the reliability and durability of joints exposed to harsh marine environments. The ability to weld high-strength steels and corrosion-resistant alloys makes hybrid processes essential for ensuring the long-term performance and safety of these installations. Furthermore, the capacity to adapt to underwater and remote-controlled welding applications expands their utility in offshore repair and maintenance, reducing the risks associated with manual welding in hazardous conditions.

Looking forward, the scope of hybrid welding extends to its integration with robotics, digital twin simulations, and AI-driven monitoring systems, supporting the transition toward fully automated and smart shipyards. Its adaptability to diverse materials, joint geometries, and environmental conditions positions hybrid welding as a cornerstone technology for achieving greater efficiency, sustainability, and reliability in both shipbuilding and offshore engineering projects.

6.2 HYBRID LASER-ARC WELDING (HLAW)

HLAW is one of the most widely adopted hybrid welding techniques in the marine and offshore industries, combining the high penetration and precision of laser welding with the filler metal adaptability and gap-bridging capability of arc welding. This integration allows the process to overcome the limitations of each method when used independently, making it highly effective for large-scale shipbuilding applications.

In HLAW, the laser beam provides deep, narrow penetration and creates a stable keyhole, while the arc welding process (commonly GMAW) supplies filler material, enhances gap tolerance, and stabilizes the molten pool. The result is a high-quality weld with reduced porosity, excellent mechanical properties, and minimal distortion, even in thick sections.

For shipbuilding, HLAW offers several advantages: it enables welding of thick steel plates in a single pass, reducing production time; it ensures consistency in long seam welds, critical for hull and deck assembly; and it significantly lowers rework by minimizing heat input and distortion. In addition, the process shows strong adaptability for welding high-strength steels and aluminum alloys, materials increasingly used in modern ship and offshore structure designs.

The marine industry is also exploring the integration of HLAW with robotic systems and digital twins, allowing for precise control, automated seam tracking, and real-time monitoring. This makes HLAW a cornerstone of advanced shipyard automation strategies, aligning with the goals of efficiency, quality, and sustainability in next-generation shipbuilding.

6.2.1 Principle and Process Mechanism

The principle of HLAW lies in the simultaneous application of a high-energy laser beam and an arc welding process (commonly GMAW) to create a synergistic effect that enhances welding performance. The laser beam provides deep penetration and forms a narrow keyhole in the workpiece, while the arc process contributes filler material, gap-bridging ability, and improved pool stability. Together, they overcome the limitations of using either process individually, resulting in higher welding efficiency and superior joint quality.

The process mechanism of HLAW can be described in three stages.

6.2.1.1 Laser Keyhole Formation

The focused laser beam rapidly melts and vaporizes the material, creating a deep, narrow keyhole. This ensures high penetration and reduced heat input compared to conventional arc welding.

6.2.1.2 Arc Welding Contribution

The arc welding process operates adjacent to or coaxially with the laser beam. It adds filler metal to the molten pool, stabilizes the weld bead, and increases tolerance to joint gaps and misalignments.

6.2.1.3 Synergistic Interaction

The simultaneous action of the laser and arc produces a wider, deeper, and more stable weld pool. This leads to reduced porosity, lower distortion, and improved mechanical properties of the joint.

By combining the precision and penetration of laser welding with the flexibility and adaptability of arc welding, HLAW achieves high travel speeds, reduced number

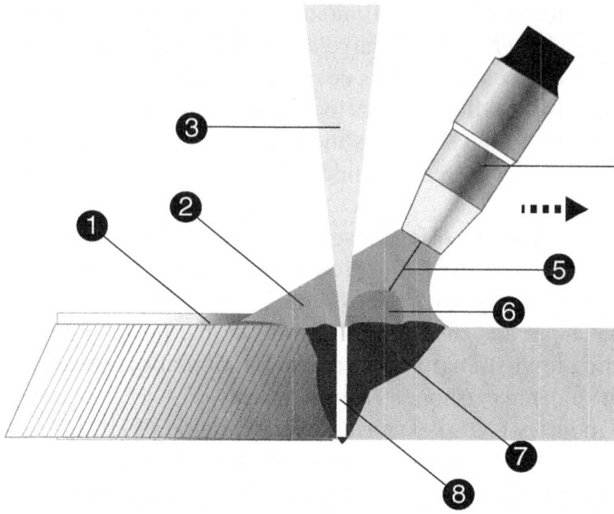

FIGURE 6.1 Hybrid laser-arc welding (HLAW) uses a wire arc welding process with a laser beam. The laser provides concentrated energy for high penetration, while the arc helps overcome fit-up challenges. Process elements include (i) newly formed weld bead, (ii) inert shielding gas, (iii) focused laser beam, (iv) GMAW gun, (v) wire electrode, (vi) electric arc, (vii) molten pool, and (viii) the keyhole. (https://www.thefabricator.com/thefabricator/article/laserwelding/hybrid-laser-arc-welding-takes-on-heavy-transportation.)

of passes, and better metallurgical characteristics, making it particularly suitable for welding thick steel plates and long seams in shipbuilding and offshore structures.

Figure 6.1 illustrates the principle of HLAW, which combines a laser beam with a wire arc welding process to achieve both high penetration and improved adaptability. The laser beam (3) creates a narrow, deep keyhole (8) in the material, ensuring efficient energy concentration and penetration, while the GMAW gun (4) delivers the wire electrode (5) that melts under the electric arc (6) to provide filler metal and stabilize the molten pool (7). The inert shielding gas (2) protects the weld zone from atmospheric contamination, allowing the formation of a clean and defect-free weld bead (1). This hybrid interaction results in a stable molten pool where both laser and arc energies combine, ensuring strong metallurgical bonding and effective gap bridging. By integrating the precision of laser welding with the flexibility of arc welding, HLAW produces high-quality welds with reduced distortion and improved efficiency, making it particularly suited for shipbuilding and offshore structures.

6.3 UNDERWATER WELDING TECHNIQUES

Underwater welding is a specialized process employed to repair, maintain, and modify marine and offshore structures without the need to bring them to dry docks. It is an essential technique for ship hull restoration, offshore oil and gas platform maintenance, pipeline repair, and renewable energy installations such as offshore wind turbines. The ability to perform welding directly underwater saves both time and cost, ensuring minimal disruption to marine operations while extending the service life of critical assets.

The techniques used in underwater welding are broadly classified into wet welding and dry (habitat) welding. In wet welding, the diver performs welding operations directly in the water using waterproof electrodes, making it a cost-effective and rapid solution for temporary or emergency repairs. However, issues such as rapid quenching, porosity, hydrogen embrittlement, and limited visibility often affect the quality and durability of welds. On the other hand, dry welding involves enclosing the weld area in a specially designed pressurized chamber (habitat) where water is displaced by gas, creating a dry and controlled environment similar to surface welding. Although this technique provides superior weld quality and is suitable for critical and long-term repairs, it requires more time, higher costs, and sophisticated equipment.

By selecting the appropriate technique based on structural requirements, environmental conditions, and cost considerations, underwater welding continues to serve as a vital enabler of efficient marine maintenance and repair in both commercial and defense sectors.

Figure 6.2 shows a diver-welder performing underwater welding on a marine structure. The image highlights the practical execution of wet welding, where the welder operates directly in the water using specialized waterproof electrodes and protective gear. The diver is equipped with a helmet for breathing support, gloves for insulation, and communication lines for surface coordination. Electrical cables and gas hoses are connected to supply power and ensure safety during the operation. The bright arc seen in the figure demonstrates the electrode generating intense heat to create a molten weld pool, which fuses the damaged section of the structure.

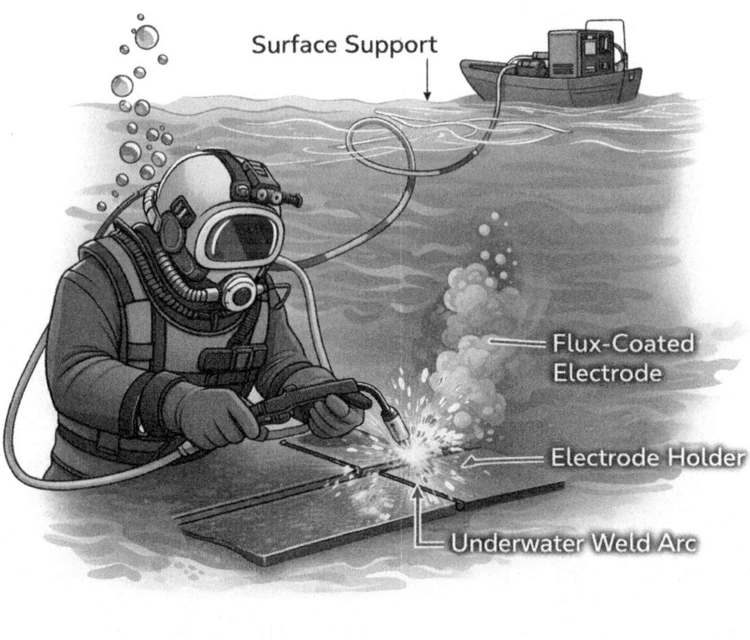

FIGURE 6.2 Underwater welding. (https://en.wikipedia.org/wiki/Hyperbaric_welding.)

This process allows for in situ repairs of pipelines, ship hulls, and offshore structures without removing them from the water, saving both time and cost. However, as illustrated, the underwater environment imposes challenges such as reduced visibility, rapid cooling of the weld due to water quenching, and potential risks of hydrogen embrittlement and porosity in the weld metal. Despite these challenges, underwater welding remains a vital technique for maintaining marine assets and extending the service life of critical infrastructure.

6.3.1 Wet Welding

Wet welding is one of the most widely practiced underwater welding methods, where the welder performs the operation directly in the water using specially designed waterproof electrodes or consumables. The process is carried out by diver welders equipped with protective suits, helmets, and breathing apparatus, allowing them to work at various depths. The most common techniques include shielded metal arc welding (SMAW), flux-cored arc welding (FCAW), and occasionally GTAW with proper shielding arrangements. Among these, SMAW is the predominant process due to its simplicity, portability, and reliability in wet conditions. Figure 6.3 depicts a diver practicing underwater welding in a controlled training pool environment. The diver is equipped with a full-face diving helmet that provides breathing gas, communication systems, and protection against water pressure. The welder is holding a waterproof electrode, producing an arc that generates the heat necessary to fuse the metal workpiece. Training pools like this are used to simulate real underwater conditions, allowing divers to practice electrode handling, arc stability, and weld bead formation without the risks associated with open-water environments (Rowe and Liu 2001).

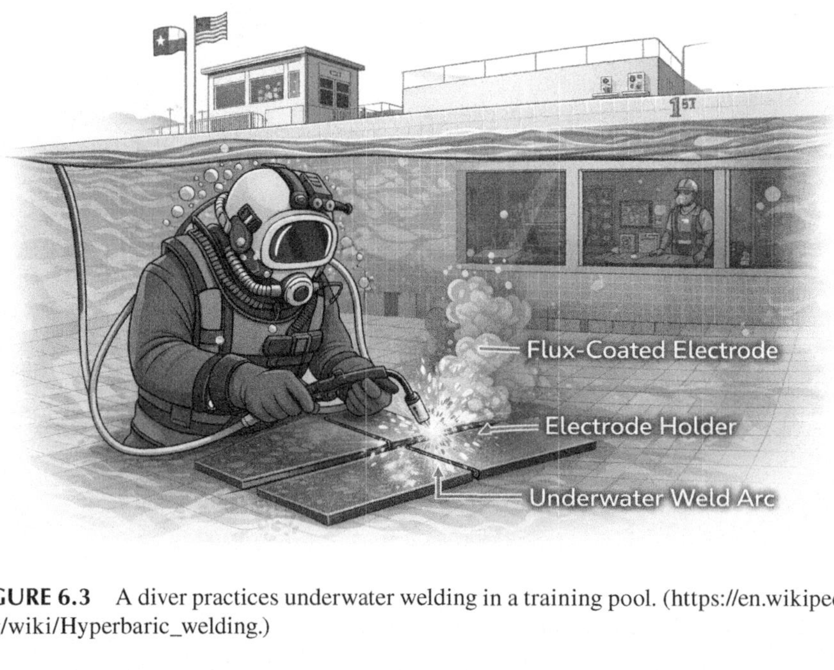

FIGURE 6.3 A diver practices underwater welding in a training pool. (https://en.wikipedia.org/wiki/Hyperbaric_welding.)

Such training is crucial because underwater welding requires not only technical knowledge of welding processes like SMAW or FCAW but also strong diving skills and the ability to work in low-visibility, high-pressure conditions. By practicing in controlled settings, diver welders can develop precision, enhance safety awareness, and prepare for actual marine maintenance and repair operations in ship hulls, offshore platforms, and underwater pipelines.

6.3.1.1 Process Overview (SMAW, FCAW, and Other Methods)

Shielded Metal Arc Welding (SMAW): The most widely used method for wet welding, employing waterproof stick electrodes that are easy to handle and suitable for underwater conditions.

Flux-Cored Arc Welding (FCAW): Provides higher deposition rates and can improve productivity but requires more advanced equipment and control.

Other Methods (e.g., GTAW): Used for specialized applications requiring high precision, though they are less common due to the difficulty in shielding the arc underwater.

6.3.1.2 Advantages and Limitations

Wet welding offers several advantages, including low equipment cost, ease of setup, flexibility for emergency repairs, and the ability to carry out in situ maintenance without lifting the structure from the water. However, it has notable limitations such as rapid quenching of the weld due to direct water contact, which increases hardness and brittleness. Issues like porosity, hydrogen embrittlement, poor visibility, and limited depth capability also reduce weld quality and long-term reliability.

6.3.1.3 Safety Considerations and Operator Skill Requirements

Wet welding poses significant safety risks, including the potential for electric shock, exposure to high pressure, entanglement of hoses and cables, and limited visibility for divers. Therefore, strict safety protocols must be followed, with surface monitoring teams supporting the diver. Operator skill is critical, as wet welding requires not only advanced diving training but also strong welding expertise to manage unstable arc conditions, electrode manipulation, and defect control. Certified diver welders must undergo rigorous training to ensure safe and effective underwater operations.

6.3.2 Dry Welding (Habitat Welding)

Dry welding, also known as habitat welding, is a specialized underwater welding technique where the welding operation is carried out in a sealed, pressurized chamber that creates a dry environment around the work area. This chamber, called a habitat, is filled with a breathable gas mixture (commonly helium and oxygen), displacing the surrounding water and enabling welders to work as if they were on the surface (Fleury and Schofield 1979).

6.3.2.1 Principles and Setup

The principle of habitat welding relies on enclosing the damaged section with a pressure-resistant chamber that maintains an internal atmosphere at the same pressure

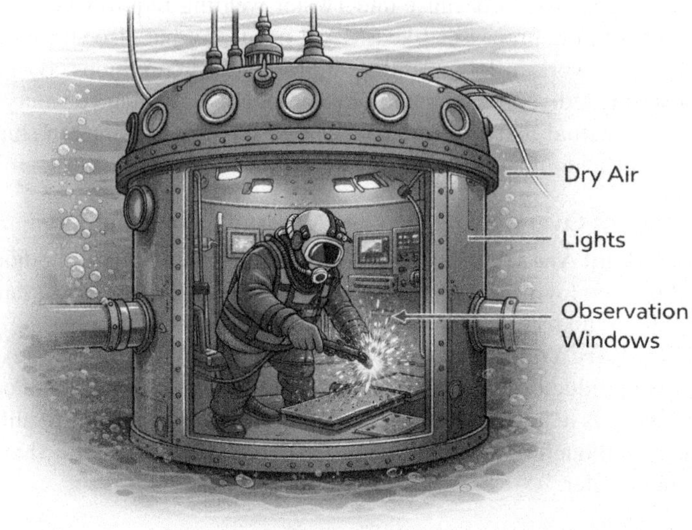

FIGURE 6.4 Underwater welding habitat for dry hyperbaric welding. (https://en.wikipedia.org/wiki/Hyperbaric_welding.)

as the surrounding water. Welders equipped with diving suits and helmets enter the chamber to perform welding using conventional processes such as SMAW, GTAW, or GMAW. The controlled dry environment improves visibility, stability of the arc, and accessibility to the weld zone, while also preventing direct contact of molten metal with water. Figure 6.4 illustrates an underwater welding habitat designed for dry hyperbaric welding, a technique in which a sealed, pressurized chamber is placed around the area requiring repair. The chamber, known as the habitat, is supplied with a breathable gas mixture (commonly helium-oxygen) and maintains an internal pressure equal to the surrounding water to prevent flooding. Inside the habitat, water is displaced, creating a dry working environment that allows welders to use conventional processes such as SMAW, GMAW, or GTAW with full visibility and control.

The habitat provides space for the welder, welding equipment, and tools, while also integrating life-support systems and communication lines connected to the surface. By isolating the weld zone from direct contact with water, this setup ensures superior weld quality, reduced porosity, and improved longevity compared to wet welding methods. Such hyperbaric welding habitats are essential for repairing offshore pipelines, oil and gas platforms, ship hulls, and other subsea structures that demand high-strength, permanent joints.

6.3.2.2 Applications in Offshore Platforms and Ship Repairs

Dry welding is widely applied in offshore oil and gas platforms, subsea pipelines, underwater foundations, and ship hulls where permanent, high-strength repairs are required. It is especially important in deepwater repairs and critical components where wet welding cannot deliver the desired reliability or mechanical performance. Naval operations also use habitat welding for emergency and structural repair of vessels.

6.3.2.3 Benefits in Weld Quality and Longevity

Habitat welding offers significant advantages in terms of weld integrity and service life. Since the process is performed in a dry, controlled atmosphere, it minimizes hydrogen-induced cracking (HIC), porosity, and rapid quenching of the weld metal. This results in higher mechanical strength, improved metallurgical properties, and long-lasting joints capable of withstanding harsh marine environments. Though more expensive and logistically complex than wet welding, habitat welding remains the most effective solution for critical, safety-sensitive underwater repairs.

6.4 METALLURGICAL CONSIDERATIONS IN HYBRID AND UNDERWATER WELDING

Metallurgical behavior is a key factor influencing the performance and reliability of welds in both hybrid and underwater welding. In underwater welding, the surrounding water acts as a strong quenching medium, causing rapid cooling of the weld metal and HAZ. This often leads to the formation of hard and brittle microstructures such as martensite, increasing the risk of cracking and reducing overall toughness. Additionally, hydrogen absorption from moisture in the environment can result in HIC, while gas entrapment frequently causes porosity. These issues, combined with the risk of alloying element loss, contribute to reduced ductility and corrosion resistance of underwater welds.

In contrast, hybrid welding processes, such as laser-arc combinations, allow for better control of heat input, producing fine and uniform microstructures with reduced residual stresses. The controlled shielding gases used in hybrid welding improve weld pool stability, minimize porosity, and significantly reduce hydrogen-related defects. Furthermore, the precise energy delivery of hybrid techniques enables deeper penetration with less distortion, leading to improved tensile strength, fatigue resistance, and fracture toughness. While underwater welds are more susceptible to galvanic corrosion and long-term degradation, hybrid welds, especially when optimized with appropriate filler materials and shielding gases, exhibit superior mechanical performance and enhanced service life.

6.4.1 Heat-Affected Zone (HAZ) Characteristics

The HAZ is the region of the base metal adjacent to the weld that does not melt but undergoes significant microstructural and mechanical changes due to exposure to high thermal cycles. In underwater welding, the HAZ is strongly influenced by rapid cooling caused by direct contact with water, which often results in the formation of hard and brittle phases such as martensite. This increases hardness but reduces toughness, making the weld joint more susceptible to cracking and hydrogen-induced embrittlement. The narrow and sharply defined HAZ in wet welding further contributes to localized stresses and reduced durability.

In hybrid welding processes, the characteristics of the HAZ are more favorable due to better control of heat input and cooling rates. The combination of laser and arc sources provides deep penetration with a balanced energy distribution, producing a finer microstructure with reduced residual stresses. The HAZ in hybrid welds typically exhibits improved toughness, lower hardness variations, and enhanced metallurgical stability compared to underwater welds. Controlling HAZ properties is critical, as it directly affects weld integrity, mechanical performance, and the long-term service life of marine structures exposed to harsh environments.

6.4.2 Microstructural Changes in Marine Alloys

Marine alloys, such as high-strength steels, stainless steels, and aluminum alloys, undergo significant microstructural transformations during welding due to rapid heating and cooling cycles. In underwater welding, the intense quenching effect of water leads to the formation of hard and brittle phases like martensite in steels, which increases hardness but reduces ductility and fracture toughness. This rapid cooling may also cause non-uniform grain structures, segregation of alloying elements, and porosity, all of which compromise the weld's mechanical performance and resistance to cracking. For stainless steels, the high cooling rates can destabilize protective oxide layers and promote sensitization, making the alloy more prone to localized corrosion in marine environments.

In hybrid welding, such as laser-arc welding, the controlled heat input produces finer and more uniform microstructures with reduced grain coarsening. The synergy between the laser's deep penetration and the arc's filler support promotes stable weld pools and minimizes defects, leading to improved toughness, fatigue resistance, and long-term durability. Aluminum alloys, commonly used in shipbuilding for their light weight and corrosion resistance, also experience microstructural changes such as grain refinement and precipitation hardening under hybrid welding, resulting in improved strength without significant distortion. Managing these microstructural changes is crucial to ensure the mechanical reliability and corrosion resistance of welded joints in harsh marine conditions.

6.4.3 Corrosion Resistance and Protective Measures

Corrosion resistance is a critical consideration in marine welding since welded joints are continuously exposed to seawater, salt spray, and humid environments that accelerate electrochemical degradation. In underwater welding, the lack of effective shielding and rapid cooling often leads to increased porosity, microcracks, and uneven microstructures, all of which reduce corrosion resistance. Hydrogen absorption from moisture and the breakdown of passive oxide films further promote pitting and galvanic corrosion, particularly in high-strength steels and stainless steels. These effects can significantly shorten the service life of welded marine structures if not properly managed.

In contrast, hybrid welding processes provide improved corrosion resistance due to their controlled atmosphere, balanced heat input, and smoother weld profiles. The use of appropriate shielding gases, filler materials, and post-weld treatments helps maintain a stable microstructure and prevents oxidation or sensitization. To enhance corrosion protection, several protective measures are commonly applied: surface coatings (such as epoxy, polyurethane, or thermal spray coatings), cathodic protection using sacrificial anodes or impressed current systems, and post-weld heat treatment to relieve residual stresses and restore metallurgical balance. Regular inspection and maintenance, combined with corrosion monitoring sensors, further ensure the longevity and integrity of welded components in shipbuilding and offshore applications.

6.5 QUALITY ASSURANCE AND TESTING IN HYBRID MARINE WELDING

Quality assurance in hybrid marine welding is essential to ensure the structural reliability, safety, and long-term performance of welded joints used in shipbuilding and offshore applications. Since hybrid welding combines laser and arc processes, maintaining process stability and weld integrity requires strict control of parameters such as heat input, travel speed, beam alignment, and shielding gas composition. Comprehensive testing and inspection protocols are employed at every stage from pre-weld preparation to post-weld evaluation to verify compliance with international marine welding standards such as ISO, AWS D3.6, and American Bureau of Shipping (ABS) rules.

Non-destructive testing (NDT) methods play a key role in hybrid welding quality control. Techniques such as ultrasonic testing (UT), radiographic testing (RT), magnetic particle testing (MT), and liquid penetrant testing (PT) are used to detect internal and surface defects like porosity, cracks, lack of fusion, and inclusions. Advanced NDT methods, including phased array ultrasonics and real-time laser scanning, are increasingly used for precision evaluation of weld integrity in automated systems. Additionally, destructive testing such as tensile, bend, impact, and hardness tests provides detailed information on the mechanical and metallurgical properties of the welded joints.

To achieve consistent quality, hybrid welding systems are often integrated with real-time monitoring and feedback control mechanisms that track parameters like temperature, arc stability, and weld penetration using sensors and machine vision. These data-driven systems allow automatic correction of deviations, reducing the likelihood of defects and rework. Documentation of welding procedure specifications (WPS), welder qualification records, and inspection reports forms an integral part of the quality assurance process. Through a combination of standardized testing, automated monitoring, and digital quality management, hybrid marine welding ensures superior weld performance and compliance with stringent marine classification and safety standards.

6.5.1 Inspection Standards for Hybrid and Underwater Welds

Hybrid and underwater welding operations are governed by a range of international standards and classification society guidelines to ensure quality and safety. Key standards include ISO 3834 (Quality Requirements for Fusion Welding of Metallic Materials), ISO 15614-11 (Qualification of Welding Procedures for HLAW), and AWS D3.6M (Underwater Welding Code). Additionally, marine classification societies such as ABS, Lloyd's Register (LR), and Det Norske Veritas Germanischer Lloyd (DNV-GL) specify inspection and certification requirements for welds used in hull structures, subsea pipelines, and offshore platforms. These standards define acceptance criteria for visual inspection, NDT, mechanical testing, and corrosion resistance evaluation. Periodic audits and third-party inspections are often mandated to maintain compliance and traceability throughout the fabrication process.

6.5.1.1 Testing and Quality Control Methods

Both non-destructive and destructive testing techniques are used to verify the integrity of hybrid and underwater welds. NDT methods such as UT, RT, MT, and PT are applied to detect internal and surface defects, including porosity, lack of fusion, or cracks. For automated hybrid systems, phased array UT (PAUT) and 3D laser scanning offer real-time monitoring and precise flaw detection. Destructive tests such as tensile strength, bend, Charpy impact, and hardness testing are performed on sample coupons to assess mechanical performance.

Modern hybrid welding systems also incorporate sensor-based real-time monitoring, using infrared thermography, acoustic emission analysis, and optical sensors to detect anomalies during welding. This feedback control ensures parameter consistency and minimizes rework. All inspection results are documented as part of the WPS and Procedure Qualification Record (PQR) to ensure traceability and compliance with classification standards.

6.5.2 Case Studies from Shipbuilding and Offshore Repair

Case Study 6.1: Hybrid Welding in Ship Hull Fabrication

A major European shipyard implemented hybrid laser-GMAW (HLAW) for joining 8–12-mm thick high-strength steel plates in hull panel assembly. The process reduced welding passes by 60%, minimized distortion, and achieved consistent penetration, verified through ultrasonic and RT. Compliance with ISO 15614-11 and ABS Class Rules ensured acceptance for marine service.

Case Study 6.2: Dry Hyperbaric Welding in Offshore Pipeline Repair

In a North Sea offshore pipeline maintenance project, dry habitat welding was used for repairing a corroded section at 150 m depth. The controlled environment enabled defect-free welds verified by UT and radiography. The welds met the requirements of AWS D3.6M and DNV-OS-F101 standards for subsea pipeline integrity.

Case Study 6.3: Automated Quality Monitoring in Ship Deck Assembly

A Japanese shipbuilder integrated AI-based visual inspection into robotic hybrid welding systems for deck structures. Real-time monitoring of bead geometry and thermal signatures improved first-pass yield by 25%, while automated NDT ensured compliance with ISO and LR guidelines for large vessel fabrication.

6.6 FUTURE TRENDS IN HYBRID MARINE WELDING

The future of hybrid marine welding is being shaped by rapid advancements in automation, AI, digitalization, and sustainable manufacturing practices. As shipbuilding and offshore industries move toward fully digital shipyards, hybrid welding processes—particularly HLAW—are evolving into intelligent, adaptive systems capable of real-time process optimization and autonomous operation. These developments promise to further enhance welding precision, efficiency, and consistency while significantly reducing material waste and energy consumption (Table 6.1)

TABLE 6.1
Key Emerging Technologies and Trends in Hybrid Marine Welding

Emerging Technology/ Trend	Description	Key Benefits in Marine Applications
AI and machine learning integration	Real-time analysis and adaptive control of welding parameters such as heat input, wire feed rate, and arc stability using intelligent algorithms	Improved weld consistency, predictive maintenance, reduced defects, and higher productivity
Machine vision and deep learning	Use of cameras and image processing systems for seam tracking, defect detection, and automated quality assurance	Enhanced accuracy in complex weld geometries, faster inspection, and reduced manual supervision
Digital twin technology	Creation of virtual models to simulate, monitor, and optimize the hybrid welding process in real time	Reduced trial-and-error, improved planning efficiency, and integration with digital shipyard systems
Industry 4.0 connectivity	Integration of welding robots with IoT-enabled platforms, CAD/CAM systems, and production management networks	Seamless data flow, remote process monitoring, and synchronized digital manufacturing
Smart sensors and real-time monitoring	Use of temperature, acoustic, and optical sensors to collect process data for feedback and quality control	Instant anomaly detection, adaptive correction, and assurance of weld integrity
Sustainability and green manufacturing	Development of energy-efficient welding techniques, recyclable filler materials, and reduced emissions	Lower environmental impact, compliance with sustainability standards, and cost efficiency
Mobile and modular welding units	Compact, portable hybrid welding systems designed for confined or complex ship sections	Improved accessibility, flexibility, and reduced setup time in shipyards and offshore structures
Adaptive control systems	Automated control mechanisms that adjust parameters dynamically to maintain optimal welding conditions	Increased precision, minimized human error, and consistent weld quality across varying materials

One of the most promising trends is the integration of AI and machine learning (ML) into hybrid welding systems. By analyzing sensor data, thermal profiles, and real-time imaging, AI-driven algorithms can automatically adjust parameters such as laser power, arc current, and wire feed rate to maintain optimal weld conditions. This not only improves weld quality but also enables predictive maintenance, minimizing downtime and ensuring consistent production flow. Similarly, machine vision and deep learning are revolutionizing seam tracking and defect detection, allowing automated identification of irregularities that were once dependent on human inspection.

Another major direction is the adoption of digital twins and virtual simulation platforms for hybrid welding planning and process control. Digital twins replicate the welding environment and structural behavior, enabling virtual optimization of parameters before physical welding begins. This ensures accuracy, reduces trial-and-error costs, and supports data-driven decision-making throughout the production life cycle. Integration with Industry 4.0 frameworks allows hybrid welding stations to communicate with other digital systems such as CAD/CAM platforms and robotic controllers—creating a synchronized, fully automated workflow across shipyards.

Sustainability is also emerging as a central focus. Future hybrid welding systems are expected to emphasize energy efficiency, reduced emissions, and recyclable filler materials, aligning with global goals for environmentally responsible manufacturing. The development of lightweight, modular, and mobile hybrid welding units will increase accessibility in confined or complex ship structures, while adaptive control algorithms will further reduce human intervention. The technologies and trends summarized in Table 6.1 highlight the transformation of hybrid marine welding into a digitally integrated, intelligent, and sustainable manufacturing process. The convergence of AI, machine vision, and digital twin simulation ensures precision and adaptability, while Industry 4.0 connectivity enables fully automated workflows across modern shipyards. Simultaneously, sustainability-driven innovations promote energy efficiency and environmentally responsible practices. As these advancements continue to mature, hybrid marine welding will serve as a cornerstone of next-generation shipbuilding delivering superior quality, reliability, and efficiency while aligning with global goals for smart and sustainable marine manufacturing.

6.7 SUMMARY

The chapter on hybrid welding technologies for marine applications explores the integration of advanced welding processes to enhance productivity, precision, and durability in shipbuilding and offshore structures. It highlights the concept and importance of hybrid welding, emphasizing its advantages over conventional methods through improved penetration, faster speed, and superior weld quality. The discussion on HLAW details its principles, synergy between laser and arc heat sources, equipment configuration, and applications in hull and panel fabrication, demonstrating its efficiency in large-scale marine construction. The chapter also covers underwater welding techniques, comparing wet and dry (habitat) welding in terms of process mechanisms, benefits, limitations, and safety requirements for marine maintenance and repair. Metallurgical aspects such as HAZ characteristics, microstructural changes, and corrosion resistance are analyzed to ensure structural integrity and longevity of welded joints. Quality assurance and testing methods, including NDT techniques, inspection standards, and practical case studies, underscore the need for reliability and compliance with international marine codes. Finally, future trends emphasize AI-driven process control, robotic integration, digital twin simulations, and eco-friendly consumables, positioning hybrid marine welding as a cornerstone of intelligent, automated, and sustainable shipbuilding for the next generation.

6.8 REVIEW QUESTIONS

1. What is the basic concept of hybrid welding and why is it significant in the marine industry?
2. How does hybrid welding combine the advantages of different welding processes?
3. List three major benefits of hybrid welding compared to conventional welding methods.
4. What are the main challenges faced by conventional welding techniques in shipbuilding?
5. Describe the scope of hybrid welding applications in shipbuilding and offshore structures.
6. Explain the working principle of hybrid laser-arc welding (HLAW).
7. How does the laser and arc combination create a synergistic effect in HLAW?
8. What are the essential components in the equipment configuration of an HLAW setup?
9. Which parameters most influence weld quality and penetration in HLAW?
10. Discuss how HLAW improves productivity and weld consistency in ship panel and hull fabrication.
11. What are the major challenges and limitations of using HLAW in marine environments?
12. Why is underwater welding important for marine maintenance and offshore repair?
13. Differentiate between wet welding and dry (habitat) welding in terms of process and quality.
14. Which welding processes are commonly used in wet welding and why?
15. What are the major limitations of wet welding related to weld quality and safety?
16. Describe the principle and setup of dry (habitat) welding.
17. What advantages does habitat welding offer over wet welding in terms of weld longevity?
18. In what applications is dry welding most commonly used in ship repair and offshore platforms?
19. Explain the metallurgical changes that occur in the heat-affected zone (HAZ) during marine welding.

20. How do hybrid welding processes help control cooling rates and reduce brittleness?
21. What microstructural transformations commonly occur in marine alloys during underwater welding?
22. Why is corrosion resistance critical for marine welded joints, and how can it be improved?
23. What protective measures are commonly applied to enhance weld corrosion performance?
24. Which non-destructive testing (NDT) techniques are used for quality control in hybrid marine welding?
25. What are the major international standards governing inspection of hybrid and underwater welds?
26. Summarize one case study where hybrid welding was successfully applied in shipbuilding.
27. How does real-time monitoring improve the quality assurance process in hybrid welding?
28. Explain the role of AI and machine learning in the future of hybrid marine welding.
29. What sustainable and energy-efficient practices are emerging in hybrid welding technologies?
30. How will digital twins and robotic automation shape the next generation of shipbuilding and offshore fabrication?

7 Wire Arc Additive Manufacturing (WAAM) in Marine Applications

7.1 INTRODUCTION

Wire arc additive manufacturing (WAAM) is an emerging metal 3D printing technology that combines the principles of arc welding and additive manufacturing (AM) to fabricate large, complex, and high-strength metallic components directly from wire feedstock. Unlike conventional subtractive manufacturing, which involves material removal, WAAM builds parts layer by layer through controlled deposition of molten metal, enabling efficient use of materials and significant reductions in production time and cost. In recent years, WAAM has gained remarkable attention in the marine and offshore industries, where the demand for lightweight, durable, and corrosion-resistant structures is growing rapidly (Vishnukumar et al. 2021).

Moreover, the integration of WAAM with robotic systems, computer-aided design (CAD), and digital twin technologies aligns seamlessly with the concept of digital shipbuilding, allowing real-time monitoring, process optimization, and adaptive control. The technology also supports the repair and remanufacturing of worn or damaged components directly on-site, extending the operational life of marine assets while reducing logistics and downtime.

As the marine industry transitions toward smart manufacturing and sustainability, WAAM stands out as a key enabler, offering flexibility, cost-effectiveness, and design freedom. This chapter explores the principles, process mechanisms, materials, challenges, and future potential of WAAM in marine applications, emphasizing its role in revolutionizing how ships and offshore structures are designed, built, and maintained.

7.1.1 Evolution of Additive Manufacturing in the Marine Sector

The evolution of AM in the marine sector has been driven by the growing need for innovative, efficient, and sustainable production methods capable of meeting the complex design and performance demands of modern ships and offshore structures. Initially, the marine industry relied heavily on traditional manufacturing techniques such as casting, forging, and machining, which, although reliable, were material-intensive, time-consuming, and limited in design flexibility. The early adoption of AM in marine applications began with polymer-based 3D printing for prototyping and small-scale modeling, primarily used for testing hydrodynamic designs and component fitment during ship design stages. However, as metal AM technologies matured especially

DOI: 10.1201/9781003217350-7

those capable of processing high-strength and corrosion-resistant alloys, AM found broader applications in functional marine components.

The development of WAAM marked a significant milestone in this evolution. Derived from conventional arc welding processes, WAAM introduced a cost-effective and scalable method for producing large metal parts with high deposition rates. This breakthrough addressed one of the key limitations of powder-based AM, which struggled with size constraints and high material costs. By utilizing wire feedstock, WAAM enabled the fabrication of large-scale structural components such as hull sections, propeller blades, and offshore platform brackets with reduced lead times and minimal waste.

Over time, the integration of robotics, computer-aided design (CAD)/computer-aided manufacturing (CAM), and real-time process monitoring transformed WAAM into an intelligent manufacturing system aligned with the goals of Industry 4.0 and digital shipbuilding. Shipyards and offshore engineering firms began adopting hybrid manufacturing strategies that combine additive and subtractive processes, allowing near-net-shape components to be finished with precision machining. Furthermore, WAAM has become a key tool for repair and remanufacturing, enabling the restoration of damaged components directly on-site, reducing downtime and costs associated with replacement parts.

Today, the evolution of AM in the marine sector represents a paradigm shift from conventional fabrication toward flexible, data-driven, and sustainable production ecosystems. As advancements continue in sensor integration, process control, and material development, AM particularly WAAM stands poised to redefine marine engineering by enhancing performance, reducing environmental impact, and supporting the next generation of smart shipbuilding practices. Table 7.1 summarizes the key milestones in the evolution of AM within the marine sector, highlighting its transformation from basic polymer 3D printing to advanced, digitally integrated metal fabrication systems. Beginning in the 1990s with polymer-based prototyping for ship design visualization, the field progressed through the 2000s with the introduction of metal AM for small functional components. The mid-2010s marked a breakthrough with the development of WAAM, enabling large-scale metal component production with reduced cost and lead time. By the 2020s, integration with robotics, CAD/CAM systems, and digital shipbuilding tools positioned AM as a cornerstone of Industry 4.0 in marine fabrication. Looking ahead, smart, sustainable, and artificial intelligence (AI)-driven AM is expected to revolutionize shipbuilding through greater automation, adaptability, and environmental efficiency.

7.1.2 Advantages of WAAM over Conventional Marine Fabrication

WAAM offers several distinct advantages over conventional marine fabrication methods such as casting, forging, and subtractive machining. Its primary strength lies in the ability to build large, complex metal components layer by layer, eliminating the need for extensive tooling, molds, or material removal. This significantly reduces material wastage often by up to 70%–80% and enables more sustainable production practices in shipbuilding and offshore engineering. Unlike traditional methods, WAAM can produce near-net-shape components directly from digital models,

TABLE 7.1
Key Milestones in the Evolution of Additive Manufacturing in the Marine Sector

Period/Year	Development Stage	Key Achievements and Technological Advancements	Impact on Marine Industry
1990s	Introduction of polymer 3D printing	Early adoption of polymer-based 3D printing for design prototyping and model testing	Improved design visualization, rapid prototyping for hull forms and components
Early 2000s	Emergence of metal additive manufacturing	Development of metal AM processes using powder-bed fusion and direct energy deposition	Production of small-scale metal components with enhanced precision
Mid-2010s	Introduction of wire arc additive manufacturing (WAAM)	Integration of arc welding and additive manufacturing using wire feedstock	Enabled fabrication of large metal parts at lower cost and higher deposition rates
Late 2010s	Industrial adoption and large-scale applications	Use of WAAM for ship hull reinforcements, propeller blades, and offshore brackets	Reduced lead times, material waste, and cost in shipbuilding and repair
2020s	Integration with robotics and digital shipbuilding	Implementation of robotic WAAM systems with CAD/CAM, sensors, and process monitoring	Automated, precise, and repeatable manufacturing aligned with Industry 4.0
Future (beyond 2025)	Smart and sustainable additive manufacturing	Incorporation of AI, digital twins, and eco-efficient materials for intelligent production	Sustainable, adaptive, and data-driven manufacturing for next-generation shipyards

requiring only minor post-processing, which shortens production cycles and lowers overall costs.

A major advantage of WAAM is its compatibility with a wide range of marine-grade materials, including steels, aluminum, titanium, and nickel-based alloys, allowing manufacturers to tailor mechanical and corrosion-resistant properties for different applications. The process also enables high deposition rates and scalability, making it ideal for fabricating large structural elements such as propeller blades, hull reinforcements, brackets, and stiffener components that are often time-consuming and expensive to produce conventionally.

Moreover, the integration of WAAM with robotic systems and CAM enhances automation, repeatability, and precision, reducing dependency on manual labor and human error. The digital nature of the process allows for rapid design modifications, on-demand production, and easy repair of damaged components, even on-site in shipyards or offshore environments. These capabilities not only improve operational flexibility but also align with the goals of digital shipbuilding and sustainable manufacturing, where efficiency, adaptability, and resource optimization are key priorities.

In summary, WAAM surpasses traditional marine fabrication techniques through its advantages in material efficiency, cost reduction, design flexibility, and digital integration, making it a transformative technology for the next generation of shipbuilding and marine repair systems.

7.1.3 Scope of WAAM in Shipbuilding and Offshore Engineering

WAAM is rapidly emerging as a transformative technology in the shipbuilding and offshore sectors due to its ability to produce large, complex, and customized metal components with high efficiency and structural integrity. Its versatility extends across a wide range of applications from new ship construction and component fabrication to repair, retrofitting, and life extension of existing marine assets. The adaptability of WAAM to marine-grade materials such as high-strength steels, aluminum, and titanium alloys enables its use in both surface vessels and deep-sea structures, where weight reduction, corrosion resistance, and mechanical reliability are crucial.

In shipbuilding, WAAM is increasingly being used for manufacturing hull reinforcements, deck fittings, propeller blades, rudders, and structural brackets. The ability to fabricate near-net-shape components directly from digital CAD models allows shipyards to bypass lengthy casting and machining processes, drastically reducing production lead times. Additionally, WAAM enables lightweight optimization through topology design, which supports improved fuel efficiency and lower emissions, key goals for sustainable shipbuilding.

In offshore engineering, WAAM has proven valuable in fabricating and repairing platform supports, subsea connectors, riser brackets, mooring elements, and corrosion-resistant cladding for subsea pipelines. Its capability for in situ repair and remanufacturing under controlled conditions minimizes downtime and logistical costs associated with part replacement in remote marine environments. The use of robotic WAAM systems further enhances precision and repeatability, allowing for the production of large, geometrically complex parts such as transition pieces and mounting structures for offshore wind turbines.

The scope of WAAM extends beyond fabrication, and it is also central to the ongoing digital transformation of shipyards. By integrating WAAM with robotics, digital twins, and sensor-based feedback control, modern shipyards are developing automated, data-driven manufacturing ecosystems that enable real-time process monitoring, adaptive control, and predictive quality assurance. This synergy between AM and digital shipbuilding supports the creation of "smart shipyards," capable of producing customized components faster, with reduced environmental impact and greater precision.

7.2 FUNDAMENTALS OF WAAM FOR SHIPBUILDING

WAAM is an advanced form of metal AM that utilizes an electric arc as the heat source and a metal wire as the feedstock to fabricate three-dimensional components layer by layer. The process is built upon conventional arc welding technologies such as gas metal arc welding (GMAW), gas tungsten arc welding (GTAW), and plasma arc welding (PAW) but is digitally controlled and optimized for additive rather than joining applications. In shipbuilding, WAAM offers the capability to produce large, high-strength, and corrosion-resistant metal structures directly from CAD data, significantly reducing both manufacturing time and material waste compared to traditional fabrication methods.

The fundamental principle of WAAM involves controlled metal deposition in successive layers. A continuous wire feed is melted by an electric arc, and the molten material solidifies upon deposition, gradually building the desired geometry. Each layer's thickness, arc current, voltage, wire feed rate, and travel speed are carefully adjusted to achieve dimensional accuracy and consistent mechanical properties. The process is often performed using robotic manipulators or multi-axis CNC systems to ensure precise path control and repeatability, which is particularly critical in shipbuilding applications involving large and complex geometries.

WAAM's integration with digital design and automation technologies represents a key enabler for modern shipyards. Through real-time process monitoring using sensors, thermal cameras, and feedback control systems, parameters such as heat input, cooling rate, and bead geometry are continuously optimized. This ensures that the fabricated parts meet stringent marine standards for strength, toughness, and corrosion resistance. Post-processing steps such as heat treatment, machining, and surface finishing are typically applied to refine tolerances and enhance performance.

For shipbuilding, the fundamentals of WAAM provide a foundation for cost-effective, flexible, and sustainable manufacturing. Components like hull reinforcements, propeller blades, deck brackets, and structural joints can be fabricated directly within the shipyard using locally available feedstock, reducing dependency on external suppliers and long lead times. Furthermore, the ability to perform on-site or near-site AM supports rapid repair and remanufacturing, improving vessel uptime and operational readiness.

7.2.1 Principle and Working Mechanism of WAAM

The principle of WAAM is based on the controlled melting and solidification of a continuous metal wire feedstock using an electric arc as the heat source. This process is derived from conventional arc welding methods primarily GMAW, though GTAW and PAW are also employed depending on the required precision and material type. WAAM transforms these traditional joining processes into a layer-by-layer additive technique, enabling the creation of near-net-shape metallic components directly from digital models (Bhuvanesh Kumar et al. 2023).

In WAAM, a wire feeder continuously supplies the metal wire to the molten pool formed by the arc on a substrate or previously deposited layer. The localized heat from the arc melts the wire, and as it solidifies, a new layer is formed. Successive deposition of layers builds up the component's geometry as defined by the CAD model. The process typically involves multi-axis robotic systems or CNC manipulators to control torch position, travel path, and deposition orientation. The parameters governing the process such as current, voltage, wire feed rate, torch speed, and interpass temperature are precisely monitored and adjusted to maintain stable metal transfer and uniform layer thickness.

The process mechanism of WAAM can be divided into three main stages:

Deposition Stage: The electric arc melts the wire and base material to form a stable molten pool. Shielding gas (commonly argon or argon-helium mixtures) is used to prevent oxidation and contamination.

Solidification Stage: As the heat source moves forward, the molten pool solidifies, forming a metallurgically bonded layer with a characteristic weld bead geometry. The rapid cooling rate influences the microstructure, grain size, and mechanical properties.

Layer Integration Stage: Subsequent layers are deposited over the solidified ones, with controlled heat input ensuring proper fusion between layers without excessive remelting or distortion.

The WAAM process relies heavily on feedback and control systems for consistent quality. Advanced sensors, cameras, and thermographic systems are used for real-time monitoring of bead geometry, temperature distribution, and arc stability. Data from these sensors are processed through control algorithms or AI-assisted systems to dynamically adjust parameters, preventing defects such as porosity, undercutting, or incomplete fusion. Figure 7.1 illustrates the principle and process flow of WAAM. The diagram shows the key components of the process, including the wire feed system, welding torch, and electric arc that melts the wire to form the molten pool on a substrate. Shielding gas surrounds the arc to protect the molten metal from oxidation and contamination. As the molten material solidifies, it forms a single layer, and successive passes create the desired 3D geometry through controlled layer-by-layer deposition. This figure visually represents the integration of welding and AM principles, fundamental to producing large, near-net-shape metal parts for marine applications.

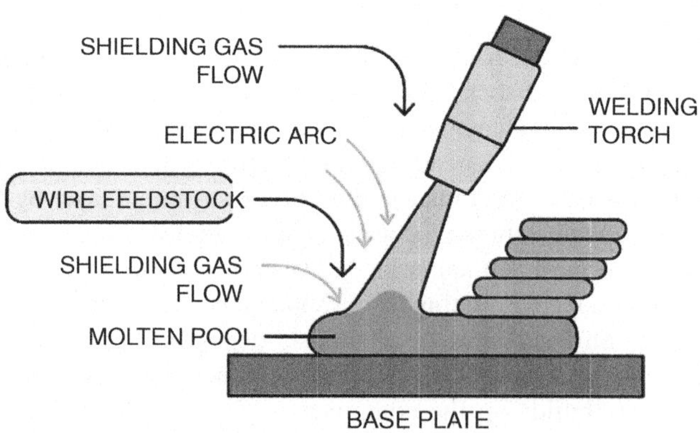

FIGURE 7.1 Principle and process flow of WAAM.

7.2.2 Comparison with Other Additive Manufacturing Processes

WAAM distinguishes itself from other AM processes primarily through its scalability, cost-effectiveness, and suitability for large metallic structures features that make it particularly advantageous for shipbuilding and offshore applications. While powder-based metal AM techniques such as selective laser melting (SLM), electron beam melting (EBM), and laser metal deposition (LMD) offer exceptional precision and surface finish, they are generally limited by build size, deposition rate, and high material cost. In contrast, WAAM uses commercial welding wire feedstock, which is more economical, readily available, and capable of producing components several meters in size.

The deposition rate of WAAM (typically 2–5 kg/hour and up to 10 kg/hour with optimized systems) far exceeds that of powder-bed processes, which are usually below 1 kg/hour. This makes WAAM particularly suitable for fabricating large marine parts such as propeller blades, hull reinforcements, brackets, and stiffeners, where high build volume and productivity are critical. Moreover, WAAM's near-net-shape manufacturing minimizes material wastage—an advantage over subtractive techniques and even some powder-based AM systems where unused powder requires recycling and handling under strict safety conditions.

However, WAAM typically results in a rougher surface finish and lower geometric resolution compared to SLM or EBM. Therefore, post-processing operations such as milling, grinding, or heat treatment are often required to achieve final dimensional accuracy and mechanical performance. On the other hand, WAAM excels in material flexibility, supporting a wide range of alloys including steels, aluminum, titanium, and nickel-based materials, many of which are difficult to process using powder-based methods due to oxidation or flowability issues.

From an environmental and economic standpoint, WAAM is also more energy-efficient and sustainable, consuming less power and generating less waste than powder-based AM systems. The ability to integrate WAAM with robotic automation, CAD/CAM, and digital twin technologies further enhances its appeal for the marine sector, where the fabrication of large, complex, and corrosion-resistant components is essential. Table 7.2 compares WAAM with other leading metal AM processes, emphasizing its superior scalability, deposition rate, and material efficiency. Unlike powder-based techniques such as SLM and EBM, WAAM's use of wire feedstock enables cost-effective fabrication of large, structural components with minimal waste, an essential advantage for shipbuilding and offshore industries. While WAAM requires post-processing to achieve fine tolerances, its compatibility with diverse alloys, robotic automation, and digital shipbuilding workflows makes it the most practical AM approach for large-scale marine applications.

7.2.3 WAAM Equipment and Setup in Marine Fabrication

The successful implementation of WAAM in marine fabrication relies on the proper configuration of specialized equipment that integrates traditional welding hardware with modern automation and digital control systems. The WAAM setup typically includes five core subsystems: the power source, wire feed system, robotic or motion

TABLE 7.2
Comparison of WAAM with Other Additive Manufacturing Processes

Parameter	Wire Arc Additive Manufacturing (WAAM)	Selective Laser Melting (SLM)	Electron Beam Melting (EBM)	Laser Metal Deposition (LMD)
Heat source	Electric arc	Laser beam	Electron beam	Laser beam
Feedstock form	Metal wire	Metal powder	Metal powder	Metal powder or wire
Deposition rate	High (2–10 kg/h)	Low (<1 kg/h)	Moderate (1–3 kg/h)	Moderate (1–5 kg/h)
Build volume	Very large (up to several meters)	Small to medium	Medium to large	Medium to large
Dimensional accuracy	Moderate	High	Moderate	High
Surface finish	Rough (requires post-processing)	Excellent	Good	Good
Material utilization efficiency	>90%	60%–80% (powder loss)	70%–85%	80%–90%
Material types	Wide range (steel, aluminum, titanium, nickel alloys)	Limited (mainly Ti, Al, and Co-based alloys)	Limited (mainly Ti and Ni-based alloys)	Wide (steel, Ti, Ni, Al alloys)
Energy consumption	Low to moderate	High	High	Moderate
Process cost	Low	High	High	Moderate
Process environment	Open atmosphere or controlled	Inert gas atmosphere	Vacuum	Inert gas atmosphere
Application suitability	Large structural components, shipbuilding, repair, offshore	Small, high-precision parts, aerospace, medical	Aerospace, turbine blades, implants	Medium-scale industrial components, repair
Automation compatibility	High (robotic, multi-axis systems)	Moderate	Moderate	High

control unit, shielding gas supply, and a build platform or substrate table. Together, these components ensure precise deposition, consistent quality, and repeatability, which are essential for large-scale marine applications.

The power source is the heart of the WAAM system, providing the stable arc energy required for melting the wire feedstock. Depending on the process variant, it may employ GMAW, GTAW, or PAW power systems. These power sources are designed to operate under controlled current, voltage, and pulse modulation to achieve consistent bead geometry and minimal spatter. The wire feed system continuously supplies the metallic wire to the weld pool at a precisely regulated rate, synchronized with the travel speed and current to maintain uniform layer buildup.

A robotic arm or multi-axis CNC gantry serves as the motion control unit, directing the torch along preprogrammed tool paths generated from CAD/CAM models. In marine fabrication, robotic WAAM setups are particularly beneficial for producing large, complex parts such as propeller blades, hull reinforcements, and stiffening structures. The integration of robotic manipulators ensures high positional accuracy and consistent layer alignment across large build envelopes, which is vital for maintaining structural integrity.

The shielding gas system, typically using argon or argon-helium mixtures, protects the molten pool from oxidation and contamination, ensuring sound metallurgical bonding. In certain marine alloys like aluminum or titanium, precise shielding is critical to prevent defects and improve surface finish. The build platform or substrate usually made of a similar material to the deposited metal provides a thermally conductive base that ensures proper adhesion of the first layer and minimizes residual stress during cooling.

Modern WAAM setups also integrate real-time monitoring and feedback control systems, which employ cameras, thermal sensors, and arc stability detectors to track process parameters such as temperature, bead height, and deposition rate. This digital feedback allows automatic adjustments to maintain consistent quality throughout the build process. Additionally, hybrid configurations combining WAAM with CNC machining heads enable in situ finishing, achieving the precision and surface finish required for marine components. In Figure 7.2, typical WAAM equipment setup for marine fabrication shows the essential components of a WAAM system. The setup includes a power source that provides arc energy, a wire feeder supplying continuous metal wire, and a robotic arm that positions the welding torch for precise material deposition. The shielding gas system protects the molten pool from oxidation, while a monitoring unit captures real-time data on arc stability and bead geometry. The substrate or base plate serves as the foundation for layer-by-layer buildup. This

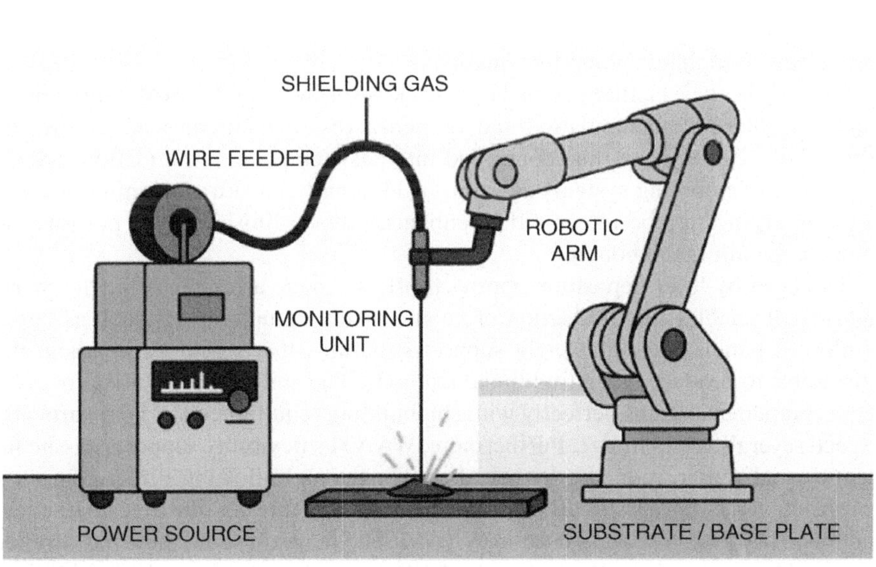

FIGURE 7.2 Typical WAAM equipment setup for marine fabrication.

configuration demonstrates how automation and control systems integrate to ensure accuracy, repeatability, and high-quality fabrication in marine manufacturing.

7.2.4 Layer-by-Layer Deposition in Large Marine Structures

The fundamental principle of WAAM lies in its layer-by-layer deposition approach, where metal is progressively built up from a digital 3D model to form large and complex components. This controlled sequential deposition enables the fabrication of near-net-shape marine structures with exceptional material efficiency and design flexibility key advantages for shipbuilding and offshore applications. Unlike conventional subtractive fabrication, where excess material is removed, WAAM adds only the necessary amount of metal in each layer, optimizing both production time and resource utilization.

In the WAAM process, each layer is deposited by melting a continuous wire feedstock through an electric arc, forming a molten pool that solidifies upon cooling. The deposition path is governed by preprogrammed tool trajectories derived from CAD/CAM models, ensuring dimensional accuracy and repeatability. The robotic or multi-axis motion system maneuvers the welding torch precisely, maintaining consistent bead geometry and interlayer fusion. After each layer is solidified, subsequent passes are deposited in a controlled sequence, gradually building the component's full geometry. The interlayer temperature is closely monitored to maintain a balance between sufficient fusion and minimal distortion, which is especially critical for large-scale marine components such as propeller blades, hull reinforcements, and stiffening ribs.

For large structures, WAAM employs multi-pass and multi-directional deposition strategies to accommodate varying geometries and load requirements. Adaptive slicing algorithms divide the digital model into optimized layers based on part thickness, ensuring uniform heat distribution and minimizing residual stresses. Real-time process control systems equipped with sensors and cameras continuously monitor arc stability, bead width, layer height, and cooling rate, automatically adjusting process parameters to maintain consistent quality.

One of the key challenges in large-scale layer-by-layer WAAM fabrication is managing thermal accumulation and residual stress, which can lead to distortion or cracking. To mitigate this, controlled interpass temperature regulation, preheating, and active cooling systems are used. Additionally, in situ machining or rolling between layers may be integrated to enhance surface finish, reduce porosity, and improve metallurgical bonding.

The layer-by-layer deposition approach offers unique advantages for the marine industry. It enables the production of geometrically complex parts such as lattice-reinforced panels and lightweight support structures that would be impractical or impossible to produce using traditional methods. The ability to scale the process for large components aligns perfectly with shipbuilding requirements, where parts often exceed several meters in size. Furthermore, WAAM's flexibility supports on-site fabrication and repair, reducing logistical challenges and allowing direct component construction in shipyards or offshore environments. Figure 7.3 illustrates the sequential metal deposition process used in WAAM. The diagram shows how the wire feed is melted by an electric arc to create a molten pool on the substrate, which solidifies to form each layer. The robotic control path guides the movement of the torch, ensuring

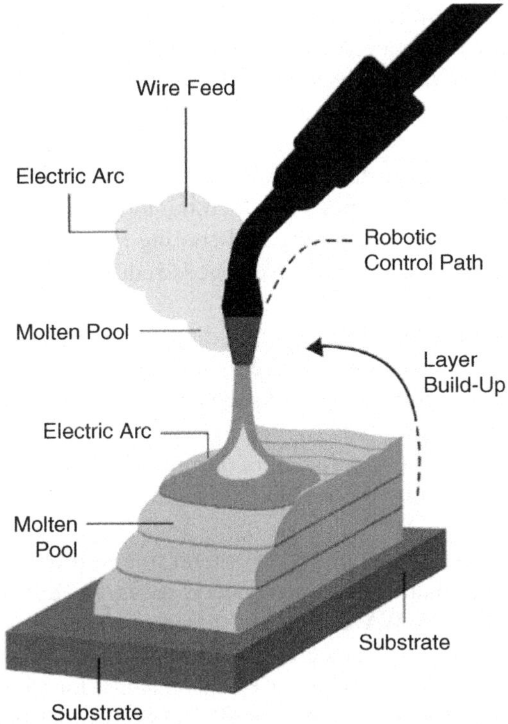

FIGURE 7.3 Layer-by-layer deposition in WAAM for large marine components.

precise alignment and uniform layer buildup. This continuous process demonstrates how WAAM fabricates large marine components with controlled geometry, strong metallurgical bonding, and high material efficiency.

7.3 WAAM PROCESS PARAMETERS AND MATERIAL CONSIDERATIONS

The quality, mechanical properties, and geometric accuracy of parts produced through WAAM depend largely on the control and optimization of process parameters, shielding conditions, and material selection. For marine applications, where corrosion resistance, strength, and durability are paramount, careful regulation of electrical, thermal, and metallurgical variables is essential to achieve consistent layer deposition and reliable structural performance.

The four primary process parameters current, voltage, travel speed, and wire feed rate determine the thermal energy input, bead geometry, and bonding quality between successive layers.

Current (A): The welding current controls the heat generation and penetration depth. Higher current increases melting efficiency and deposition rate but can cause excessive heat input, leading to distortion or wider beads.

For marine-grade steels, currents typically range between 150 and 300 A depending on wire diameter and desired layer thickness.

Voltage (V): Voltage influences arc length and stability. A longer arc produces a wider bead with less penetration, while too short an arc risks irregular melting and spatter. Maintaining a stable voltage ensures uniform bead shape and minimizes surface defects.

Travel Speed (mm/s): The speed at which the torch moves along the deposition path affects layer uniformity and cooling rate. Lower speeds increase heat accumulation, while excessive speeds reduce fusion and may lead to underfill defects.

Wire Feed Rate (m/min): The feed rate dictates the rate of material deposition. It must be balanced with current and travel speed to maintain consistent molten pool volume and avoid porosity or bead irregularities.

Optimal coordination of these parameters ensures stable metal transfer, consistent bead morphology, and minimal distortion, especially crucial when fabricating large marine components.

The shielding gas composition plays a critical role in protecting the molten pool from oxidation, contamination, and atmospheric gases, directly affecting weld quality and microstructure. In WAAM, argon, helium, or argon-helium mixtures are commonly used. Argon provides stable arc characteristics, while helium improves penetration and heat transfer, making Ar-He blends ideal for aluminum and titanium alloys used in shipbuilding. For steels, small additions of carbon dioxide (CO_2) or oxygen (O_2) (1%–5%) enhance arc stability and bed wetting but must be carefully controlled to prevent oxidation.

The wire composition must match or complement the base material to ensure good metallurgical compatibility. Marine-grade filler wires are selected based on corrosion resistance, mechanical strength, and resistance to hydrogen embrittlement. For example, austenitic stainless steels (ER308L, ER316L) and low-carbon steels (ER70S-6) are widely used in ship structures, while Al-Mg (ER5356) and Ti-6Al-4V wires are employed for lightweight or corrosion-critical applications.

WAAM supports a wide range of marine-grade materials suitable for shipbuilding and offshore structures:

Steels: Mild- and low-alloy steels are the most common materials due to their cost-effectiveness and mechanical reliability. They exhibit good weldability and are used in hull fabrication, stiffeners, and deck structures. Stainless steels (especially austenitic grades) are favored for their corrosion resistance in seawater environments.

Aluminum Alloys: Aluminum alloys (e.g., Al-Mg and Al-Mg-Si series) are increasingly used in ship superstructures and small vessels due to their lightweight and corrosion-resistant properties. WAAM enables large aluminum components to be fabricated with minimal distortion compared to conventional welding.

Titanium Alloys: Titanium alloys, particularly Ti-6Al-4V, are used in high-performance marine applications such as propellers, underwater vehicles, and offshore platforms. They offer a superior strength-to-weight ratio and corrosion resistance, though they require precise shielding and thermal control due to their sensitivity to oxidation.

These materials, combined with WAAM's adaptability, enable the fabrication, repair, and reinforcement of a broad spectrum of marine components under controlled digital environments.

Heat input is a critical factor in WAAM, influencing the cooling rate, microstructural evolution, and residual stress distribution within the deposited layers. Excessive heat input can lead to grain coarsening, distortion, and reduced mechanical properties, while insufficient heat may cause poor interlayer bonding and porosity. Therefore, maintaining a balanced energy density through precise control of current, voltage, and travel speed is vital for achieving uniform metallurgical quality.

Residual stresses develop due to the uneven thermal expansion and contraction during layer deposition, particularly in large or thick components. To mitigate this, several techniques are applied:

- Interpass temperature control, where each layer is deposited after reaching an optimal temperature range, prevents excessive accumulation of heat.
- Preheating and controlled cooling reduce thermal gradients and minimize cracking risks.
- In situ rolling or interlayer machining can refine the microstructure, relieve stress, and improve surface finish.
- Post-deposition heat treatment (PDHT) is often applied to restore ductility, homogenize the microstructure, and reduce residual stresses in high-strength marine alloys.

Effective thermal management ensures that WAAM-fabricated marine components exhibit high dimensional stability, mechanical strength, and fatigue resistance essential for long-term service in demanding maritime environments. Figure 7.4 illustrates the interrelationship between primary process parameters and their effect on weld

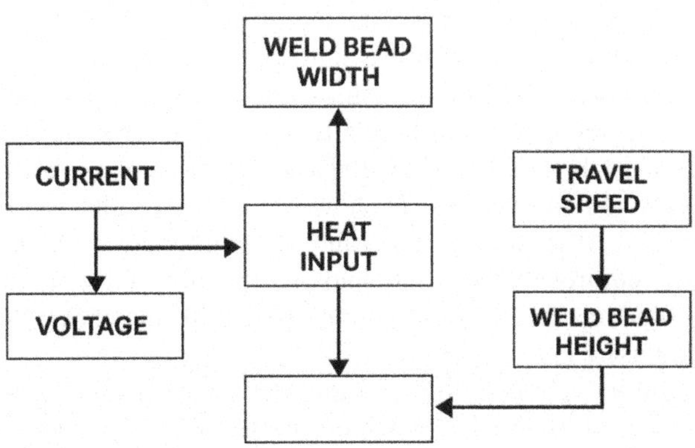

FIGURE 7.4 Influence of key WAAM process parameters on bead geometry and heat input.

bead characteristics. An increase in current and voltage elevates the overall heat input, resulting in wider bead formation and deeper penetration. Conversely, a higher travel speed reduces heat accumulation, producing narrower and higher beads with lower energy input. Proper balance among these parameters is crucial for achieving uniform layer formation, minimizing distortion, and ensuring consistent mechanical properties in WAAM-fabricated marine components.

7.4 CHALLENGES OF WAAM IN THE MARINE INDUSTRY

While WAAM offers significant potential for transforming marine fabrication through its scalability, material efficiency, and digital integration, several technical and operational challenges currently limit its widespread adoption in shipbuilding and offshore applications. These challenges are primarily related to process control, material performance, surface quality, and qualification standards, all of which must be addressed to ensure reliable, repeatable, and certified production of large marine components.

One of the foremost challenges is heat input management and residual stress control. Due to the high energy levels of arc-based processes, excessive heat accumulation can lead to distortion, warping, and microstructural inconsistencies in large builds. These thermal effects also induce residual stresses that compromise dimensional accuracy and fatigue performance, particularly in thick or complex structures. Effective mitigation requires precise control of interpass temperature, adaptive heat regulation, and PDHT, all of which add process complexity and cost.

Another major issue is surface finish and geometric accuracy. WAAM inherently produces rough surfaces and visible weld bead patterns that require extensive post-processing such as machining, grinding, or milling to meet the tight tolerances demanded by marine components. Although hybrid WAAM systems incorporating in situ machining are emerging, they increase setup requirements and machine footprint, which can be challenging in constrained shipyard environments.

Porosity, lack of fusion, and microstructural heterogeneity also pose significant metallurgical challenges. The rapid solidification and multi-pass nature of WAAM can cause defects related to gas entrapment, oxidation, or incomplete bonding between layers. For reactive materials like aluminum and titanium alloys, maintaining shielding gas purity and controlling atmospheric contamination are critical yet difficult to achieve consistently in open shipyard environments.

From a material standpoint, limited data on the long-term corrosion and fatigue behavior of WAAM-fabricated marine alloys under seawater exposure remain a concern. While steels and stainless steels show promising results, the performance of lightweight materials such as aluminum and titanium under cyclic loading and corrosive conditions requires further study and qualification. This gap in material standards currently hinders classification societies from fully certifying WAAM parts for critical structural use.

Additionally, process standardization and certification represent another major barrier. The absence of established marine-specific WAAM guidelines within international standards (e.g., ISO, AWS D3.6M, DNV, or ABS) limits its application to non-critical or experimental components. To achieve broader adoption, consistent qualification procedures, inspection protocols, and mechanical property databases must be developed through collaborative research between shipyards, academia, and classification agencies.

Operationally, implementing WAAM in shipyards requires significant investment in robotic systems, digital process monitoring, and operator training. Many shipyards still rely on manual or semi-automated welding, and transitioning to fully digital additive workflows demands a cultural and technical shift. Integration with existing CAD/CAM and digital twin systems is also challenging due to compatibility and data management issues. Table 7.3 outlines the key technical, metallurgical, and

TABLE 7.3
Summary of Key Challenges and Mitigation Strategies for WAAM in Marine Applications

Challenge	Cause/Description	Impact on Marine Components	Mitigation Strategies
Excessive heat input and residual stress	High arc energy and poor thermal control cause distortion and internal stress accumulation	Dimensional inaccuracy, warping, and reduced fatigue strength	Optimize current and voltage; use interpass temperature control, preheating, and post-deposition heat treatment (PDHT)
Surface roughness and geometric inaccuracy	Layer-by-layer weld deposition produces uneven bead profiles and visible ridges	Requires extensive machining and finishing to meet marine tolerances	Integrate hybrid WAAM-CNC systems; apply in situ milling or bead smoothing between layers
Porosity and lack of fusion	Inadequate shielding, unstable arc, or improper travel speed	Reduced mechanical strength and fatigue resistance	Maintain stable arc conditions; ensure gas purity; optimize travel speed and wire feed rate
Microstructural Inhomogeneity	Uneven cooling and reheating cycles during multi-layer deposition	Variable hardness, strength, and corrosion resistance within layers	Employ interpass temperature monitoring; use controlled cooling or active thermal management
Corrosion and fatigue uncertainty	Limited long-term testing data for WAAM-built marine alloys	Unpredictable performance under seawater exposure and cyclic loading	Conduct corrosion-fatigue studies; apply protective coatings; adopt cathodic protection for critical parts
Material-specific challenges (Al, Ti alloys)	High reactivity and oxidation risk in open environments	Formation of oxides and inclusions affecting integrity	Use enclosed chambers or local shielding; maintain high-purity argon flow
Process standardization and certification	Lack of established marine-specific WAAM standards	Limits adoption for critical ship and offshore components	Develop qualification procedures with DNV, ABS, and ISO; establish property databases and certification pathways
Operational integration in shipyards	Requirement of digital systems, robotics, and trained operators	High setup cost and steep learning curve	Implement gradual integration with existing CAD/CAM systems; provide operator training and pilot projects

operational challenges faced in implementing WAAM within the marine industry and highlights practical mitigation strategies. The primary concerns such as thermal distortion, surface roughness, porosity, and process standardization can be effectively addressed through optimized process control, hybrid manufacturing integration, and the establishment of robust qualification frameworks. By combining digital process monitoring, advanced shielding techniques, and cooperative standard development with classification agencies, WAAM can evolve from an emerging innovation to a fully certified and dependable technology for next-generation shipbuilding and offshore engineering applications.

7.5 FUTURE TRENDS OF WAAM IN THE MARINE INDUSTRY

The future of WAAM in the marine industry is being shaped by rapid advancements in automation, AI, digital integration, and material science. As shipyards increasingly adopt Industry 4.0 principles and transition toward smart manufacturing, WAAM is expected to become a cornerstone technology for sustainable, high-precision, and large-scale metal fabrication. The growing focus on eco-efficiency, cost reduction, and design flexibility positions WAAM as a vital enabler of next-generation shipbuilding and offshore production systems.

One of the most significant trends is the integration of AI and machine learning (ML) into WAAM systems for process optimization. AI-driven algorithms can analyze real-time sensor data to automatically adjust current, voltage, wire feed rate, and travel speed, ensuring consistent quality and reducing human intervention. Predictive analytics will enable defect forecasting and adaptive control, minimize rework, and improve component reliability. Machine vision and deep learning-based monitoring systems are also being developed for automated bead geometry assessment, layer tracking, and defect detection, enhancing precision in complex marine components.

The adoption of digital twin technology represents another transformative trend. Digital twins create a virtual replica of the WAAM process and component geometry, allowing engineers to simulate thermal behavior, material flow, and stress evolution before actual fabrication. This predictive modeling capability enables virtual validation of process parameters, leading to optimized energy use, improved deposition accuracy, and reduced experimental costs. When integrated with real-time data acquisition, digital twins will form the backbone of fully connected smart shipyards, enabling continuous improvement through data feedback loops.

Advancements in hybrid manufacturing systems are also reshaping the future of WAAM. Combining additive and subtractive processes in a single setup allows in situ machining or rolling between layers, producing components with near-finished tolerances and improved surface quality. Such hybrid systems are ideal for marine fabrication, where large structures require both high deposition rates and precision finishing. The integration of multi-material deposition, including graded and composite structures, is emerging as a promising avenue for producing lightweight yet high-strength components that meet marine durability standards.

Another critical direction is material innovation and sustainability. Research is expanding into high-strength low-alloy (HSLA) steels, aluminum-lithium alloys, and corrosion-resistant titanium and nickel-based alloys tailored for WAAM processing.

The development of recyclable wire feedstock and low-carbon energy sources aligns WAAM with global decarbonization goals in shipbuilding. Furthermore, on-site or near-site additive repair systems will enable shipyards to perform real-time component restoration, drastically reducing downtime, logistics, and environmental impact.

In the longer term, the marine industry envisions fully automated WAAM production cells integrated into shipyards capable of autonomous fabrication, inspection, and quality certification through embedded AI and digital inspection systems. Such facilities will embody the concept of "smart additive shipbuilding," where design, manufacturing, and maintenance are interconnected in a continuous digital ecosystem.

The future of WAAM in the marine industry lies in its intelligent, hybrid, and sustainable evolution. With advancements in automation, digital twin integration, AI-driven process control, and material development, WAAM is poised to revolutionize marine manufacturing. It will enable the creation of complex, lightweight, and corrosion-resistant structures with unprecedented efficiency ushering in a new era of digital, adaptive, and environmentally responsible shipbuilding. Table 7.4 outlines the key technological directions shaping the future of WAAM in the marine industry. The integration of AI-driven control, digital twin simulations, and

TABLE 7.4
Emerging Trends and Technological Directions of WAAM in the Marine Industry

Emerging Trend	Enabling Technologies	Key Focus/Advancement	Expected Benefits for Marine Applications
AI and machine learning integration	Smart sensors, neural networks, data analytics	Real-time parameter optimization, defect prediction, adaptive control	Improved weld consistency, reduced defects, and higher process automation
Digital twin and virtual simulation	Process modeling, thermal simulations, IoT connectivity	Virtual replication of WAAM operations for predictive control and design validation	Reduced trial-and-error, lower material waste, and enhanced accuracy
Hybrid additive-subtractive systems	Integrated CNC machining and robotic WAAM platforms	In situ machining, surface finishing, and interlayer stress relief	Near-finished surface quality, minimized distortion, and faster production cycles
Advanced materials for WAAM	Alloy design, metallurgy, wire development	Development of high-strength, corrosion-resistant marine alloys (HSLA, Ti, Ni)	Lightweight and durable marine components with enhanced performance
Multi-material and functionally graded deposition	Multi-wire feed systems, thermal gradient control	Graded composition for corrosion protection or tailored strength	Custom material properties, extended component life in harsh environments

(Continued)

TABLE 7.4 *(Continued)*
Emerging Trends and Technological Directions of WAAM in the Marine Industry

Emerging Trend	Enabling Technologies	Key Focus/Advancement	Expected Benefits for Marine Applications
Automation and robotic fabrication cells	Autonomous robotic arms, process control algorithms	Fully automated WAAM systems for shipyard integration	Continuous, high-volume production with reduced labor dependency
Sustainable manufacturing initiatives	Renewable energy integration, eco-efficient materials	Low-carbon wire feedstock and energy optimization	Reduced environmental impact and alignment with green shipbuilding goals
On-site additive repair systems	Portable WAAM units, real-time scanning, and deposition	In situ component repair and remanufacturing	Lower maintenance cost, reduced downtime, and logistical flexibility

robotic automation will enable real-time decision-making and fully automated production environments in shipyards. Meanwhile, material innovations and hybrid additive-subtractive systems will improve component quality, efficiency, and sustainability. Collectively, these emerging trends are steering WAAM toward becoming an integral part of smart, data-driven, and environmentally conscious shipbuilding ecosystems, capable of producing high-performance marine structures with unmatched precision and adaptability.

7.6 CASE STUDY: WIRE ARC ADDITIVE MANUFACTURING OF SS308L USING CMT-BASED ROBOTIC METAL INERT GAS (MIG) WELDING

This case study presents the fabrication and characterization of an SS308L stainless steel component using cold metal transfer (CMT)-based WAAM integrated with a robotic metal inert gas (MIG) welding system. The investigation emphasized microstructural evolution, mechanical performance, and thermal behavior, highlighting the process's potential for marine structural fabrication and repair applications.

The experimental setup, as shown in Figure 7.5, utilized a KUKA 6-axis robotic manipulator fitted with a Fronius TP500i welding power source and a WF25i REEL R wire feeder, controlled through an automated teach pendant. The SS308L-T1 flux-cored wire served as the feedstock material, while a shielding gas mixture of argon (Ar) and 2.5% CO_2 provided arc stability and minimized oxidation. The optimized process parameters, current: 133 A, voltage: 19 V, wire feed rate: 5 m/min, and torch travel speed: 0.368 m/min, yielded consistent bead geometry with minimal porosity and spatter.

A 3D wall structure ($149 \times 22 \times 125\,mm^3$) was fabricated layer by layer through a preprogrammed robotic path. A K-type thermocouple embedded in the substrate monitored temperature fluctuations during successive bead depositions, enabling

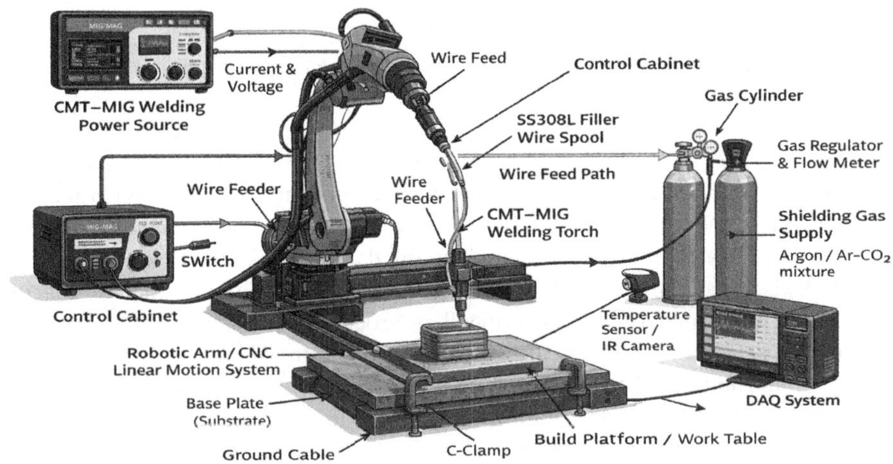

FIGURE 7.5 Experimental setup of CMT + MIG-based WAAM system for SS308L fabrication. (Haldar et al. 2024.)

analysis of heat transfer, layer reheating, and cooling trends. Recorded data revealed an initial rise in substrate temperature with increasing layer count, followed by gradual stabilization due to enhanced convective cooling at higher build heights.

Microstructural characterization revealed a dual-phase composition of γ-austenite dendrites containing residual δ-ferrite in vermicular, skeletal, and lathy morphologies. The Cr eq/Ni eq ratio (≈1.72) indicated a ferrite austenite (F/A) solidification mode, ensuring hot cracking resistance and corrosion stability. Along the build direction, a transition from fine columnar grains (near the substrate) to coarser equated grains (toward the top) was observed, primarily due to varying thermal gradients and cooling rates.

Mechanical testing confirmed directional anisotropy. The horizontal tensile specimens exhibited higher ultimate tensile strength (542 MPa) than vertical specimens (520 MPa) due to grain boundary alignment and favorable load transfer along the deposition plane. Fractographic analysis displayed ductile fracture behavior, featuring micro-dimpled surfaces and void coalescence, typical of sound metallurgical bonding. Microhardness varied between 172 and 197 $HV_{0.1}$, with higher values near the substrate owing to refined microstructure and ferrite enrichment. PDHT (1,200°C for 4 hours) transformed dendritic microstructures into equiaxed austenitic grains, reducing hardness (~142 $HV_{0.1}$) due to grain coarsening and ferrite dissolution (Haldar et al. 2024).

From a marine engineering perspective, SS308L's excellent corrosion resistance, toughness, and weldability make it ideal for offshore and underwater applications. CMT-based WAAM offers significant promise for in situ manufacturing, structural refurbishment, and repair of stainless steel components such as ship hull reinforcements, propeller shafts, ballast tank fixtures, and underwater pipeline sections. The robotic precision and thermal control inherent in CMT-WAAM ensure dimensional accuracy, low defect rates, and enhanced surface integrity, critical for demanding marine environments.

7.7 SUMMARY

The chapter provides a concise overview of how WAAM is revolutionizing shipbuilding and offshore engineering through large-scale, efficient, and digitally controlled metal fabrication. It traces the evolution of AM from polymer-based prototyping to advanced metal processes, emphasizing WAAM's advantages over conventional fabrication in terms of material efficiency, design flexibility, and cost reduction. The fundamentals of WAAM, including its principle, working mechanism, equipment setup, and layer-by-layer deposition, are explained alongside comparisons with other AM methods, establishing WAAM as the most practical solution for large marine structures. The chapter further examines key process parameters (current, voltage, wire feed rate, travel speed), shielding gas effects, and marine-grade materials such as steels, aluminum, and titanium, while addressing heat input and residual stress management essential for maintaining structural integrity. Practical applications such as propeller repair, hull fabrication, and offshore component restoration highlight WAAM's operational versatility. The discussion on challenges covers issues like surface roughness, porosity, process control, and certification gaps that must be resolved for broader industrial adoption. Finally, the chapter explores future trends, including AI-driven optimization, hybrid WAAM-CNC systems, digital twin integration, sustainable material development, and the emergence of fully automated smart shipyards, positioning WAAM as a cornerstone technology for next-generation, intelligent, and eco-efficient marine manufacturing.

7.8 REVIEW QUESTIONS

1. What is wire arc additive manufacturing (WAAM), and how does it differ from traditional metal fabrication methods?
2. Describe the evolution of additive manufacturing in the marine sector from polymer-based to metal-based processes.
3. What key limitations of conventional shipbuilding does WAAM aim to overcome?
4. List three major advantages of WAAM over conventional marine fabrication techniques.
5. Discuss the scope of WAAM in shipbuilding and offshore engineering with examples of typical applications.
6. Explain the fundamental working principle of the WAAM process.
7. What is the role of the electric arc in WAAM, and how does it contribute to layer formation?
8. Describe the layer-by-layer deposition process and its importance in fabricating large marine structures.
9. Compare WAAM with other additive manufacturing processes such as SLM, EBM, and LMD.
10. Identify the main components of a typical WAAM setup used in marine fabrication.
11. How do current, voltage, travel speed, and wire feed rate affect the WAAM process outcome?

12. Why is the control of heat input critical in WAAM, and what methods are used to regulate it?
13. Discuss the influence of shielding gas composition on the quality of WAAM deposits.
14. What are the key characteristics that make an alloy suitable for WAAM in marine applications?
15. Compare the use of steels, aluminum, and titanium alloys in WAAM for shipbuilding applications.
16. How does interpass temperature control influence the microstructure and mechanical properties of WAAM parts?
17. Explain the causes and effects of residual stress formation in WAAM-built components.
18. What techniques are commonly used to reduce or manage residual stresses during WAAM fabrication?
19. Describe how WAAM is used for the repair of damaged or worn marine components.
20. What advantages does WAAM offer in the fabrication of large ship sections and offshore structures?
21. How can WAAM be applied for on-site or remote fabrication within shipyards?
22. Discuss the role of WAAM in propeller blade repair and performance restoration.
23. In what ways does WAAM contribute to the sustainability and life cycle management of marine assets?
24. What are the primary challenges faced in implementing WAAM in the marine industry?
25. How does excessive heat input affect dimensional accuracy and surface quality?
26. Explain the causes of porosity and lack of fusion in WAAM deposits and how these can be mitigated.
27. Why is process certification and standardization a major barrier for industrial adoption of WAAM?
28. How can shipyards integrate WAAM into existing digital manufacturing workflows?
29. Discuss how AI and machine learning are expected to enhance the WAAM process in the future.
30. What are the key technological directions driving the future of WAAM in marine manufacturing, including hybrid systems, digital twins, and sustainable practices?

8 Welding for Offshore and Subsea Structures

8.1 INTRODUCTION

Welding serves as the backbone of fabrication and maintenance in the offshore and subsea industries, where structural integrity, durability, and reliability are critical to ensuring safe and long-term operation in some of the world's harshest environments. Offshore platforms, subsea pipelines, risers, and support structures are continuously subjected to extreme conditions such as high hydrostatic pressure, saltwater corrosion, strong currents, and fluctuating temperatures. These demanding service conditions make welding technology not merely a joining method but a vital engineering discipline that directly determines the safety, performance, and lifespan of marine infrastructure (Gyasi 2019).

In offshore engineering, welding is extensively employed in the construction of fixed platforms, floating production systems, jackets, decks, subsea manifolds, and pressure vessels. Similarly, in subsea operations, welding is used for pipeline tie-ins, repair, and component joining in deep and ultra-deep waters where human access is limited. The complexity of these environments requires specialized welding processes, including underwater wet and dry welding, hyperbaric welding, and advanced automated techniques, each tailored to overcome challenges of pressure, accessibility, and metallurgical behavior under submerged conditions.

The evolution of welding for offshore and subsea structures has been driven by technological advancements in arc welding, friction stir welding, laser hybrid processes, and automated robotic systems, coupled with improvements in weld metallurgy and non-destructive testing (NDT). These innovations have enabled the fabrication of large, high-strength structures that meet stringent international standards such as AWS D3.6M, ISO 15618, and DNV-OS-C401. In addition, the growing integration of digital monitoring, remote welding systems, and artificial intelligence (AI)-based inspection is transforming how subsea welding is performed, allowing for real-time quality control and predictive maintenance.

However, welding in offshore and subsea environments also presents significant challenges, including hydrogen embrittlement, porosity, microcracking, residual stress, and corrosion fatigue. The selection of appropriate materials, filler metals, and welding techniques must therefore be optimized to maintain weld integrity under cyclic loading and corrosive seawater exposure. Furthermore, stringent qualification procedures and inspection standards are essential to ensure compliance with safety and reliability requirements.

This chapter introduces the principles, techniques, and challenges associated with welding for offshore and subsea structures. It explores conventional and advanced

DOI: 10.1201/9781003217350-8

welding processes, metallurgical considerations, underwater welding technologies, inspection standards, and emerging trends such as robotic and remotely operated welding systems. By bridging theory and practical application, the chapter provides a comprehensive understanding of how welding continues to enable the safe, efficient, and sustainable development of offshore energy infrastructure across oil, gas, and renewable sectors.

8.1.1 Importance of Welding in Offshore and Subsea Engineering

Welding plays an indispensable role in offshore and subsea engineering, serving as the fundamental process for fabricating, assembling, and maintaining the vast range of structures that operate in marine environments. From towering offshore platforms and floating production systems to subsea pipelines and manifolds, welding ensures the structural integrity, leak tightness, and durability of components that must withstand enormous hydrostatic pressures, corrosive seawater exposure, and dynamic mechanical loads over decades of service. Unlike conventional land-based fabrication, offshore welding must meet exceptionally high standards of precision and reliability, as even minor weld imperfections can lead to catastrophic failures in remote or inaccessible underwater locations.

In offshore engineering, welding is central to the construction of critical load-bearing structures such as jacket frames, decks, legs, and risers. These welded joints are designed to sustain cyclic stresses from wind, waves, and currents, as well as the weight of drilling and production equipment. Similarly, in subsea systems, welding ensures the integrity of pipelines, flowlines, and pressure vessels, which transport hydrocarbons and other fluids under high-pressure and high-temperature conditions. The ability of welds to maintain mechanical strength, ductility, and corrosion resistance under these environments directly influences the safety, performance, and lifespan of offshore installations.

The importance of welding also extends to maintenance and repair operations in offshore and subsea sectors. Over time, structures experience fatigue cracking, corrosion damage, and wear due to continuous exposure to the marine environment. Welding enables on-site and underwater repair, often using advanced techniques such as hyperbaric dry welding, wet welding, and remotely operated robotic systems, which allow for restoration of damaged areas without complete system shutdown or retrieval. This capability significantly reduces downtime, operational costs, and safety risks associated with offshore maintenance.

Moreover, welding provides the flexibility to join a wide range of materials used in marine construction, including carbon steels, low-alloy steels, stainless steels, and corrosion-resistant alloys (CRAs). Each material presents unique challenges related to heat input, hydrogen absorption, and metallurgical transformations, requiring careful control of welding parameters and consumable selection. Through optimized welding procedures and post-weld treatments, engineers ensure that the mechanical and metallurgical continuity across joints meets the stringent requirements set by classification societies such as DNV, ABS, and ISO.

In modern offshore engineering, welding is not just a fabrication tool, but it is a strategic enabler of innovation and safety. The integration of automation, sensors, and real-time monitoring is transforming welding into a data-driven process that supports digital shipyards and smart offshore construction. As the industry moves toward deepwater and ultra-deepwater developments, the demand for advanced welding technologies, such as laser hybrid welding, friction stir welding, and additive repair, continues to grow.

8.1.2 Types of Offshore Structures (Fixed, Floating, Subsea)

Offshore and subsea engineering encompasses a wide variety of structures, each requiring specialized welding techniques to meet the stringent mechanical, corrosion, and safety requirements imposed by harsh marine environments. The welding approach, materials used, and quality control measures depend on the type, function, and operational conditions of each structure. Broadly, these can be classified into fixed platforms, floating systems, subsea structures, pipelines, and auxiliary components, all of which rely heavily on precision welding for their fabrication, assembly, and long-term service performance.

8.1.2.1 Fixed Offshore Platforms

Fixed platforms, such as jacket-type structures, are permanently anchored to the seabed using steel piles and serve as the primary base for drilling, production, and processing operations. These massive welded frameworks, typically made of high-strength low-alloy (HSLA) steels, require thousands of meters of welded joints that must endure cyclic stresses from ocean currents, wind, and wave loading. Welding in these

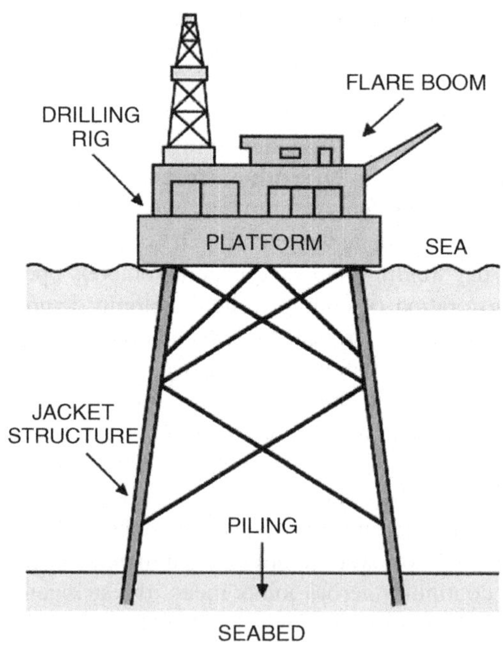

FIGURE 8.1 Fixed offshore platforms.

platforms involves extensive use of shielded metal arc welding (SMAW), flux-cored arc welding (FCAW), and submerged arc welding (SAW) to ensure deep penetration, low defect rates, and uniform mechanical properties across joints. Figure 8.1 illustrates the structural configuration of a typical jacket-type offshore platform used for oil and gas production. The diagram highlights key components, including the drilling rig, flare boom, main platform deck, jacket structure, and piling system, which are anchored into the seabed. The jacket framework, constructed from welded tubular steel members, provides stability against wave, wind, and current loads. Welded joints within the jacket and deck structures are crucial for ensuring high fatigue resistance and load-bearing capability. This figure demonstrates how welding underpins the overall strength, integrity, and longevity of fixed offshore platforms operating in harsh marine environments.

8.1.2.2 Floating Offshore Structures (FOS)

Floating structures, including floating production storage and offloading (FPSO) units, tension leg platforms (TLPs), and semi-submersibles, rely on welded steel hulls and mooring systems designed to flex and absorb dynamic ocean forces. Welding in these applications must ensure fatigue resistance and watertight integrity, particularly at stress concentration points such as joints and corners. Specialized processes such as gas metal arc welding (GMAW) and automated robotic welding are employed to achieve consistent weld quality and repeatability in complex curved geometries. The welding of cryogenic and corrosion-resistant materials is also critical for structures involved in LNG storage or transportation. Figure 8.2 illustrates the main categories of floating platforms used in deepwater offshore operations, including the TLP, semi-submersible platform, and spar platform. Each design is optimized to maintain stability and functionality under varying ocean depths and environmental conditions. The TLP is anchored to the seabed by vertical tendons providing high vertical stiffness and minimal heave motion, suitable for moderate to deep waters. The semi-submersible platform achieves buoyancy through partially submerged pontoons, offering excellent stability in rough seas and flexibility for drilling and production. The spar platform,

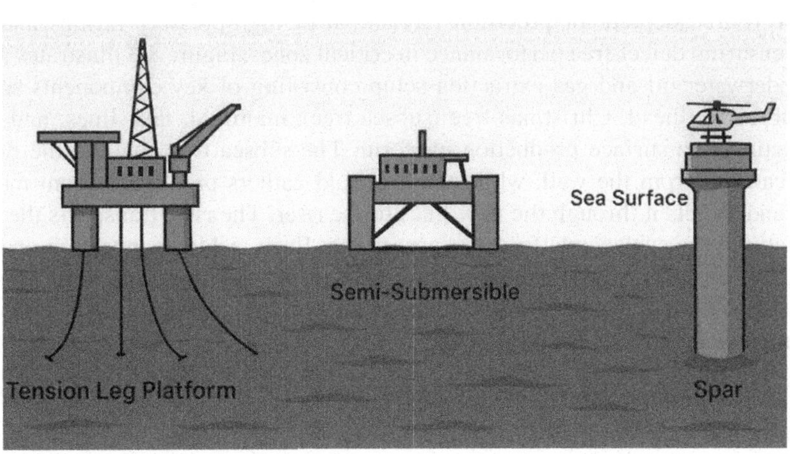

FIGURE 8.2 Floating offshore structures (FOS).

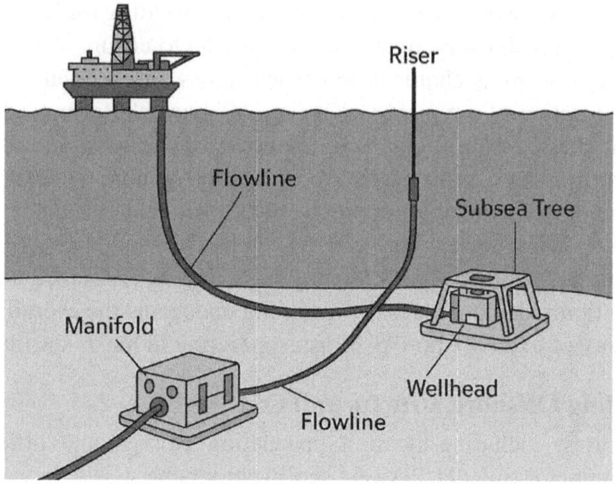

FIGURE 8.3 Subsea production systems.

characterized by a tall cylindrical hull extending deep below the surface, is designed for ultra-deepwater operations, providing superior motion damping and load distribution. Together, these structures represent the evolution of floating offshore engineering, where advanced welding and fabrication technologies ensure safety, durability, and longevity in demanding marine environments.

8.1.2.3 Subsea Production Systems

Subsea structures include manifolds, templates, risers, jumpers, and wellhead assemblies, which operate at depths exceeding several thousand meters. These components are exposed to high external pressures, requiring welds with superior toughness, ductility, and crack resistance. Hyperbaric welding techniques, both dry (habitat) and wet, are used for joining and repairing these components directly on the seabed. Additionally, narrow-gap welding and orbital TIG (gas tungsten arc welding [GTAW]) are adopted for precision fabrication of high-pressure fittings and flow lines, ensuring defect-free performance in critical zones. Figure 8.3 illustrates a typical underwater oil and gas extraction setup consisting of key components such as the subsea wellhead, Christmas tree (subsea tree), manifold, flow lines, and risers connecting to a surface production platform. The subsea tree controls the flow of hydrocarbons from the well, while the manifold gathers production from multiple wells and directs it through the flow lines to the riser. The riser transports the fluids vertically to the surface platform for processing. Each welded connection and joint in this system must withstand high pressure, temperature variations, and corrosive seawater conditions, underscoring the importance of precision welding and inspection in subsea production engineering.

8.1.2.4 Subsea and Offshore Pipelines

Pipelines and risers form the lifelines of offshore operations, transporting oil, gas, and fluids between subsea wells and surface facilities. These long, continuous structures require thousands of circumferential welds, which must withstand internal

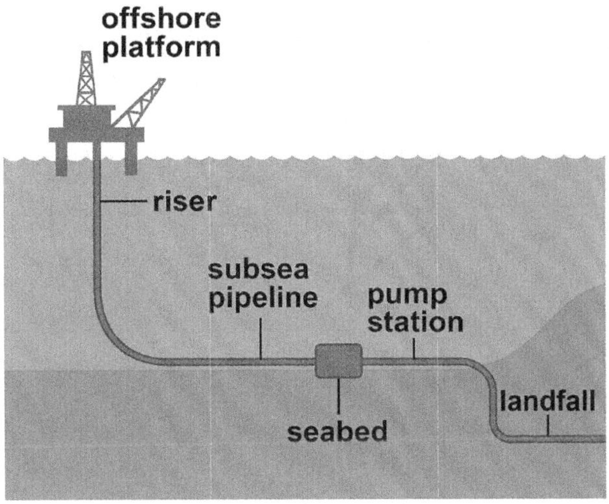

FIGURE 8.4 Subsea and offshore pipelines.

pressure, bending stresses, and external corrosion. Advanced pipeline welding methods such as mechanized GMAW, SAW, and orbital welding are employed to achieve high productivity and consistency. Automated welding systems with real-time monitoring ensure weld integrity even under challenging field conditions, often supported by in-line inspection and NDT. Figure 8.4 illustrates a typical subsea pipeline system used for transporting oil and gas between offshore platforms and onshore terminals. The diagram shows an offshore platform connected to a subsea pipeline through a riser, which transfers fluids vertically to or from the seabed. The pipeline runs along the seabed, incorporating key components such as pump or booster stations, pipeline end manifolds (PLEMs), and a landfall connection for onshore delivery. Each section of the pipeline is joined through precision welding to ensure leak-tight performance under high-pressure and corrosive marine conditions. This figure highlights how welded subsea pipeline networks form the critical link in offshore production and transportation infrastructure.

8.1.2.5 Auxiliary and Support Structures

Auxiliary offshore components such as flare towers, cranes, helidecks, riser clamps, and support frames also depend heavily on welding for fabrication and installation. While these are secondary structures, their failure can lead to severe operational and safety risks. Welding processes for these components prioritize ease of access, repairability, and corrosion protection, often using FCAW and GMAW with anti-corrosive coatings and post-weld surface treatments. Figure 8.5 illustrates various secondary components essential for offshore platform functionality, including a submerged buoy, jacket foundation, subsea template, and mooring line. These structures play crucial roles in ensuring platform stability, equipment support, and load transfer to the seabed. The jacket foundation provides the main structural base for topside facilities, while subsea templates guide well placement and subsea installations. Buoys and mooring lines assist in maintaining platform position and stability

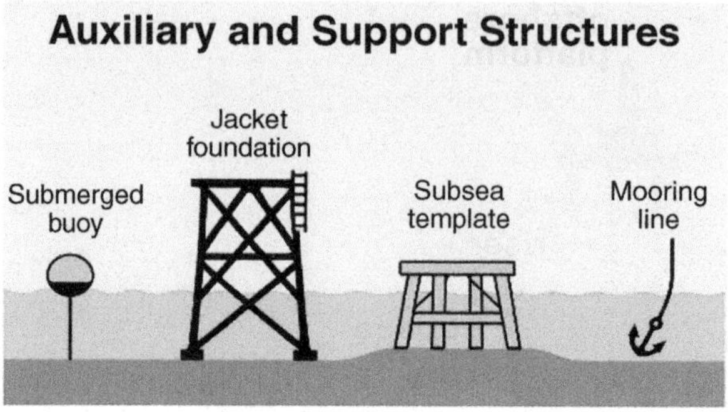

FIGURE 8.5 Auxiliary and support structures.

under dynamic sea conditions. Each component relies heavily on precise welding and fabrication to ensure durability, strength, and corrosion resistance in the marine environment.

8.1.3 Role of Welding in Structural Integrity and Safety

Welding is the cornerstone of structural integrity and safety in offshore and subsea engineering, where the performance of critical joints determines the overall reliability and lifespan of marine structures. Offshore installations operate under some of the most demanding environmental and mechanical conditions on earth exposed to cyclic loading, high hydrostatic pressure, corrosion, and dynamic wave action, making the quality of welded connections fundamental to ensuring operational safety. The ability of a structure to resist failure depends not only on its design but also on the metallurgical soundness, mechanical strength, and defect-free nature of its welds.

In offshore construction, welded joints form the load-transmitting links that unify structural members into a single, continuous framework. For example, in jacket platforms and subsea manifolds, the welds must sustain fluctuating stresses from wave and wind forces, as well as axial and bending loads from heavy equipment. Poorly executed welds or improper heat control can lead to localized weaknesses, initiating cracks, fatigue failures, or brittle fractures, especially in high-strength steels used for deepwater structures. Therefore, achieving consistent weld penetration, fusion, and microstructural uniformity is crucial for long-term structural performance.

The safety implications of welding in offshore environments are profound. Structural or pipeline failure due to weld defects can lead to catastrophic consequences, including oil spills, environmental pollution, equipment loss, and human casualties. To mitigate such risks, offshore welding practices adhere to stringent standards such as AWS D3.6M, ISO 15618, DNV-ST-F101, and ABS Rules, which define weld quality requirements, inspection procedures, and qualification of welding personnel. NDT methods such as ultrasonic testing (UT), radiography (RT), magnetic particle testing (MT), and visual inspection (VT) are routinely applied to detect flaws before structures are commissioned.

Moreover, welding plays a vital role in maintaining fatigue resistance and fracture toughness. Offshore structures experience millions of stress cycles over their operational life, and any discontinuity at a weld toe or root can act as a fatigue initiation site. Advanced welding techniques such as narrow-gap welding, friction stir welding, and automated GMAW help minimize such discontinuities, ensuring improved stress distribution and extended fatigue life. Post-weld treatments, including peening, grinding, and stress relief heat treatment (SRHT), are also employed to enhance residual stress profiles and prevent crack propagation.

In subsea operations, where repairs are often performed by divers or remotely operated vehicles (ROVs), welding integrity becomes even more critical. Hyperbaric and dry habitat welding techniques ensure precise control of the welding environment, allowing strong, defect-free joints even under high ambient pressures. The reliability of these welds directly affects the safety of subsea pipelines, manifolds, and risers, which must remain leak-proof under extreme conditions.

Ultimately, welding acts as both a structural and safety assurance mechanism in offshore and subsea systems. Its role extends beyond simple joining, and it encompasses the entire life cycle of the structure, from fabrication and assembly to inspection, maintenance, and repair. By ensuring metallurgical continuity, mechanical strength, and resistance to degradation, welding upholds the structural integrity and operational safety that form the foundation of modern offshore engineering.

8.2 CHALLENGES IN WELDING OFFSHORE RIGS AND PLATFORMS

Welding on offshore rigs and platforms presents some of the most complex challenges in modern engineering due to the harsh environmental conditions, demanding safety requirements, and logistical constraints associated with marine operations. Offshore welding must deliver consistent quality, high strength, and long-term durability under extreme and unpredictable oceanic conditions. The combination of environmental, metallurgical, and operational challenges makes offshore welding distinct from land-based fabrication, requiring specialized processes, advanced materials, and stringent inspection protocols.

8.2.1 Environmental Challenges

One of the foremost difficulties in offshore welding arises from harsh marine environments. Welders and automated systems must operate under high humidity, salt-laden air, fluctuating temperatures, and strong wind conditions—all of which can compromise weld quality. Saltwater contamination leads to oxidation, porosity, and corrosion at the weld interface, while moisture increases the risk of hydrogen-induced cracking (HIC). In subsea or splash-zone welding, pressure, visibility, and limited access further complicate welding operations. Additionally, waves and vessel motion create instability, making precision welding and arc control more difficult to achieve in offshore conditions than in controlled workshop environments.

8.2.2 Metallurgical and Material-Related Challenges

The materials used in offshore rigs and platforms—primarily HSLA steels, duplex stainless steels, and CRAs—pose significant metallurgical challenges. These materials are prone to hydrogen embrittlement, cracking, and heat-affected zone (HAZ) softening when not welded under controlled parameters. Managing thermal gradients during multi-pass welding is crucial, as improper heat input can alter the microstructure and reduce fracture toughness. The HAZ hardness and residual stresses must be carefully controlled through precise preheating, interpass temperature regulation, and post-weld heat treatment (PWHT). Furthermore, weld consumable compatibility with base materials is critical to prevent galvanic corrosion, especially when dissimilar metals are joined in structural assemblies.

8.2.3 Accessibility and Positional Welding

Offshore rigs and platforms are vast, complex structures, often requiring welding in confined, elevated, or underwater positions. Limited accessibility to joints, combined with the necessity of out-of-position welding (vertical, overhead), increases difficulty and operator fatigue. Welding in tight spaces or near live equipment also presents safety hazards due to fire risk, restricted movement, and limited ventilation. Robotic or remotely operated welding systems can mitigate some challenges, but their deployment offshore requires additional setup time, specialized calibration, and environmental protection systems.

8.2.4 Safety and Human Factors

Welding in offshore environments exposes personnel to significant risks, including electric shock, burns, fire hazards, and exposure to toxic fumes. In addition, offshore operations are subject to rigorous safety regulations requiring protective gear, grounding systems, and continuous monitoring. Fatigue from long shifts, vibration, and adverse weather can affect welder concentration and precision, increasing the likelihood of human error. Emergency response and evacuation during accidents or equipment failures are also more complicated in offshore conditions, emphasizing the need for highly trained and certified welders capable of maintaining quality under stress.

8.2.5 Equipment and Process Control Limitations

Unlike controlled workshop environments, offshore platforms have limited access to advanced equipment, power supply, and environmental control systems. Equipment must be compact, portable, and resistant to saltwater corrosion. Maintaining stable arc characteristics in windy or wet conditions is challenging, particularly for GMAW or FCAW processes. In underwater welding, the arc stability and shielding gas retention are heavily influenced by water currents and depth, necessitating specialized electrodes and habitat chambers. Real-time process monitoring is often limited, increasing the risk of undetected defects.

8.2.6 Inspection and Quality Assurance Challenges

Offshore weld inspection is difficult due to restricted accessibility and environmental interference. NDT methods such as RT and UT may be constrained by space, radiation safety concerns, or surface irregularities. Moreover, subsea weld inspection requires specialized ROVs or divers, increasing cost and complexity. Continuous real-time monitoring and digital weld traceability are increasingly being introduced to overcome these limitations, yet the technology is still evolving for full offshore deployment.

8.2.7 Economic and Logistical Constraints

Offshore welding operations are expensive and time-sensitive, as delays in fabrication or repair can lead to substantial production losses. Mobilization of welding crews, habitat systems, and inspection teams to remote offshore locations adds significant logistical overhead. Adverse weather conditions, rough seas, or equipment malfunction can halt operations, extending project timelines. These factors necessitate robust planning, redundancy in equipment, and adaptive scheduling to ensure efficiency and safety.

8.2.8 Mitigation Strategies and Technological Advancements in Offshore Welding

Overcoming the complex challenges associated with welding on offshore rigs and platforms requires a combination of strategic process control, advanced materials, environmental adaptation, and digital technology integration. Continuous innovation in welding methodologies, equipment design, and inspection systems has made it possible to improve weld quality, safety, and efficiency under demanding marine conditions. The following mitigation strategies and technological advancements are being implemented to enhance the reliability and productivity of offshore welding operations.

8.2.8.1 Advanced Welding Processes and Automation

The adoption of automated and mechanized welding systems has significantly improved the consistency and repeatability of offshore welds. Processes such as GMAW, FCAW, and SAW are now combined with robotic automation to achieve higher precision and deposition rates. In underwater environments, remote-controlled and robotic welding systems are replacing manual operations, enabling welds to be performed safely at great depths. Techniques such as narrow-gap welding, tandem arc welding, and pulsed GMAW are also used to control heat input, minimize distortion, and improve penetration in thick-walled components.

8.2.8.2 Environmental Control and Habitat Welding

To mitigate environmental effects such as wind, humidity, and saltwater exposure, welding operations are increasingly performed within controlled environments or dry habitats. In subsea applications, hyperbaric welding habitats provide a dry, pressurized chamber around the work area, allowing welders or remotely operated tools to perform high-quality welds at depth. For topside operations, wind shields, humidity barriers, and local enclosures help maintain stable arc conditions and protect the weld zone from contamination. These environmental control measures are crucial for reducing porosity, oxidation, and hydrogen absorption during welding.

8.2.8.3 Preheating, Interpass Temperature Control, and Post-Weld Treatments

Effective thermal management plays a key role in minimizing residual stresses, cracking, and distortion. Preheating reduces thermal gradients and slows cooling rates, preventing HIC. Maintaining optimal interpass temperature ensures uniform microstructure formation between weld layers. Post-weld treatments such as PWHT, stress relief annealing, and peening further improve ductility, relieve internal stresses, and enhance fatigue resistance, particularly critical for HSLA steels used in offshore platforms.

8.2.8.4 Material and Consumable Optimization

Selecting the right base materials and filler metals is essential for corrosion resistance and weld compatibility. The use of low-hydrogen electrodes, flux-cored wires with controlled alloy content, and corrosion-resistant consumables minimizes hydrogen embrittlement and galvanic corrosion. For subsea applications, nickel-based alloys, duplex stainless steels, and titanium alloys are gaining popularity due to their superior mechanical and corrosion properties. Consumables are also tailored to match the weld environment, with specialized coatings and moisture-resistant electrodes designed for humid or underwater conditions.

8.2.8.5 Non-Destructive Testing (NDT) and Real-Time Monitoring

Advancements in inspection technology have made real-time weld monitoring and digital inspection a practical solution for offshore quality assurance (QA). Techniques such as phased array UT (PAUT), digital RT, and eddy current testing provide precise defect detection even in inaccessible areas. The integration of sensors and cameras into welding systems allows for live monitoring of bead geometry, temperature, and arc stability. Data-driven inspection ensures that defects are identified immediately, reducing rework and ensuring compliance with classification standards.

8.2.8.6 Artificial Intelligence (AI) and Digital Twin Integration

AI and machine learning (ML) are revolutionizing offshore welding through predictive analytics and process optimization. AI algorithms can process data from sensors and cameras to predict defect formation, suggest parameter adjustments, and ensure process consistency in real time. The use of digital twin technology, a virtual replica of the welding process, enables simulation of heat flow, distortion, and material behavior before actual welding, allowing for fine-tuning of parameters and minimizing experimental errors. These digital tools contribute to greater process reliability, cost reduction, and continuous improvement.

8.2.8.7 Corrosion Protection and Coating Systems

Since corrosion is one of the primary threats to offshore structures, the integration of protective coatings and cathodic protection (CP) systems with welded components is vital. Welded joints are typically coated with zinc-rich primers, epoxy layers, or thermal spray coatings to prevent seawater intrusion. For subsea pipelines and risers, fusion-bonded epoxy (FBE) and polyurethane coatings are applied after welding to enhance resistance against biofouling and mechanical damage.

8.2.8.8 Skill Development and Safety Enhancements

The reliability of offshore welding also depends on the expertise of personnel. Continuous training and certification programs ensure that welders are familiar with the latest processes, standards, and safety protocols. Simulated welding environments and virtual reality (VR) training systems are now being used to train operators for high-risk offshore conditions. Enhanced personal protective equipment (PPE), arc shielding systems, and fire-resistant enclosures further improve safety in offshore welding environments.

8.2.8.9 Sustainable and Eco-Efficient Practices

Modern offshore welding strategies are increasingly focusing on sustainability. Energy-efficient power sources, eco-friendly shielding gases, and recyclable filler materials are being adopted to reduce the environmental footprint. Automation also contributes to waste minimization and optimized resource use, aligning offshore fabrication with global sustainability and carbon reduction targets.

8.3 WELDING MATERIALS FOR OFFSHORE APPLICATIONS

The selection of appropriate welding materials is one of the most critical aspects of offshore fabrication and repair, as the performance of welded joints directly affects the structural integrity, corrosion resistance, and safety of offshore and subsea installations. Offshore environments impose unique challenges such as high salinity, cyclic loading, temperature fluctuations, hydrostatic pressure, and microbial corrosion, all of which demand the use of materials with superior mechanical and metallurgical characteristics. The choice of base metals, filler materials, and consumables must therefore ensure strength compatibility, weldability, and long-term durability in aggressive marine environments. Figure 8.6 illustrates the decision-making process used in selecting suitable materials for offshore applications. The framework connects design requirements, environmental factors, and mechanical performance to the selection of base materials and filler materials. Design parameters such as load capacity, structure type, and service life determine the primary material class, while environmental conditions like salinity, temperature, and pressure guide corrosion resistance and coating choices. Mechanical

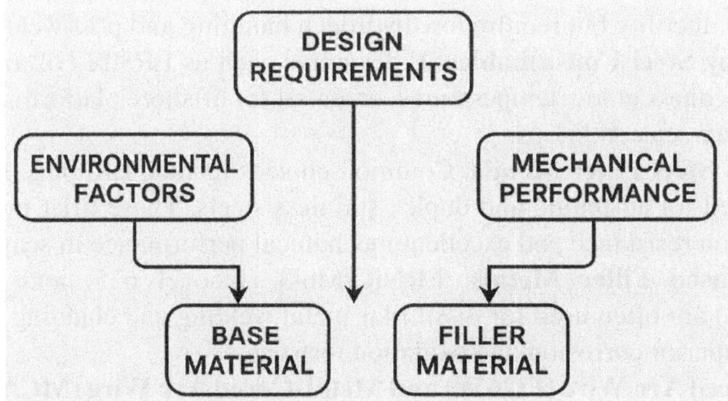

FIGURE 8.6 Material selection framework for offshore welding.

performance, including tensile strength, toughness, and fatigue resistance, ensures the final selection meets operational safety and durability criteria. This systematic approach ensures optimized material pairing for reliable offshore welding performance.

8.3.1 Base Materials Used in Offshore Structures

Offshore structures such as platforms, risers, pipelines, and subsea systems are primarily constructed using carbon steels, HSLA steels, stainless steels, and CRAs.

Carbon Steels: Widely used for jackets, decks, and topside structures due to their cost-effectiveness and mechanical strength. Typical grades include API 5L, ASTM A36, and A516, which offer excellent weldability. However, carbon steels require coatings or CP systems to mitigate corrosion in marine environments.

High-Strength Low-Alloy (HSLA) Steels: Commonly used in deepwater and heavy-load-bearing applications, HSLA steels such as ASTM A572 and A709 provide superior yield strength, toughness, and fatigue resistance while maintaining good weldability. Preheating and controlled heat input are essential to prevent hydrogen cracking in these steels.

Stainless Steels: Austenitic (304L, 316L) and duplex (2205, 2507) stainless steels are used where corrosion resistance is paramount, such as in subsea manifolds and flowlines. Duplex grades combine the strength of ferritic steels with the corrosion resistance of austenitic types, making them ideal for seawater-exposed components.

Nickel-Based and Titanium Alloys: Employed in high-performance applications like risers, manifolds, and subsea connectors where exceptional resistance to pitting, crevice corrosion, and hydrogen embrittlement is required. Alloys such as Inconel 625 and Ti-6Al-4V are used for extreme pressure and temperature conditions.

8.3.2 Filler Metals and Welding Consumables

Filler materials must be compatible with the base metal and capable of producing a weld deposit that meets or exceeds the required mechanical and corrosion properties.

Carbon Steel Filler Metals: Electrodes such as E7018 (SMAW) and wires like ER70S-6 (GMAW) are common for structural welding. These provide high tensile strength and ductility but require low-hydrogen handling and post-weld protection.

Low-Alloy Steel Consumables: Filler wires such as ER80S-Ni2 or E10018-D2 enhance toughness at low temperatures, essential for offshore platforms operating in Arctic or deep-sea conditions.

Stainless Steel Filler Metals: Common choices include ER308L, ER316L, and ER2209, used for austenitic and duplex stainless steels. These filler metals provide high corrosion resistance and excellent mechanical performance in seawater.

Nickel-Based Filler Metals: ERNiCrMo-3 (Inconel 625) and ERNiFeCr-2 (Inconel 718) are often used for dissimilar metal welding and cladding applications, providing superior corrosion and oxidation resistance.

Flux-Cored Arc Wire (FCAW) and Metal-Cored Arc Wire (MCAW): Favored for offshore fabrication due to high deposition rates, positional welding capabilities, and tolerance to environmental variations.

8.3.3 Corrosion and Hydrogen Control

Corrosion and hydrogen embrittlement are among the most serious concerns in offshore welding. The high chloride concentration in seawater accelerates pitting, crevice corrosion, and stress corrosion cracking (SCC). To counteract these effects, materials with high chromium, molybdenum, and nickel content are preferred. Low-hydrogen electrodes and controlled preheating are used to minimize hydrogen diffusion, while PWHT helps reduce residual stresses and HIC risks.

Protective coatings such as zinc-rich primers, epoxy-based systems, or thermally sprayed aluminum (TSA) are commonly applied after welding to prevent corrosion initiation. In subsea structures, CP systems using sacrificial anodes or impressed current are implemented in conjunction with corrosion-resistant weld metals.

8.3.4 Metallurgical Compatibility and Heat-Affected Zone (HAZ) Control

The HAZ in welded joints is often the weakest region due to metallurgical transformations caused by rapid heating and cooling. Selection of welding materials with compatible thermal expansion and solidification characteristics helps prevent cracking and hardness variations. For HSLA steels and duplex stainless steels, maintaining precise heat input and interpass temperature is essential to preserve phase balance and prevent the formation of brittle microstructures such as martensite or sigma phases.

8.3.5 Material Standards and Qualification Requirements

All materials and consumables used in offshore welding must conform to stringent international standards such as AWS D1.1, AWS D3.6M, ISO 15614, DNV-ST-F101, and API 1104. Qualification testing ensures that welds meet specific criteria for tensile strength, impact toughness, corrosion resistance, and fatigue performance. The procedure qualification record (PQR) and welder performance qualification (WPQ) verify that both the materials and personnel meet regulatory and design requirements for offshore operations.

8.3.6 Emerging Trends in Offshore Welding Materials

Recent advancements in metallurgy have led to the development of super duplex stainless steels, high-nitrogen alloys, and high-entropy materials designed for superior resistance to marine corrosion and fatigue. Additive manufacturing techniques such as wire arc additive manufacturing (WAAM) are being used to deposit corrosion-resistant layers on structural components, reducing material waste and repair time. The ongoing integration of AI-driven material selection and predictive corrosion modeling is also helping engineers design more durable offshore systems with tailored material properties.

Table 8.1 summarizes the commonly used base materials, filler metals, and welding processes applied in offshore fabrication and subsea systems. The table highlights how material selection depends on the operating environment, mechanical requirements, and corrosion resistance of the structure. Carbon and HSLA steels dominate

TABLE 8.1
Common Welding Materials and Filler Metals Used in Offshore Applications

Base Material	Typical Filler Metal/ Electrode	Applicable Welding Process	Key Properties/ Advantages	Typical Offshore Applications
Carbon steel (ASTM A36, API 5L Gr B)	E7018, ER70S-6	SMAW, GMAW	Excellent weldability, high ductility, economical; requires corrosion protection	Jackets, decks, topside modules
High-strength low-alloy (HSLA) steel (ASTM A572, A709)	E8018-B2, ER80S-Ni2	SMAW, FCAW	High yield strength, good impact toughness, suitable for cold climates	Legs, braces, load-bearing frames
Austenitic stainless steel (304L, 316L)	ER308L, ER316L	GTAW, GMAW	High corrosion resistance, good ductility, non-magnetic	Process piping, handrails, tanks
Duplex stainless steel (2205, 2507)	ER2209	GTAW, GMAW, SAW	High strength, excellent chloride-induced SCC resistance	Subsea manifolds, flowlines, risers
Nickel-based alloy (Inconel 625, 718)	ERNiCrMo-3, ERNiFeCr-2	GTAW, GMAW	Superior corrosion and oxidation resistance; suitable for dissimilar metal welding	Cladding, risers, wellhead components
Titanium alloy (Ti-6Al-4V)	ERTi-5	GTAW	Lightweight, excellent corrosion, and fatigue resistance	Deepwater connectors, ROV components
Corrosion-resistant alloy (CRA) overlay	ERNiCrMo-4, ER309L	GMAW, FCAW, SAW	Provides protective coating against seawater corrosion and wear	Pipeline internal cladding, flange surfaces
Flux-cored filler wire (for offshore fabrication)	E71T-1C/M, E81T-Ni1	FCAW	High deposition rate, suitable for all positions, good toughness	Field fabrication, repair welding
Metal-cored wire	E70C-6M	MCAW	Excellent arc stability and productivity, reduced spatter	Automated welding of large sections

primary structural fabrication, while duplex stainless steels and nickel alloys are preferred for subsea and high-corrosion environments. The integration of flux-cored and metal-cored wires enhances productivity and weld consistency under offshore conditions. Overall, proper pairing of base and filler materials ensures the long-term strength, safety, and corrosion durability essential for offshore platforms, pipelines, and subsea assemblies.

8.4 HYPERBARIC WELDING

Hyperbaric welding is a specialized process used to join or repair metallic components in underwater and deep-sea environments, where conventional welding methods cannot be applied due to hydrostatic pressure and restricted accessibility. The term hyperbaric refers to the elevated pressure conditions under which the welding is performed. This technique is widely employed in offshore oil and gas infrastructure, subsea pipelines, platform legs, and marine vessel repairs, where the ability to perform reliable welds at depth is critical for maintaining structural integrity and operational safety.

Hyperbaric welding is broadly classified into two categories: dry hyperbaric welding and wet welding. In dry hyperbaric welding, the weld area is enclosed within a pressurized habitat or chamber from which water is displaced by an inert gas (typically helium–argon mixtures). This creates a controlled dry environment that allows welders or remotely operated tools to perform high-quality welds under pressure. In contrast, wet welding is performed directly in the water, where the electrode and arc are exposed to the surrounding medium. Although wet welding offers simplicity and speed, it produces welds of lower quality due to rapid quenching and hydrogen embrittlement.

The key advantage of hyperbaric dry welding is its ability to maintain a stable arc and controlled atmosphere, minimizing oxidation, porosity, and hydrogen contamination. The chamber pressure is equalized with the surrounding hydrostatic pressure to prevent structural collapse and allow safe operation. Within the chamber, welders use modified GTAW, GMAW, or FCAW techniques, depending on the material and joint configuration. The process parameters, such as arc voltage, current, gas composition, and electrode feed rate, must be precisely adjusted to account for increased pressure and reduced arc length stability.

From a metallurgical standpoint, hyperbaric conditions influence arc behavior, heat transfer, and cooling rates, which affect the weld's microstructure and mechanical properties. Increased ambient pressure enhances arc constriction and stability but also accelerates heat dissipation, resulting in finer grain structures. However, excessive cooling can lead to martensitic transformation in carbon steels and increased hardness in the HAZ, potentially reducing ductility and toughness. To counteract these effects, preheating, controlled interpass temperatures, and PWHT are often applied after the welding operation.

Hyperbaric welding systems consist of key components, including the welding habitat (pressure chamber), gas control and life-support system, power supply unit, communication interface, and remote operation controls. The habitat is designed to accommodate welders or robotic systems while maintaining visibility, ventilation, and pressure stability. For deep-sea operations, ROVs equipped with robotic manipulators are increasingly used to perform automated hyperbaric welding, eliminating the need for divers and enhancing safety at depths beyond 300 m.

Despite its advantages, hyperbaric welding presents several challenges. Maintaining the sealing integrity of the habitat, ensuring pressure equalization, and preventing gas contamination require precise engineering control. Furthermore, diver fatigue, limited workspace, and the logistical complexity of habitat deployment make this process cost-intensive. However, its unmatched ability to deliver high-quality, durable welds in deep underwater environments makes it indispensable for critical offshore maintenance and subsea infrastructure projects.

In recent years, the development of robotic and AI-assisted hyperbaric welding systems has revolutionized the process. These innovations allow for real-time monitoring of arc characteristics, automated control of weld parameters, and adaptive path correction using machine vision and sensors. Combined with digital twin simulations and data-driven process optimization, modern hyperbaric welding has become more efficient, predictable, and sustainable, aligning with the growing trend of autonomous underwater fabrication and repair in the offshore industry.

8.4.1 Principles of Hyperbaric Welding

The principles of hyperbaric welding are rooted in the control of environmental and pressure conditions that enable stable arc welding operations underwater or at elevated ambient pressures. The process is designed to replicate, as closely as possible, the conditions of surface welding within a submerged environment, ensuring that welds retain the mechanical integrity, metallurgical quality, and corrosion resistance required for offshore and subsea structures.

At its core, hyperbaric welding operates on the same physical principles as arc welding the creation of an electric arc between an electrode and the workpiece generates heat sufficient to melt and fuse metals. However, the key difference lies in the ambient pressure and atmospheric control under which the welding occurs. As the depth of the operation increases, hydrostatic pressure rises (approximately 1 bar for every 10 m of depth), which significantly influences arc stability, heat transfer, gas behavior, and cooling rates. To maintain control, a pressurized habitat or chamber is used to isolate the weld area from direct water contact and provide a controlled gaseous environment for welding.

The welding habitat a sealed chamber placed around the work area is filled with a gas mixture (commonly helium–argon) at a pressure equal to the surrounding seawater. This prevents water ingress while ensuring diver or equipment safety. Inside the chamber, welding is performed using modified versions of conventional processes such as GTAW, GMAW, or FCAW. These methods are chosen for their ability to maintain a stable arc and high deposition rate under varying pressure conditions.

Under hyperbaric conditions, pressure and gas composition directly affect the arc's electrical and thermal characteristics. Increased pressure results in a shorter, narrower, and more constricted arc, leading to higher current density and improved penetration. However, excessive constriction can cause spatter or instability. The ionization potential of the shielding gas also changes under pressure, influencing arc ignition and energy distribution. Helium, for example, requires a higher voltage for arc initiation but offers better heat transfer, while argon provides easier arc starting and smoother metal transfer. Therefore, helium–argon mixtures are typically used to balance arc stability and heat input.

The heat transfer and cooling behavior of hyperbaric welds differ markedly from surface welding. At higher pressures, increased gas density enhances convective heat loss, causing faster cooling rates and smaller molten pools. This can refine grain structure, improving strength, but may also lead to brittleness and higher residual stresses if not properly controlled. Consequently, precise management of current, voltage, and interpass temperature is critical to prevent the formation of undesirable microstructures, such as martensite or carbide precipitates, in steels and alloys.

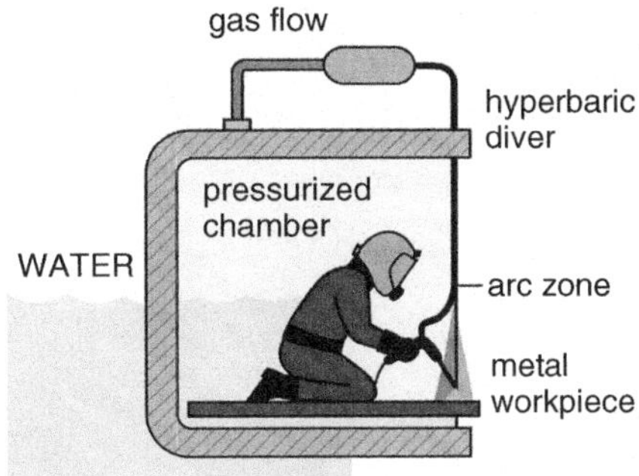

FIGURE 8.7 Principle of hyperbaric welding operation.

The metallurgical principles of hyperbaric welding focus on maintaining weldment toughness, minimizing hydrogen absorption, and avoiding porosity. Hydrogen, introduced through moisture or decomposition of hydrocarbons, becomes more soluble under pressure and can cause HIC during cooling. To mitigate this, dry habitats, low-hydrogen consumables, and preheating techniques are employed. Additionally, continuous gas circulation removes contaminants and maintains consistent oxygen levels inside the habitat.

In modern hyperbaric systems, remote sensors and feedback control are integrated to monitor key parameters such as pressure, gas composition, temperature, and arc voltage in real time. These data are used to regulate the welding environment dynamically, ensuring process consistency. Digital twin simulations and computational fluid dynamics (CFD) models are also increasingly utilized to predict thermal behavior and optimize chamber design before actual operation. Figure 8.7 illustrates the working concept of dry hyperbaric welding. The diagram shows a pressurized chamber attached to an underwater structure, where a diver or robotic arm performs welding in a dry, gas-filled environment. The gas flow system (typically a helium–argon mixture) maintains a stable atmosphere inside the chamber by displacing water and equalizing pressure with the surrounding seawater. The arc zone is shown between the electrode and the metal workpiece, where fusion occurs under controlled pressure. This setup allows for high-quality, defect-free welds in deep-sea environments by stabilizing the arc, preventing hydrogen absorption, and minimizing contamination.

8.4.2 Wet Hyperbaric Welding (Direct Seawater Environment)

Wet hyperbaric welding, often referred to simply as underwater wet welding, is a process in which the welding operation is performed directly in the surrounding seawater, without isolating the weld zone within a dry chamber. The arc, electrode, and molten weld pool are completely exposed to the underwater environment. This method is primarily used for rapid, temporary, or emergency repairs on offshore and subsea structures where the deployment of a dry habitat system is impractical or uneconomical. Despite its limitations in weld quality compared to dry hyperbaric welding, wet welding remains a vital technique due to its simplicity, cost-effectiveness, and quick setup (Figure 8.8).

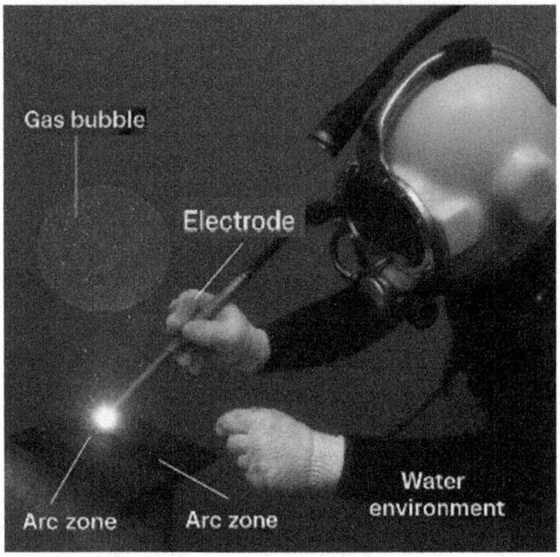

FIGURE 8.8 Wet hyperbaric welding process (direct seawater environment).

In wet hyperbaric welding, the diver performs welding using specially designed waterproof electrodes and a constant current power supply connected through a waterproof umbilical cable. The process is predominantly carried out using SMAW, as this method is more tolerant of unstable environments and does not require complex gas shielding systems. The electrode coating plays a crucial role, and it must provide arc stability, deoxidation, and shielding against rapid quenching by the surrounding water. The most commonly used electrodes include E7014- and E6013-type waterproof variants, which contain flux compositions modified to minimize hydrogen absorption.

The main challenge in wet hyperbaric welding is the interaction between the electric arc and seawater. When the arc is struck, it generates a small cavity or "gas bubble" around the arc zone due to vaporization of water and decomposition of electrode coatings. This bubble temporarily shields the molten pool from direct water contact, but the constant turbulence and bubble instability make the arc highly erratic. This leads to fluctuations in heat input, resulting in incomplete fusion, porosity, and slag inclusions. Furthermore, rapid cooling of the molten metal in water can produce hard and brittle microstructures, increasing susceptibility to HIC and reducing impact toughness in the weld and HAZ.

Another critical concern in wet hyperbaric welding is hydrogen embrittlement. The decomposition of water (H_2O) under high temperatures releases hydrogen and oxygen gases, some of which dissolve into the molten weld pool. Upon solidification, this trapped hydrogen can cause delayed cracking, particularly in high-strength steels. To minimize this effect, low-hydrogen electrodes, preheating techniques, and controlled interpass temperature are employed where possible. However, in practical underwater conditions, precise temperature control is challenging due to the continuous heat loss to seawater.

Wet hyperbaric welding is typically performed at depths up to 300 m, although most operations occur within 30–50 m where divers can safely operate. The arc stability and visibility decrease with increasing depth due to higher ambient pressure, water density, and reduced gas bubble retention. To counter these effects, constant current DC power sources with remote voltage control and closed-circuit video monitoring systems are used. Diver welders receive training in manual dexterity, buoyancy control, and electrode manipulation to compensate for unstable conditions and limited visibility.

Despite its drawbacks, wet hyperbaric welding offers significant operational flexibility for in situ repairs of subsea pipelines, platform legs, ship hulls, and structural braces. It allows for quick deployment without extensive setup, making it ideal for temporary fixes or pre-stabilization welds before permanent dry habitat repairs are conducted. Modern advancements such as the use of programmable power supplies, specialized flux coatings, and robotic-assisted underwater welders are gradually improving the consistency and mechanical performance of wet welds.

From a metallurgical standpoint, wet hyperbaric welds exhibit fine-grained microstructures due to rapid cooling, but they often require post-weld inspection and reinforcement through surface grinding, overlay welding, or coating application. NDT methods, including magnetic particle inspection (MPI) and UT, are used to detect defects, although inspection under water remains technically challenging.

8.4.3 Dry Hyperbaric Welding (Habitat Welding)

Dry hyperbaric welding, also known as habitat welding, is an advanced underwater welding process that provides a controlled dry environment around the weld zone, allowing high-quality welding operations to be performed at significant depths. Unlike wet welding—where the arc and molten pool are directly exposed to seawater—dry hyperbaric welding completely isolates the weld area within a sealed and pressurized chamber (habitat). This chamber is filled with an inert gas mixture, typically helium and argon, and is maintained at a pressure equal to the surrounding seawater to prevent water ingress. The result is a stable, contamination-free welding environment that enables precision welding similar in quality to that achieved on the surface.

The principle of habitat welding involves enclosing the area to be welded in a pressure-tight chamber fitted with access ports, gas supply lines, lighting, and control systems. The chamber is lowered and clamped around the workpiece on the seabed or structure, and the internal pressure is gradually increased to match the external hydrostatic pressure. Water inside the chamber is displaced by the gas mixture, creating a dry, breathable workspace for the welder or robotic system. Welding is then performed using modified GTAW, GMAW, or FCAW, depending on material type, joint configuration, and required mechanical properties.

Inside the habitat, temperature, humidity, and gas composition are carefully regulated. Helium is often used as a base gas due to its low density and inertness, which minimizes contamination, while argon may be added to stabilize the arc. The use of helium, however, increases thermal conductivity and heat loss, necessitating adjustments in current and voltage settings to maintain adequate penetration. Operators must also account for the pressure-dependent behavior of the arc, which becomes more constricted and shorter at higher depths, influencing bead shape and heat input.

Dry hyperbaric welding offers significant metallurgical advantages over wet welding. The dry environment prevents hydrogen diffusion from seawater, thus minimizing HIC and porosity formation. Controlled atmosphere and temperature ensure consistent cooling rates, reducing the formation of brittle microstructures in the weld metal and HAZ. The absence of water contact also allows for the use of high-quality filler metals and gas shielding systems, resulting in improved ductility, toughness, and fatigue life of welded joints—critical for subsea structures exposed to cyclic loading and pressure variations.

A standard habitat welding system includes the following:

- **Welding Chamber (Habitat)**: A sealed, pressurized enclosure with access hatches and observation ports.
- **Gas Control and Life-Support Unit**: Supplies the gas mixture, maintains internal pressure, and removes contaminants.
- **Power and Control Systems**: Provide regulated current, voltage, and remote monitoring of arc stability.
- **Communication and Safety Equipment**: Allow continuous communication between divers and surface operators.

In practice, dry hyperbaric welding can be performed either by diver welders inside the habitat or by remotely operated robotic systems. Diver welders typically operate at depths up to 300m, while robotic or automated hyperbaric systems are used for greater depths to eliminate human exposure to high-pressure environments. These remotely operated systems use precision robotic arms equipped with sensors, cameras, and digital feedback control for automated welding in confined or hazardous locations.

The applications of habitat welding are extensive, particularly in the offshore oil and gas industry and subsea infrastructure maintenance. It is used for pipeline tie-ins, riser repairs, platform leg restoration, and structural reinforcements. Due to its high weld integrity, dry hyperbaric welding is the preferred method for critical load-bearing joints and permanent underwater repairs, where mechanical reliability and long service life are mandatory.

However, habitat welding also involves several challenges and limitations. The setup is complex and costly, requiring specialized chambers, gas control systems, and trained personnel. The logistics of deploying and positioning the habitat in deep water can be time-consuming. Additionally, gas composition, humidity, and pressure must be continuously monitored to prevent condensation, oxidation, or weld porosity. Diver fatigue and restricted maneuverability within the habitat further complicate manual welding tasks.

Despite these challenges, technological advancements are rapidly transforming dry hyperbaric welding into a digitally controlled, automated process. The integration of real-time monitoring sensors, AI-based parameter optimization, and digital twin simulations allows engineers to predict thermal behavior and adjust process variables dynamically. Hybrid systems, combining robotic automation with supervised human control, are emerging as a reliable solution for deepwater welding tasks, ensuring both efficiency and safety. Figure 8.9 illustrates a pressurized

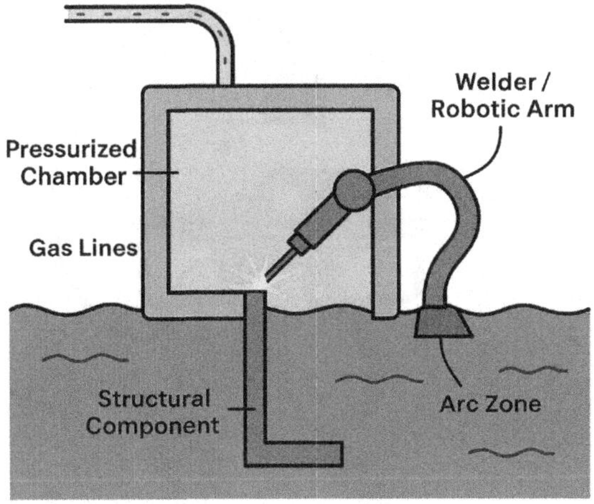

FIGURE 8.9 Dry hyperbaric welding (habitat welding) setup.

underwater chamber used for performing welding in a dry environment at depth. The diagram shows the pressurized chamber attached to a structural component, with gas lines supplying the helium–argon mixture to displace water and maintain internal pressure. A welder or robotic arm operates inside the habitat, performing welding at the arc zone in a controlled atmosphere free from seawater contamination. This setup allows for precise control of arc stability, gas composition, and temperature, enabling high-quality, defect-free welds suitable for critical offshore and subsea applications.

8.5 DEEP-SEA WELDING TECHNIQUES

Deep-sea welding techniques encompass advanced welding processes and technologies designed to perform joining and repair operations at extreme depths, typically below 300m, where hydrostatic pressure, low temperature, limited visibility, and accessibility make conventional underwater welding impractical. As the offshore energy industry continues to expand into deeper waters for oil, gas, and renewable energy extraction, the demand for reliable and automated deep-sea welding systems has increased significantly. These techniques focus on maintaining weld integrity, operator safety, and equipment reliability under severe hydrodynamic and pressure conditions, ensuring that critical subsea structures such as pipelines, risers, and manifolds can be constructed and repaired efficiently. Figure 8.10 shows the deep-sea robotic welding system that illustrates an ROV performing an automated welding operation on a subsea pipeline. The ROV is equipped with a robotic arm, welding torch, and multi-camera vision system for precise alignment and real-time monitoring of the weld zone. Sensors and power supply lines provide feedback and energy to maintain optimal welding conditions under high-pressure environments. The subsea pipeline joint is shown as the primary work area, where the robotic arm executes welding with stability and precision, ensuring high-quality joints in deep-sea fabrication and repair operations.

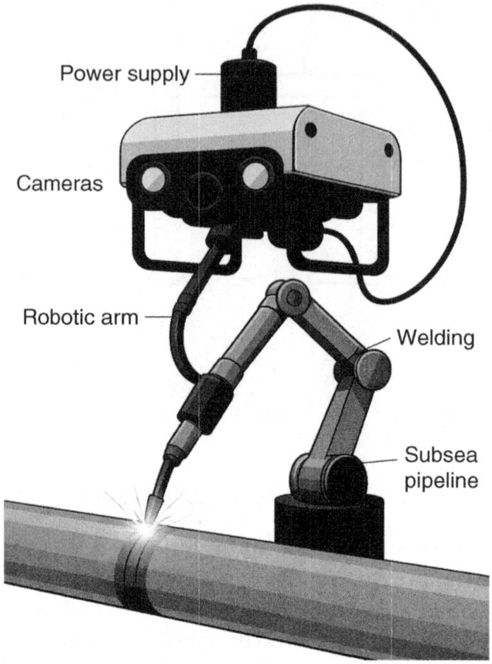

FIGURE 8.10 Deep-sea robotic welding system.

8.5.1 Challenges in Deep-Sea Welding

Welding at great depths presents unique technical and environmental challenges. Hydrostatic pressure increases by approximately 1 MPa for every 100 m of depth, causing significant effects on the welding arc, gas shielding, and molten metal behavior. Elevated pressure compresses the plasma arc, altering its heat distribution and increasing current density, which can destabilize the arc or reduce penetration. The low temperatures of deep seawater accelerate cooling, leading to rapid solidification, high hardness, and potentially brittle microstructures in the HAZ. Visibility is also severely reduced, complicating manual or VT of weld quality. These factors make conventional diver-assisted welding unfeasible beyond 300 m, necessitating the use of remotely operated or automated systems.

8.5.2 Robotic and Remotely Operated Welding Systems

Modern deep-sea welding heavily relies on ROVs and autonomous underwater vehicles (AUVs) equipped with robotic manipulators, cameras, and welding tools. These robotic systems perform GTAW, GMAW, or laser-based welding under controlled conditions without human divers. ROVs can operate at depths exceeding 3,000 m, performing tasks such as pipeline tie-ins, crack repair, and structural reinforcements. High-definition cameras, sonar guidance, and laser scanning ensure precise alignment and weld positioning even in turbid environments.

Advanced robotic systems utilize AI-based control algorithms to adjust welding parameters in real time, compensating for environmental fluctuations. Sensors continuously monitor arc voltage, current, temperature, and bead geometry, feeding data into ML models that optimize weld quality. These systems are integrated with digital twin platforms, allowing operators at the surface to simulate, monitor, and control underwater welding processes virtually, ensuring high repeatability and minimal human intervention.

8.5.3 Deepwater Habitat and Pressure-Controlled Systems

For operations where permanent or critical welds are required, deepwater dry habitat systems are deployed. These pressure-balanced enclosures create localized dry environments around the weld joint, similar to shallow-water hyperbaric chambers but capable of withstanding extreme pressures. Inside the habitat, automated welding robots perform the weld under precisely regulated temperature, humidity, and gas conditions. These systems enable high-quality metallurgical performance comparable to onshore welding and are primarily used for deepwater pipeline installations and riser repairs.

Pressure-controlled habitats are equipped with remote gas regulation, video feedback, and data acquisition systems, allowing surface operators to oversee every aspect of the process. The internal gas environment—typically a helium-based mixture—is adjusted to ensure arc stability and minimal hydrogen absorption. The ability to control environmental variables makes this approach ideal for critical, long-term repairs in deep-sea infrastructure.

8.5.4 Laser and Hybrid Welding for Deep-Sea Applications

Recent developments in laser-based and hybrid welding technologies have opened new possibilities for deep-sea applications. Fiber laser welding and laser-arc hybrid welding (HLAW) can produce deep, narrow welds with minimal heat input and high precision, even in pressurized environments. These processes can be integrated into ROVs or fixed subsea workstations, offering excellent control over weld penetration and minimizing distortion. The high energy density of laser beams allows operation through transparent windows or controlled gas chambers, reducing the need for large-scale habitats.

HLAW combines the deep penetration of lasers with the filler addition capability of arc welding, resulting in improved gap bridging, faster travel speeds, and enhanced mechanical strength. These systems are ideal for automated pipeline and structural welds, especially where access is limited or component geometry is complex.

8.6 QUALITY CONTROL AND INSPECTION IN OFFSHORE WELDING

Quality control (QC) and inspection are critical aspects of offshore welding, ensuring that welded joints meet stringent mechanical, metallurgical, and safety requirements under demanding marine conditions. Offshore structures operate in highly corrosive and dynamic environments, where weld failures can lead to catastrophic consequences such as oil spills, structural collapse, or human and environmental hazards. Therefore, maintaining strict QA and control measures throughout the design, fabrication, installation, and maintenance phases is essential to guarantee structural integrity, reliability, and long service life.

QC in offshore welding involves a systematic approach encompassing process qualification, material testing, weld monitoring, and non-destructive evaluation (NDE). It begins with defining the welding procedure specification (WPS), which outlines approved parameters for welding current, voltage, preheat temperature, and filler material. This is validated through PQRs, where sample welds are tested to verify compliance with mechanical and metallurgical criteria. Additionally, WPQ ensures that welding personnel possess the necessary skill and competency to perform welds under offshore conditions.

8.6.1 Standards and Codes for Offshore Welding Quality

Offshore welding is governed by international standards established by organizations such as

- **AWS D1.1/D3.6M**: Structural and underwater welding codes.
- **ASME Section IX**: Qualification standards for welding and brazing procedures.
- **ISO 15614, ISO 3834**: Qualification and quality requirements for welding processes.
- **DNV-ST-F101/DNV-OS-C401**: Subsea pipeline and offshore structural welding standards.
- **API 1104/ABS Rules**: Specifications for pipeline and offshore platform welding.

These standards define acceptable limits for weld geometry, discontinuities, and mechanical properties, ensuring uniformity and global compliance. Certification from recognized authorities such as Lloyd's Register, Bureau Veritas, or ABS further validates the weld quality for offshore deployment.

8.6.2 Welding Process Control and Monitoring

During offshore fabrication, maintaining consistent welding parameters is essential to prevent defects such as porosity, undercutting, lack of fusion, or hydrogen cracking. Continuous process monitoring is performed using advanced tools that record real-time data for current, voltage, arc length, and travel speed. In automated or robotic systems, embedded sensors, cameras, and thermal imaging are used for tracking bead geometry and arc stability. Any deviation from the pre-qualified parameters triggers alerts for corrective action.

Modern digital quality systems use data logging and ML algorithms to detect trends and predict potential weld issues before they occur. Integration with digital twin platforms allows engineers to simulate and validate welding operations virtually, ensuring optimal process control even in deepwater or high-risk environments.

8.6.3 Non-Destructive Testing (NDT) and Inspection Methods

Inspection of offshore welds is primarily carried out using NDT techniques that assess internal and surface defects without damaging the component. The choice of method depends on the weld type, accessibility, and environmental conditions:

In subsea environments, ROV-mounted NDT systems are increasingly used to perform remote inspections, reducing the need for divers and enhancing safety. Real-time imaging and acoustic mapping technologies provide visual verification of weld condition and surface roughness even in low-visibility waters.

8.6.4 Metallurgical and Mechanical Testing

In addition to NDT, destructive testing is conducted during the qualification stage to validate the weld's structural performance. Tests include

- **Tensile testing** for assessing weld strength.
- **Bend tests** for ductility and fusion quality.
- **Charpy impact tests** for toughness at subzero temperatures.
- **Hardness mapping** to evaluate HAZ properties.
- **Macro/microstructural analysis** to observe grain structure, inclusions, and phase distribution.

These results are correlated with process parameters to establish relationships between welding conditions and mechanical performance, ensuring long-term reliability under cyclic loading and corrosion.

8.6.5 Corrosion and Fatigue Inspection

Offshore structures are continuously exposed to seawater, leading to corrosion, pitting, and fatigue cracking in welded joints. Periodic inspection using corrosion monitoring sensors, ultrasonic thickness gauges, and CP assessment is essential to prevent structural degradation. For fatigue-prone joints, acoustic emission and strain monitoring systems detect crack initiation and growth in real time, enabling early intervention before failure occurs.

8.6.6 Digital Quality Assurance and Predictive Maintenance

The evolution of Industry 4.0 technologies is transforming offshore welding inspection into a digital process. Integrated IoT-enabled weld monitoring systems collect continuous data during fabrication and service, feeding it into centralized databases for quality analytics. AI-driven defect recognition uses machine vision to classify discontinuities from radiographic or ultrasonic images with high accuracy. Predictive models based on digital twins allow maintenance scheduling based on actual weld performance rather than fixed inspection intervals, optimizing operational efficiency.

Figure 8.11 illustrates the systematic workflow followed in ensuring weld integrity in offshore fabrication. The process begins with WPS qualification, followed by the

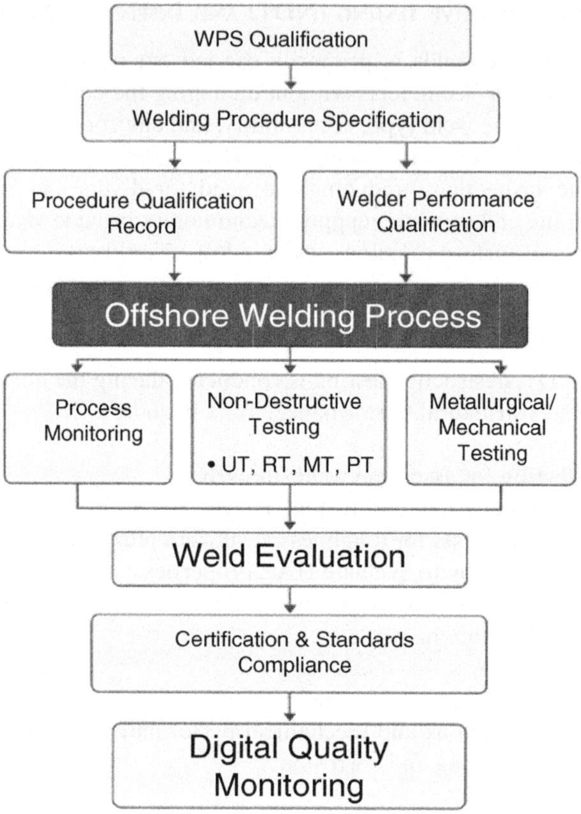

FIGURE 8.11 Offshore weld quality control and inspection framework.

development of the WPS and validation through PQRs and WPQ. During the offshore welding process, parameters are continuously monitored, and welds undergo NDT (UT, RT, MT, penetrant testing [PT]) and mechanical/metallurgical testing. The results feed into weld evaluation and subsequent certification and standard compliance, concluding with digital quality monitoring for traceability and predictive maintenance in modern offshore operations.

Table 8.2 provides a concise overview of the major inspection and testing techniques used in offshore welding to ensure the integrity, reliability, and safety of welded structures under marine conditions. Each method serves a specific purpose ranging from surface flaw detection (VT, PT, MT) to internal defect identification (RT, UT, PAUT) and is selected based on the material type, weld geometry, and environmental exposure. Advanced techniques such as PAUT and acoustic emission monitoring (AEM) enable real-time, high-resolution assessment of welds even in subsea environments. Together, these inspection methods form the foundation of the offshore QC framework, ensuring that all welds meet international standards and performance expectations for deep-sea and offshore applications.

TABLE 8.2
Summary of Key Inspection and Testing Methods in Offshore Welding

Inspection/ Testing Method	Principle/Technique	Purpose	Typical Offshore Application
Visual testing (VT)	Direct visual observation using magnifiers, borescopes, or ROV-mounted cameras	Detects surface flaws, cracks, and undercuts	Weld seams, joints, platform braces
Magnetic particle testing (MT)	Uses magnetic fields and iron particles to locate surface and near-surface defects	Identifies discontinuities in ferromagnetic materials	Structural steel members, nodes
Dye penetrant testing (PT)	Liquid penetrant highlights surface-breaking flaws under UV or visible light	Detects cracks, porosity, or surface defects	Stainless steel and non-ferrous components
Radiographic testing (RT)	X-rays or gamma rays reveal internal voids, inclusions, and lack of fusion	Evaluates internal weld quality in critical components	Pipelines, risers, pressure vessels
Ultrasonic testing (UT)	High-frequency sound waves reflect from internal flaws or boundaries	Measures thickness and detects subsurface defects	Weld joints, hull plating, deep-sea structures
Phased array ultrasonic testing (PAUT)	Multi-angle ultrasonic scanning for 3D flaw imaging	Provides detailed volumetric inspection	Subsea pipelines, complex weld geometries
Eddy current testing (ECT)	Induced electromagnetic fields detect surface and near-surface discontinuities	Used for corrosion and coating inspection	Conductive alloys, hull surfaces
Acoustic emission monitoring (AEM)	Captures transient elastic waves from crack growth during loading	Real-time monitoring of fatigue and structural integrity	Offshore jackets, dynamic risers
Hardness and tensile testing	Measures hardness and tensile strength of welds and HAZ	Verifies mechanical strength and ductility	Qualification and quality verification
Charpy impact testing	Measures energy absorbed in fracture at subzero temperatures	Ensures toughness for Arctic and deepwater operations	Cryogenic or cold-climate offshore platforms

8.7 FUTURE TRENDS IN OFFSHORE AND SUBSEA WELDING

The future of offshore and subsea welding is being shaped by rapid advancements in automation, AI, digital simulation, and material science, aimed at improving safety, productivity, and weld quality in increasingly challenging marine environments. As offshore energy exploration extends into ultra-deep waters and as the offshore renewable energy sector (such as floating wind farms and subsea power systems) expands, welding technologies must evolve to operate autonomously, intelligently, and sustainably under extreme conditions. The emerging trends focus on enhancing precision, reducing human risk, and ensuring long-term structural integrity in subsea applications. Figure 8.12 illustrates the integration of next-generation technologies that are transforming marine welding. The diagram highlights five major innovation pillars: robotic and autonomous welding, AI and digital twin technology,

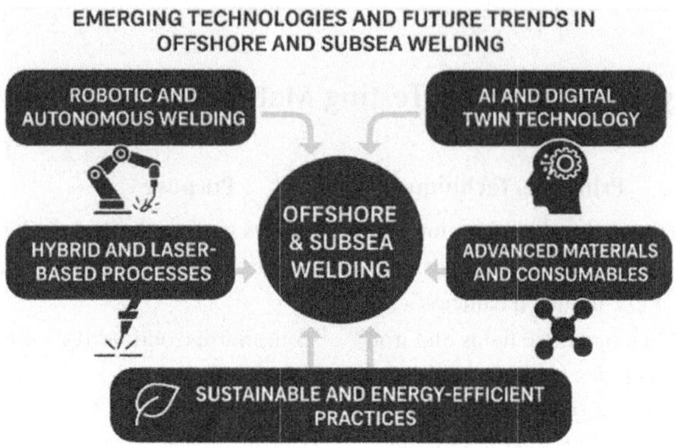

FIGURE 8.12 Emerging technologies and future trends in offshore and subsea welding.

hybrid and laser-based processes, advanced materials and consumables, and sustainable energy-efficient practices, all converging toward enhancing offshore and subsea welding capabilities. These advancements collectively improve precision, QA, and environmental performance, setting the foundation for intelligent, automated, and sustainable offshore fabrication systems.

8.7.1 Robotic and Autonomous Welding Systems

One of the most significant trends in subsea fabrication is the shift toward robotic and autonomous welding systems. Modern ROVs and AUVs equipped with multi-axis robotic arms, sensors, and vision systems are replacing manual diver-assisted welding. These systems can perform complex welding, inspection, and maintenance tasks at depths beyond 3,000 m, eliminating human exposure to hazardous environments.

Robotic welders are now integrated with AI-based control algorithms that adjust welding parameters such as current, voltage, and wire feed rate in real time based on sensor feedback. This adaptive control ensures consistent weld quality even under fluctuating environmental conditions like temperature, pressure, and water currents. Combined with machine vision and laser scanning, autonomous systems can precisely locate joints, track seams, and execute welds with micron-level accuracy, significantly improving efficiency in subsea fabrication.

8.7.2 Integration of Artificial Intelligence (AI) and Digital Twin Technology

The integration of AI, ML, and digital twin simulation is transforming the planning, execution, and QA of offshore welding operations. Digital twin's virtual replicas of physical systems allow engineers to simulate welding processes, predict thermal behavior, and identify potential defects before field implementation. By combining real-time sensor data with predictive analytics, AI algorithms can optimize weld parameters dynamically, reducing rework and improving structural reliability.

These intelligent systems also enable predictive maintenance, where AI models analyze weld performance and environmental exposure to forecast degradation or

FIGURE 8.13 Integration of artificial intelligence (AI) and digital twin technology in offshore welding.

corrosion over time. As a result, offshore operators can schedule targeted maintenance before failures occur, extending the lifespan of subsea assets and minimizing downtime. Figure 8.13 illustrates how real-time data from a welding operation are transmitted to a digital twin model for continuous process optimization. Sensors collect information such as current, voltage, temperature, and weld geometry, which is analyzed by an AI system in the cloud. The digital twin simulates and predicts weld behavior, feeding optimization data back to the physical welding system. This closed-loop feedback enables enhanced control, predictive maintenance, and defect prevention, resulting in superior weld quality and reliability in offshore operations.

8.7.3 Hybrid and Laser-Based Welding Advancements

Future offshore welding systems are expected to leverage hybrid processes, such as HLAW and plasma–MIG hybrid techniques, to combine the high penetration of laser beams with the filler adaptability of arc welding. These methods reduce heat input, distortion, and residual stress, while enabling high-speed, high-precision welding suitable for thick-section marine components and pressure vessels.

Laser-based welding systems, when integrated into ROVs or modular subsea habitats, provide deep penetration with minimal contamination risk. The development of fiber-delivered laser systems has made it possible to transmit high-energy laser beams through long optical fibers to underwater work zones, enabling remote, high-precision welding at extreme depths. Figure 8.14 illustrates modern developments combining laser and arc-based technologies for offshore and subsea applications. The figure highlights key components such as high-power laser sources, arc torches, robotic manipulators, and advanced process controls. The synergy between laser and arc heat sources enables deep penetration, reduced distortion, and improved deposition efficiency. Integration with robotic automation and AI-driven control systems ensures precise parameter optimization, making hybrid laser-arc welding an essential advancement for high-strength, corrosion-resistant marine fabrication.

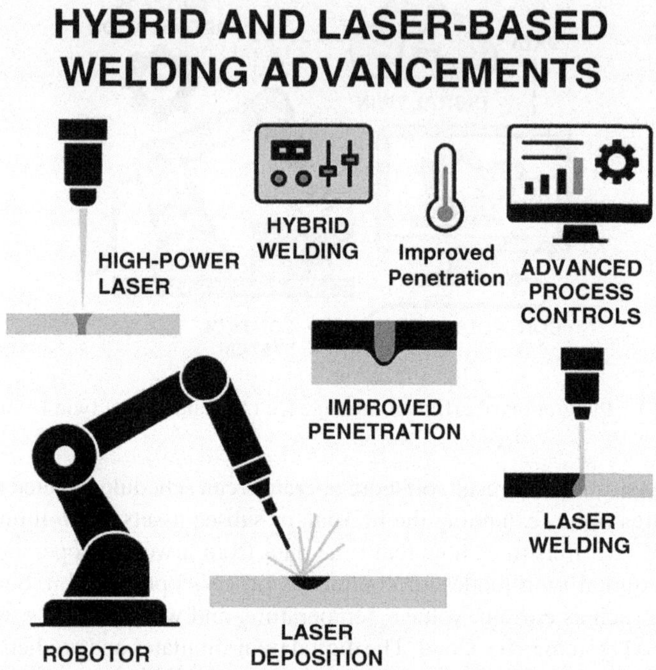

FIGURE 8.14 Hybrid and laser-based welding advancements.

8.7.4 Advanced Materials and Smart Consumables

The evolution of marine-grade alloys and consumables is another defining trend in the future of subsea welding. The introduction of high-entropy alloys (HEAs), super duplex stainless steels, and nickel-titanium composites is improving corrosion resistance, fatigue performance, and weldability under deep-sea conditions. Similarly, smart consumable electrodes and filler wires embedded with sensors or alloying agents are being developed to monitor temperature, detect hydrogen concentration, and control solidification during welding.

The application of additive manufacturing techniques, such as WAAM, for subsea component repair is also emerging. WAAM can rebuild worn or damaged sections directly on-site, minimizing the need for replacement and reducing downtime for offshore platforms.

8.7.5 Sustainable and Energy-Efficient Welding Practices

Sustainability is becoming a major driver of innovation in offshore welding. Future welding systems are expected to operate with eco-efficient power sources, reduced emissions, and recyclable shielding gases. The integration of renewable energy (e.g., offshore wind or wave power) into subsea construction platforms is being explored to reduce carbon footprints. Additionally, cold metal transfer (CMT) and low-energy pulse welding are gaining attention for minimizing energy consumption and thermal distortion while maintaining high weld integrity.

8.7.6 Integration with Offshore Renewable Energy and Deepwater Infrastructure

As the global energy industry transitions toward renewable sources, the role of subsea welding will expand to include offshore wind turbine foundations, tidal energy converters, and subsea power transmission systems. The ability to perform autonomous welding and inspection at depth will be crucial for building and maintaining these complex systems. Similarly, in the oil and gas sector, new subsea tie-backs and high-pressure, high-temperature (HPHT) systems will require advanced welding technologies capable of maintaining reliability under extreme operational stresses.

8.8 SUMMARY

The chapter provides a comprehensive overview of the critical role of welding in ensuring the safety, reliability, and longevity of marine and deep-sea infrastructure. It explores how welding underpins the fabrication and maintenance of offshore platforms, subsea pipelines, risers, and structural components exposed to extreme hydrostatic pressure, corrosion, and cyclic loading. The chapter highlights the types of offshore structures, principles of hyperbaric (dry and wet) welding, and emerging deep-sea and robotic welding techniques that enable precision and safety in challenging underwater environments. Metallurgical aspects such as HAZ control, hydrogen embrittlement, and corrosion resistance are discussed alongside stringent QC and inspection frameworks that ensure compliance with global standards like AWS, DNV, and ISO. The discussion concludes with insights into future trends, including the integration of AI, digital twin simulations, robotic automation, hybrid laser-arc processes, and sustainable welding practices, which are revolutionizing offshore fabrication toward safer, smarter, and more efficient next-generation marine engineering.

8.9 REVIEW QUESTIONS

1. Define offshore and subsea welding and explain why it is critical in marine engineering.
2. What are the main challenges encountered during welding operations in offshore environments?
3. Differentiate between fixed and floating offshore structures in terms of design and welding requirements.
4. Explain the role of welding in maintaining structural integrity and safety of offshore platforms.
5. What are the key factors affecting weld quality in subsea applications?
6. Describe the main environmental challenges that influence offshore welding performance.
7. List and explain the most commonly used welding processes for offshore fabrication.
8. What is hyperbaric welding, and how does it differ from conventional underwater welding?
9. Compare wet hyperbaric welding and dry hyperbaric welding in terms of process control, weld quality, and application.

10. Describe the principle of operation in dry hyperbaric (habitat) welding.
11. What are the main equipment components of a hyperbaric welding system?
12. How does pressure affect the arc behavior and cooling rate in underwater welding?
13. What are the metallurgical issues commonly observed in subsea welds, and how can they be mitigated?
14. Discuss the role of preheating, interpass temperature control, and post-weld heat treatment in offshore welding.
15. Explain the significance of hydrogen embrittlement and its prevention in deep-sea welding operations.
16. What materials and filler metals are commonly used for offshore structures, and why?
17. How do duplex stainless steels and nickel-based alloys contribute to corrosion resistance in subsea applications?
18. Describe the importance of quality control and inspection in offshore welding operations.
19. List five non-destructive testing (NDT) methods commonly used for inspecting offshore welds and briefly describe their principles.
20. How does phased array ultrasonic testing (PAUT) improve upon conventional ultrasonic testing for subsea weld inspection?
21. Explain the role of standards and codes (e.g., AWS, DNV, ISO, API) in regulating offshore welding practices.
22. Discuss the impact of automation and robotics on offshore welding productivity and safety.
23. What are deep-sea welding techniques, and how do robotic systems contribute to their success?
24. Describe the use of artificial intelligence (AI) and digital twin technology in optimizing offshore welding processes.
25. How do hybrid and laser-based welding techniques enhance offshore and subsea fabrication?
26. Explain how additive manufacturing methods like wire arc additive manufacturing (WAAM) can support offshore repair applications.
27. What are the primary advantages of using robotic and autonomous welding systems for deepwater projects?
28. Identify common types of weld defects found in offshore structures and suggest inspection methods for detecting them.
29. Discuss the importance of sustainability and energy efficiency in future offshore welding practices.
30. Predict how the integration of smart sensors, AI-based monitoring, and digital quality assurance will redefine the future of offshore welding.

9 Non-Destructive Testing (NDT) for Marine Welds

9.1 INTRODUCTION

In the marine and offshore industries, where structures are continuously exposed to extreme environmental conditions, the integrity of welded joints determines not only the performance and longevity of assets but also the safety of operations. Welds in ships, offshore platforms, subsea pipelines, and underwater components must endure high mechanical stresses, corrosive seawater, pressure variations, and dynamic loading throughout their service life. Any undetected flaw, such as cracks, porosity, lack of fusion, or inclusions, can propagate rapidly under these conditions, leading to catastrophic structural failures. To prevent such occurrences, the marine sector relies extensively on non-destructive testing (NDT) techniques, which enable the detection and evaluation of surface and subsurface discontinuities without damaging or altering the welded component (Tang and Yu 2020).

NDT plays a crucial role in the quality assurance (QA), maintenance, and life cycle management of marine welded structures. It forms an integral part of both fabrication-stage inspection and in-service monitoring, ensuring that welds comply with international standards such as American Welding Society (AWS) D3.6M, ISO 17635, DNV-ST-F101, and American Society of Mechanical Engineers (ASME) Section V. Unlike destructive testing methods that require sample removal, NDT provides a cost-effective and reliable means of assessing weld integrity while keeping the component in operation. The adoption of advanced NDT technologies in marine engineering has been driven by the need for precision, real-time data acquisition, and remote inspection capability, especially in subsea and confined environments where human accessibility is limited.

Traditional NDT methods such as visual testing (VT), magnetic particle testing (MT), dye penetrant testing (PT), ultrasonic testing (UT), and radiographic testing (RT) continue to serve as the backbone of marine weld inspection. However, recent advancements in phased array ultrasonic testing (PAUT), acoustic emission (AE) analysis, time-of-flight diffraction (TOFD), and digital radiography (DR) have significantly enhanced defect detection accuracy, depth resolution, and automation. Moreover, the integration of robotic platforms, remotely operated vehicles (ROVs), and artificial intelligence (AI)-based image recognition systems now enables continuous inspection of underwater welds at extreme depths, improving safety and reducing downtime.

In the modern marine context, NDT is evolving from a purely diagnostic function into a predictive and data-driven discipline. By leveraging digital twins, sensor fusion, and machine learning (ML) algorithms, NDT not only detects flaws but also

DOI: 10.1201/9781003217350-9

predicts degradation trends, allowing for condition-based maintenance (CBM) and proactive asset management. This transformation aligns with the broader vision of smart shipyards and digital offshore infrastructure, where inspection, welding, and repair processes are seamlessly integrated through intelligent automation.

In summary, this chapter provides a comprehensive overview of NDT for marine welds, discussing the principles, techniques, equipment, and applications of major NDT methods used in shipbuilding and offshore engineering. It also highlights the emerging trends in robotic, digital, and AI-enhanced inspection systems, emphasizing their role in ensuring structural reliability, operational safety, and sustainability in the global marine industry.

9.1.1 Importance of NDT in Marine Welding Quality Assurance

NDT plays a pivotal role in ensuring the QA of welded joints in marine and offshore structures. In shipbuilding, offshore platform construction, and subsea fabrication, welds serve as the primary load-transferring elements of the structure. The harsh operating environment characterized by high salinity, cyclic stresses, and extreme pressure variations makes even minor weld discontinuities potentially catastrophic. Therefore, NDT is indispensable for verifying that each weld meets design, material, and structural integrity requirements before the asset enters service.

The primary purpose of NDT in marine welding QA is to detect surface and subsurface flaws such as cracks, inclusions, porosity, incomplete fusion, or undercuts without impairing the weld or base material. By identifying these defects early in the fabrication stage, rework can be performed efficiently, minimizing costly in-service failures. NDT also ensures that welding processes remain within qualified procedure parameters, maintaining consistency and repeatability across different sections of the structure.

Moreover, NDT supports continuous quality monitoring, where periodic inspection throughout the operational life of a vessel or offshore installation ensures long-term weld performance. Techniques like UT and PAUT are widely used for in-depth inspection, while radiography (RT) and MT provide accurate detection of internal and surface discontinuities. In essence, NDT is the backbone of marine welding QA, providing quantitative and visual evidence that welds conform to the highest safety and reliability standards.

9.1.2 Role of NDT in Safety, Reliability, and Regulatory Compliance

The safety and reliability of marine and offshore structures depend fundamentally on the soundness of welded joints, and NDT serves as the primary verification tool for achieving compliance with international standards and classification requirements. Offshore platforms, ships, subsea pipelines, and floating systems are all governed by strict codes established by organizations such as AWS, American Bureau of Shipping (ABS), Det Norske Veritas (DNV), Lloyd's Register (LR), and ISO/IEC. Compliance with these standards is mandatory not only for fabrication approval but also for periodic certification during service.

From a safety perspective, NDT minimizes the risk of structural failure by ensuring defect-free welds capable of withstanding high loads, vibrations, and corrosive marine environments. Defects such as cracks or lack of fusion can act as initiation points for fatigue failure or stress corrosion cracking (SCC) conditions that often remain undetected until failure occurs. By employing advanced inspection methods such as DR, TOFD, or acoustic emission monitoring (AEM), engineers can identify early-stage damage, enabling preventive maintenance before critical failures arise.

In terms of reliability, NDT contributes to life cycle management of marine structures by establishing baseline data during commissioning and tracking changes in weld condition over time. This historical record aids in predicting the remaining service life and scheduling maintenance or repair operations more effectively. Furthermore, modern NDT systems integrate automated data logging and digital reporting, ensuring full traceability and transparency during regulatory audits.

Therefore, NDT is not only a tool for defect detection but also a strategic assurance mechanism for achieving compliance, enhancing structural resilience, and safeguarding lives, assets, and the marine environment.

9.1.3 Challenges of NDT in Marine Environments

Conducting NDT in marine and offshore environments poses significant challenges due to environmental, technical, and operational constraints that are often absent in land-based inspection. The most fundamental challenge arises from limited accessibility and visibility in underwater or confined areas. Welds located on subsea pipelines, platform legs, or internal tank structures are difficult to inspect, requiring specialized tools, ROVs, or diver-assisted equipment.

Environmental factors such as high hydrostatic pressure, seawater salinity, marine growth, and low temperatures can interfere with both inspection accuracy and equipment functionality. For instance, ultrasonic waves are affected by water temperature and salinity, while RT becomes impractical in deep-sea environments due to radiation scattering and shielding limitations. Additionally, corrosion, biofouling, and coating layers may obscure defects or alter signal reflections, complicating interpretation.

Operational challenges also stem from time constraints and safety risks associated with underwater inspection. Diver exposure time is limited, and equipment deployment at great depths requires precise coordination and logistics. Maintenance windows for offshore installations are narrow, demanding rapid yet reliable inspection solutions. Moreover, data interpretation in underwater NDT is complex, as acoustic noise and environmental variability can introduce errors or false readings.

To overcome these challenges, modern NDT practices in marine settings increasingly employ robotic automation, advanced sensors, and AI-assisted data analytics. ROV-mounted ultrasonic and phased array probes, for example, enable high-resolution mapping of subsea welds without direct human intervention. Similarly, digital imaging and acoustic tomography improve flaw detection accuracy in murky or inaccessible environments. Despite these advancements, the unique conditions of the marine domain continue to demand specialized procedures, skilled operators, and rigorous calibration to ensure accurate and reliable NDT results.

9.2 OVERVIEW OF NDT METHODS FOR MARINE APPLICATIONS

NDT methods form the foundation of inspection, maintenance, and QA programs across all marine and offshore structures. Given the harsh and unpredictable conditions of the marine environment characterized by corrosive seawater, high pressure, biofouling, and cyclic mechanical stresses, it is critical that welds and structural materials are inspected using techniques capable of identifying surface and subsurface defects without impairing the component's function or integrity.

NDT methods for marine applications can be broadly categorized into surface inspection methods, volumetric inspection methods, and advanced digital and automated techniques. Each category serves a distinct purpose depending on the weld type, material composition, accessibility, and inspection environment (in-air, underwater, or confined spaces). Figure 9.1 illustrates the primary categories of NDT used in marine and offshore inspection. The diagram classifies NDT techniques into surface, subsurface, volumetric, and advanced digital methods. Each category represents a distinct level of inspection depth and application ranging from surface evaluation (VT and PT) to internal flaw detection (UT and RT) and digital/automated approaches (infrared thermography [IRT], AE, and robotic NDT). This classification emphasizes how modern inspection strategies integrate traditional and advanced methods to ensure comprehensive weld integrity assessment across all marine environments.

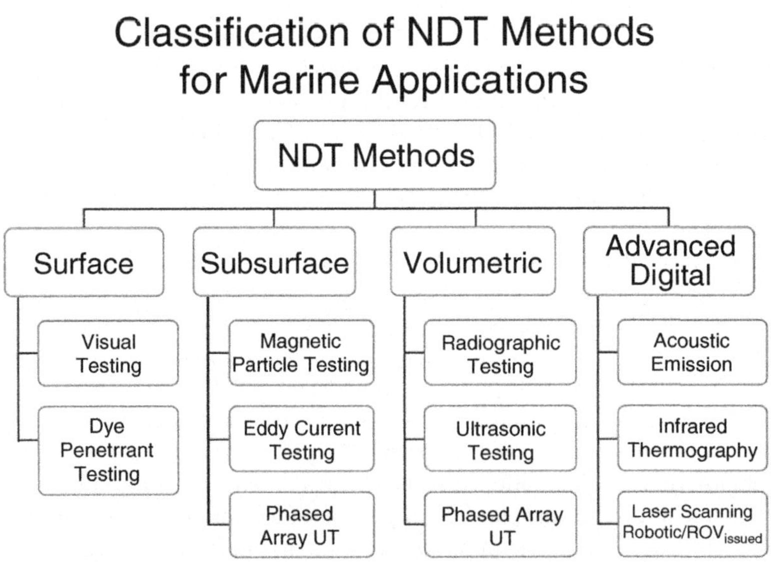

FIGURE 9.1 Classification of NDT methods for marine applications.

9.2.1 Surface vs Volumetric Testing Techniques

9.2.1.1 Surface Testing Techniques

Surface inspection techniques are employed to detect discontinuities that appear on or near the surface of welds, such as cracks, porosity, undercuts, and surface-breaking flaws. The most widely used methods include the following:

Visual Testing (VT): The simplest yet most essential NDT technique, VT involves direct or remote visual examination of weld surfaces using magnifiers, borescopes, or remotely operated video systems. In marine applications, ROV-mounted cameras and digital imaging systems enable remote visual inspection (RVI) of subsea welds, detecting corrosion, misalignment, or weld irregularities.

Dye Penetrant Testing (PT): This method is used to locate open-to-surface defects in non-porous materials. The process involves applying a liquid dye to the surface, allowing it to penetrate flaws, and then using a developer to draw out the dye for visibility under white or ultraviolet (UV) light. PT is commonly used for stainless steel, aluminum, and non-ferrous marine components, especially in ship hulls and propeller assemblies.

Magnetic Particle Testing (MT): MT detects surface and near-surface discontinuities in ferromagnetic materials by magnetizing the weld area and applying fine magnetic particles. The presence of a defect distorts the magnetic field, causing particle accumulation that visually indicates flaw locations. In marine structures, MT is effective for platform braces, risers, and steel hull structures but requires surface cleaning and is unsuitable for underwater environments.

9.2.1.2 Volumetric Inspection Methods

Volumetric methods provide a deeper assessment of the internal integrity of welds and structural materials, detecting defects that are invisible on the surface. These methods are critical for ensuring the mechanical reliability of pressure-containing components, pipelines, and load-bearing joints in offshore systems.

Radiographic Testing (RT): RT uses X-rays or gamma rays to produce images of internal flaws within welds. Variations in material density due to inclusions, voids, or cracks appear as dark regions on the radiograph. This method provides a permanent visual record of weld quality and is extensively used in pipeline welding and pressure vessel fabrication. However, RT poses radiation safety challenges and is difficult to apply underwater.

Ultrasonic Testing (UT): UT employs high-frequency sound waves transmitted into the material to detect internal discontinuities. Reflected echoes from flaws or interfaces are displayed on a screen, allowing accurate measurement of defect size and depth. Contact and immersion probes are used depending on accessibility. UT is one of the most widely adopted NDT methods in offshore pipeline inspection, ship hull evaluation, and structural integrity assessment.

Phased Array Ultrasonic Testing (PAUT): An advanced form of UT, PAUT uses multiple ultrasonic transducers that can be electronically controlled to steer and focus the beam at various angles. This allows for three-dimensional imaging and precise flaw characterization. PAUT is invaluable for complex weld geometries and automated underwater inspections, often integrated with robotic systems.

Time-of-Flight Diffraction (TOFD): TOFD measures the time difference between ultrasonic signals diffracted from crack tips, enabling highly accurate depth sizing of defects. It is often used in conjunction with PAUT for critical subsea pipeline welds, ensuring compliance with high-integrity standards.

9.2.2 Advanced and Specialized NDT Methods

Modern marine engineering has witnessed the integration of specialized and digital NDT techniques that offer greater precision, automation, and real-time data processing essential for inspection in deepwater or inaccessible environments.

Acoustic Emission Testing (AET): This technique detects transient elastic waves generated by active crack growth or material deformation during loading. In marine applications, AET is used for fatigue monitoring of offshore platforms, risers, and pressure hulls, allowing early detection of crack initiation before failure occurs.

Eddy Current Testing (ECT): Based on electromagnetic induction, ECT detects surface and near-surface defects in conductive materials. It is effective for corrosion mapping, crack detection in non-ferrous alloys, and inspection of coatings on ship structures or underwater components.

Infrared Thermography (IRT): IRT uses infrared cameras to detect temperature variations caused by subsurface defects, delamination, or corrosion under coatings (CUCs). It is a fast, non-contact method ideal for large-area surveys of hulls, decks, and composite materials.

Laser and Optical Scanning: These digital imaging methods create high-resolution 3D maps of weld profiles and structural surfaces, enabling deformation tracking, crack growth analysis, and digital documentation for predictive maintenance.

9.2.3 Underwater and Remote NDT Techniques

The unique challenges of subsea environments restricted visibility, pressure effects, and limited accessibility have led to the development of remote and automated NDT systems. ROV- and AUV-mounted NDT tools combine ultrasonic, visual, and magnetic sensors for integrated inspection. These systems can operate autonomously, transmitting real-time data to surface control units for analysis. Additionally, DR and underwater acoustic tomography are emerging as promising methods for volumetric flaw detection at extreme depths.

Table 9.1 summarizes the principal NDT methods utilized in the inspection of marine welds, outlining their operating principles, detection capabilities, and application areas across shipbuilding, offshore, and subsea engineering. Surface-level inspection techniques like VT and PT are ideal for detecting surface-breaking flaws, while MT enhances detection in ferromagnetic components. For volumetric assessment, methods such as UT, PAUT, and RT provide detailed insights into internal discontinuities and weld integrity.

TABLE 9.1
Summary of Major NDT Methods for Marine Weld Inspection

NDT Method	Principle of Operation	Detection Capability	Typical Marine Applications
Visual testing (VT)	Direct or remote optical examination using magnifiers, cameras, or ROVs	Detects surface discontinuities such as cracks, porosity, undercuts, or misalignment	Ship hull inspection, weld seams on platforms, structural joints
Dye penetrant testing (PT)	Capillary action draws dye into surface-breaking flaws, revealed with developer under white/UV light	Detects surface cracks, pinholes, and porosity in non-porous materials	Stainless steel and aluminum ship components, propeller blades
Magnetic particle testing (MT)	Magnetic field applied to ferromagnetic materials; magnetic particles reveal flux leakage at defect sites	Detects surface and near-surface cracks and inclusions	Platform braces, risers, and structural steel welds
Radiographic testing (RT)	X-rays or gamma rays pass through material to form images of internal discontinuities on film or digital detector	Detects internal defects such as porosity, inclusions, lack of fusion, and cracks	Offshore pipeline welds, pressure vessels, critical load joints
Ultrasonic testing (UT)	High-frequency sound waves reflect from internal flaws; echoes analyzed for flaw size and location	Detects internal discontinuities, delamination, and thickness loss	Subsea pipeline inspection, ship hull plating, offshore tanks
Phased array ultrasonic testing (PAUT)	Electronically controlled ultrasonic beam scans at multiple angles for 3D flaw imaging	Provides volumetric flaw detection with high accuracy and depth profiling	Complex weld geometries, deep-sea joints, robotic underwater inspection
Time-of-flight diffraction (TOFD)	Measures diffraction time of ultrasonic waves from defect tips to calculate flaw depth	Accurate sizing and characterization of cracks and inclusions	High-integrity welds in offshore pipelines and subsea manifolds
Eddy current testing (ECT)	Induced electromagnetic fields detect discontinuities by changes in electrical conductivity	Detects surface and near-surface cracks and corrosion in conductive materials	Aluminum hulls, propellers, and corrosion detection on coatings

(Continued)

TABLE 9.1 *(Continued)*
Summary of Major NDT Methods for Marine Weld Inspection

NDT Method	Principle of Operation	Detection Capability	Typical Marine Applications
Acoustic emission testing (AET)	Detects elastic waves produced by active crack growth or material deformation	Monitors dynamic defect growth, fatigue, or leak detection	Fatigue monitoring of offshore platforms, pressure hulls, risers
Infrared thermography (IRT)	Detects heat patterns caused by subsurface defects using infrared imaging	Identifies corrosion, delamination, and coating degradation	Hull and deck surveys, composite structures, corrosion mapping
Laser and optical scanning	Uses laser triangulation or structured light for 3D imaging of weld surfaces	Detects distortion, dimensional deviations, and surface cracks	Digital weld profiling, shipyard inspection, deformation analysis
Digital radiography (DR)	Uses digital detectors instead of film for real-time radiographic imaging	Detects internal weld flaws with improved image resolution	In-service inspection of offshore and subsea welds
Robotic/ ROV-based NDT	Integration of UT, PAUT, or visual sensors into remotely operated vehicles	Enables automated inspection in inaccessible or deep-sea environments	Subsea pipelines, platform legs, remote structural inspection

Advanced digital techniques TOFD, ECT, IRT, and laser scanning offer high-resolution imaging and real-time defect mapping, which are invaluable for critical and inaccessible areas of marine structures. Furthermore, robotic and ROV-based NDT systems represent the latest evolution in offshore inspection, allowing for autonomous and deepwater operation with minimal human intervention. Collectively, these techniques form a comprehensive inspection toolkit ensuring the structural reliability, safety, and regulatory compliance of marine welded structures throughout their service life.

9.3 ULTRASONIC TESTING (UT)

UT is one of the most widely adopted volumetric NDT techniques used in the marine and offshore industries to evaluate the internal integrity of welds and structural materials. It operates by transmitting high-frequency sound waves into the material and analyzing the reflected signals (echoes) from internal discontinuities such as cracks, porosity, inclusions, lack of fusion, or laminations. UT offers excellent penetration depth, high accuracy, and portability, making it indispensable for shipbuilding, offshore platform fabrication, subsea pipelines, and pressure hull inspection.

9.3.1 Principles of Ultrasonic Inspection

The principle of ultrasonic inspection is based on the propagation of high-frequency acoustic waves (typically 0.5–20 MHz) through a material. When these waves encounter an interface between materials of differing acoustic impedance, such as a crack, void, or inclusion, part of the wave energy is reflected back to the transducer, while the remainder continues to propagate. By measuring the time of flight and amplitude of the reflected signal, the position, size, and orientation of internal flaws can be determined.

A standard UT setup consists of a transducer (probe), pulse-receiver unit, couplant, and display system (A-scan or digital interface). The transducer converts electrical energy into mechanical vibrations (sound waves) and vice versa. A couplant usually water, oil, or gel is applied to eliminate air gaps between the probe and surface, ensuring efficient sound transmission. The display system shows the amplitude of echoes against time, where each peak corresponds to an interface or discontinuity within the weld.

In marine applications, UT can be performed using two principal methods:

Contact Testing: The probe is directly placed on the structure, commonly used for in-air inspection of ship hulls or tanks.

Immersion Testing: The component is submerged in water, and ultrasonic energy is transmitted through the liquid medium ideal for underwater or subsea weld inspection using ROV-mounted transducers.

The angle-beam technique is frequently employed for weld testing, where the transducer introduces the sound wave at an oblique angle to detect planar flaws like lack of fusion or cracks aligned parallel to the weld surface. Figure 9.2 illustrates how ultrasonic waves propagate through a material during testing. The transducer emits high-frequency sound waves that travel through a couplant into the workpiece. When the sound waves encounter a defect or boundary, part of the energy reflects back to the transducer, producing an echo on the display. The figure shows both contact testing, where the probe is placed directly on the surface with a thin couplant layer, and

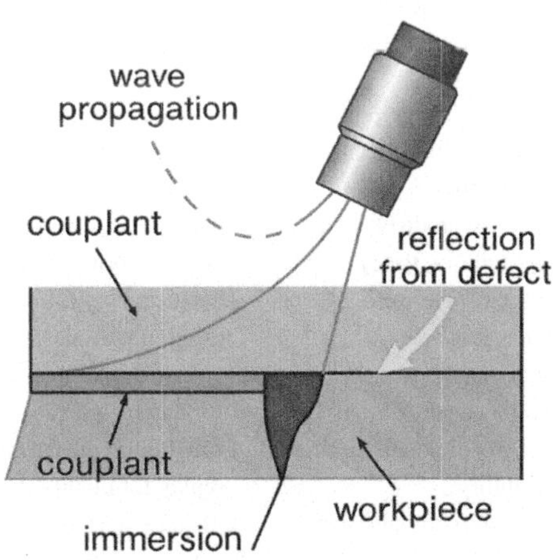

FIGURE 9.2 Principle of ultrasonic inspection.

immersion testing, where sound travels through a water medium ideal for underwater or subsea weld inspection in marine applications.

9.3.2 Conventional UT vs. Phased Array UT in Marine Welds

While conventional UT remains widely used, the evolution of technology has introduced PAUT as a superior alternative, offering greater precision, automation, and data visualization particularly beneficial in complex marine structures. Table 9.2 compares the key characteristics, capabilities, and applications of conventional UT and PAUT as applied to marine weld inspection. The comparison highlights differences in principle of operation, data visualization, coverage, automation potential, and inspection accuracy. While conventional UT remains effective for routine inspections, PAUT provides advanced imaging, greater flaw characterization accuracy, and full digital integration, making it ideal for complex geometries and automated subsea inspection systems in modern shipbuilding and offshore engineering.

In marine weld inspection, PAUT has largely replaced conventional UT in critical applications, especially where high structural reliability is required. The ability to visualize the weld cross section in real time and automatically record results ensures traceable and repeatable inspection, reducing operator dependency. Furthermore, PAUT can be deployed using ROV- or crawler-based systems for automated underwater inspection, making it ideal for subsea pipelines, riser welds, and offshore structural nodes.

TABLE 9.2
Comparison between Conventional Ultrasonic Testing (UT) and Phased Array Ultrasonic Testing (PAUT) for Marine Weld Inspection

Aspect	Conventional UT	Phased Array UT (PAUT)
Principle	Uses a single transducer to emit and receive sound waves at a fixed angle	Uses multiple small transducer elements electronically controlled to steer, focus, and scan beams at multiple angles
Coverage	Limited to discrete points; requires multiple scans for full coverage	Provides wide area coverage and 3D volumetric imaging from a single probe position
Data representation	Simple A-scan display (amplitude vs. time)	Offers A-scan, B-scan, and C-scan images for cross-sectional and top-view analysis
Accuracy	High for simple weld geometries	High accuracy for complex geometries and dissimilar joints
Automation capability	Manual and operator-dependent	Easily integrated with robotic or ROV systems for automated subsea inspection
Recording and data storage	Manual recordkeeping; limited traceability	Fully digital data acquisition, storage, and replay for quality documentation
Marine application	Routine ship hull and tank inspection	Offshore platform welds, pipeline joints, risers, and deepwater infrastructure

9.3.3 Advantages, Limitations, and Applications in Shipbuilding

Advantages:

High Sensitivity and Accuracy: Capable of detecting very small internal flaws and determining their exact depth and orientation.
Excellent Penetration: Can inspect thick marine-grade steels and alloys used in hulls and offshore platforms.
Immediate Results: Provides instant feedback through digital displays for on-site decision-making.
Non-Hazardous and Portable: Unlike radiography, UT involves no radiation hazards and can be used in confined shipboard environments.
Digital Integration: PAUT allows 3D imaging, automated scanning, and digital recordkeeping for QA.
Adaptability for Underwater Use: Immersion and ROV-based UT systems are highly effective for in situ subsea inspections.

Limitations:

Surface Accessibility: Requires smooth, clean surfaces and good coupling for accurate results.
Operator Skill Dependence: Interpretation of signals demands high expertise, especially in complex geometries.
Calibration Requirements: Must be carefully calibrated with reference blocks for accurate flaw sizing.
Difficulties with Coarse-Grained Materials: Rough or anisotropic microstructures (e.g., cast steels) can scatter sound waves.
Limited Inspection of Very Small Welds: Beam size and geometry may restrict testing of narrow or fillet welds.

Applications in Shipbuilding and Offshore Structures:

Ship Hulls and Bulkheads: Detection of internal cracks, corrosion, and delamination in thick steel plates.
Offshore Pipelines and Risers: Verification of weld integrity and flaw depth sizing in high-pressure systems.
Ballast Tanks and Pressure Vessels: Monitoring wall thickness and detecting CUCs.
Subsea Structures: Automated inspection of critical joints using PAUT-equipped ROVs or magnetic crawlers.
Propulsion and Steering Components: Evaluation of propeller shafts, rudders, and thrust blocks for fatigue cracks

9.4 RADIOGRAPHIC TESTING (RT)

RT is one of the most reliable and established volumetric NDT techniques used for detecting internal discontinuities in welds and structural components in shipyards, offshore platforms, and subsea fabrication. It employs X-rays or gamma rays to penetrate the material and produce a permanent image (radiograph) that reveals internal flaws such as porosity, slag inclusions, lack of fusion, cracks, or incomplete penetration. Because it provides a visual record of internal structure, RT is widely regarded as a benchmark inspection method in marine and offshore QA programs.

9.4.1 Principle of Radiographic Testing (RT)

The principle of RT is based on the differential absorption of ionizing radiation as it passes through a material. When a beam of X-rays (generated electrically) or gamma rays (from radioactive isotopes like Iridium-192 or Cobalt-60) is directed at a weld, denser areas of the material absorb more radiation, while regions with internal flaws allow greater transmission. A detector such as a photographic film, phosphor imaging plate, or digital sensor captures the transmitted radiation, producing a two-dimensional image that displays internal variations as contrast differences.

Defects like voids or cracks appear as dark regions (due to higher radiation transmission), whereas dense inclusions appear lighter. By analyzing the resulting radiograph, inspectors can identify the type, size, and location of internal discontinuities.

Modern DR and computed radiography (CR) have largely replaced traditional film techniques in marine applications, allowing for real-time image acquisition, enhanced contrast, and digital storage, which significantly improve efficiency and traceability.

Figure 9.3 illustrates how X-rays or gamma rays are projected from a radiation source through a welded material to a detector or radiographic film. When radiation passes through the weld, internal discontinuities such as voids or inclusions allow greater transmission, appearing as dark regions on the image, while denser areas absorb more radiation, creating lighter zones. This contrast enables identification of internal flaws within the weld. The figure highlights the key components X-ray or

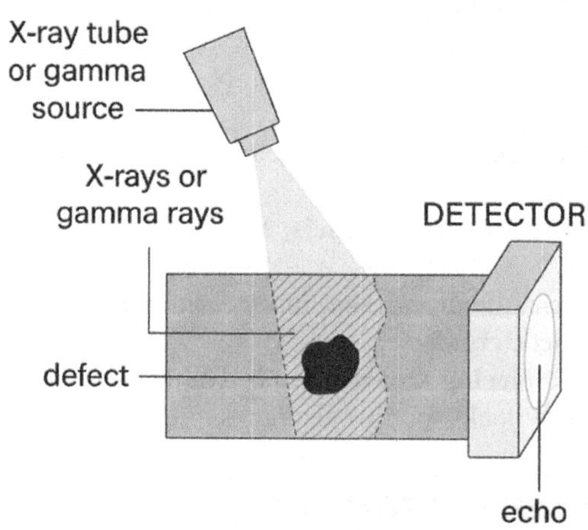

FIGURE 9.3 Principle of radiographic testing (RT).

gamma source, weld joint, defect, and detector representing the core concept of RT as used in shipyard and offshore weld inspection.

9.4.2 Applications in Shipyard and Offshore Use

RT is extensively applied across the shipbuilding and offshore sectors, particularly for critical weld joints and pressure-retaining components where internal integrity is paramount. Its ability to detect both volumetric and planar defects makes it invaluable for ensuring structural soundness and compliance with classification standards.

Key Applications Include:

Shipbuilding:

Inspection of hull and deck welds, particularly in high-stress regions such as bulkheads and longitudinal seams.

Examination of boilers, pressure vessels, and pipelines for internal defects or corrosion.

Verification of welds in propulsion systems, rudders, and structural frames.

Offshore Platforms and Pipelines:

Inspection of pipeline circumferential welds for porosity, lack of fusion, or inclusions before subsea deployment.

Quality control of riser welds, pressure manifolds, and subsea connectors exposed to dynamic loading and high pressure.

Assessment of critical offshore joints such as flare booms, jacket nodes, and load-bearing members.

Subsea and Remote Inspection:

Application of gamma radiography using compact radioactive sources for on-site inspection at offshore installations, where X-ray equipment may be impractical.

Digital radiography using crawler-mounted or ROV-based imaging systems for remote inspection in confined or submerged environments.

9.4.3 Advantages of Radiographic Testing (RT)

RT offers several key advantages that make it an essential method for weld QA in marine and offshore operations:

Internal Defect Detection: Capable of identifying internal flaws such as inclusions, porosity, cracks, and incomplete fusion that are invisible to surface inspection methods.

Permanent Record: Produces a visual image (film or digital) that serves as verifiable documentation of weld quality and inspection traceability.

Wide Material Applicability: Suitable for metals, alloys, composites, and non-metallic materials used in shipbuilding and offshore engineering.

High Sensitivity: Detects small volumetric defects and provides clear contrast differentiation.

Digital Integration: Modern DR systems enable instant image capture, enhanced image processing, and cloud-based data storage.

Non-Contact Method: Effective for inspecting components where physical access is limited, including in-service pressure systems or subsea welds.

9.4.4 Limitations of Radiographic Testing (RT)

Despite its effectiveness, RT presents several limitations that must be managed, particularly in marine environments:

Safety Hazards: Involves the use of ionizing radiation, requiring strict safety procedures, shielding, and certified operators.

Access and Geometry Constraints: Requires access to both sides of the component, which can be difficult in confined shipyard areas or underwater environments.

Thickness Limitations: Effectiveness decreases with increasing material thickness, requiring higher radiation energy or longer exposure times.

Interpretation Complexity: Radiographs provide only 2D images of 3D structures; defect orientation may be difficult to assess accurately.

Environmental Restrictions: Radiation safety regulations and environmental conditions may limit its use during active shipyard or offshore operations.

Cost and Time: Film-based RT can be time-consuming and expensive due to setup, exposure, and development processes (partially mitigated by digital systems).

9.5 MAGNETIC PARTICLE TESTING (MT)

MT is a widely used surface and near-surface NDT technique applied primarily to ferromagnetic materials such as carbon steel and low-alloy steels commonly found in ships, offshore platforms, and subsea structures. It operates on the principle of magnetic flux leakage (MFL), where magnetic fields are distorted by discontinuities such as cracks or inclusions near the material's surface. These field disruptions attract magnetic particles, forming a visible indication of the defect's size, shape, and orientation.

MT is particularly valued in marine applications for its speed, reliability, and ability to detect fine surface-breaking flaws, making it essential in both fabrication-stage inspection and in-service maintenance of welded components exposed to dynamic loads and harsh environments.

9.5.1 Principle of Magnetic Flux Leakage

The principle behind MT relies on magnetizing the material and observing how magnetic flux behaves in the presence of a discontinuity. When a magnetic field is applied to a ferromagnetic material, the field lines (flux) pass uniformly through the body. However, if a flaw such as a crack, inclusion, or void exists near the surface, it interrupts the magnetic field, causing the flux to leak out of the material. This phenomenon is known as MFL.

To visualize this leakage, fine magnetic particles either dry powder or suspended in a liquid carrier are applied to the surface. These particles accumulate at leakage sites due to the localized magnetic attraction, forming a visible indication of the defect. The pattern and density of particle buildup reveal details about the defect's size, shape, and orientation.

Two magnetization methods are commonly used in MT:

Longitudinal Magnetization: Field lines run parallel to the material's length; used to detect transverse cracks.

Circular Magnetization: Field lines flow circumferentially around the component; used to detect longitudinal cracks.

MT can be performed using alternating current (AC) for detecting surface flaws or direct current (DC) for identifying subsurface discontinuities up to a few millimeters deep. UV-visible fluorescent particles are often used in low-light conditions, such as ship tanks or offshore maintenance sites to enhance detection accuracy. Figure 9.4 illustrates how a magnetic field interacts with a surface crack in a ferromagnetic material. The magnetic field lines flow uniformly through the component until they encounter a surface discontinuity, where MFL occurs. Iron particles, either dry or

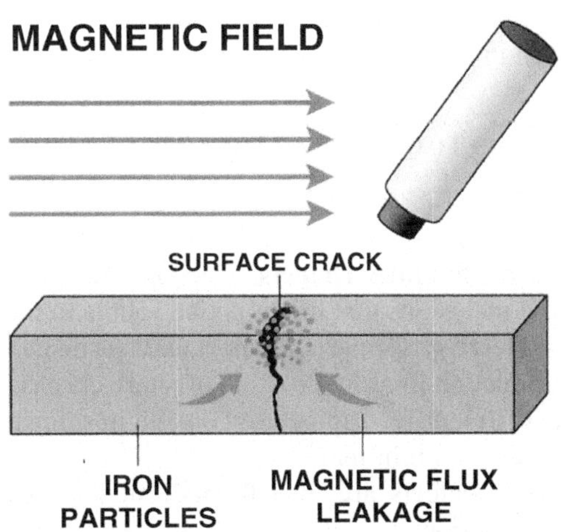

FIGURE 9.4 Principle of magnetic particle testing (MT).

suspended in liquid, accumulate at the leakage site, visually revealing the defect's location and orientation. This principle forms the foundation of MT, enabling rapid and reliable detection of surface and near-surface flaws in ferromagnetic marine components.

9.5.2 Application in Ferromagnetic Marine Materials

MT is extensively applied in the shipbuilding and offshore industries, where ferromagnetic materials dominate structural and welded assemblies. Its ability to detect early-stage cracking or surface discontinuities contributes significantly to preventive maintenance and structural reliability.

Common Marine Applications Include:

Ship Hull Structures: Inspection of longitudinal and transverse welds for fatigue or thermal cracking.

Offshore Platforms: Examination of jacket legs, riser clamps, and node connections subjected to cyclic loading and seawater corrosion.

Pressure Vessels and Storage Tanks: Detection of weld defects and stress corrosion cracks in carbon steel tanks used for fuel, water, or compressed gas storage.

Cranes, Winches, and Lifting Equipment: Checking weld joints and critical components for surface cracks before deployment.

Subsea Equipment: Inspection of ferromagnetic pipeline flanges, valves, and anchors in maintenance dry docks or shallow-water environments.

Propulsion Systems: Evaluation of rudder linkages, shaft supports, and gear housings exposed to repetitive mechanical stresses.

MT is especially effective for complex geometries where welds intersect or where visual access is limited, as the magnetic particles can reveal discontinuities even in partially hidden zones.

9.5.3 Advantages and Limitations in Shipyard and Offshore Use

Advantages:

High Sensitivity for Surface Defects: Capable of detecting fine surface cracks, laps, and inclusions that might not be visible to the naked eye.

Quick and Cost-Effective: Simple equipment and fast inspection make it ideal for shipyard production lines and routine offshore checks.

Portable and Field-Friendly: Can be used on-site in shipyards, dry docks, or offshore platforms with minimal setup.

Immediate Results: Defects are visually indicated in real time, allowing prompt corrective action.

Effective on Complex Shapes: Works well on curved or irregular welds, fillets, and nodes.

Adaptable to Various Lighting Conditions: Fluorescent particle testing enhances defect visibility under UV illumination, which is useful in confined or poorly lit marine spaces.

Limitations:

Restricted to Ferromagnetic Materials: Not applicable to non-ferrous metals such as aluminum, copper, or stainless steels commonly used in marine environments.
Surface Preparation Required: The inspection area must be clean and free from paint, rust, or marine deposits for reliable results.
Limited Penetration Depth: Detects only surface and near-surface flaws (typically up to 2–3 mm deep).
Demagnetization Requirement: Components may retain residual magnetism after inspection, requiring demagnetization to prevent interference in service.
Environmental Sensitivity: Offshore humidity, seawater exposure, or strong currents can affect particle distribution and visibility.
Operator Skill Dependence: Accurate interpretation of indications requires experience, especially in complex weld geometries.

9.6 OTHER RELEVANT NDT METHODS IN MARINE WELDING

In addition to primary NDT techniques such as UT, RT, and MT, several complementary inspection methods play vital roles in ensuring weld integrity in marine and offshore applications. These methods enhance the detection of specific defect types, provide material-specific testing capabilities, and support continuous structural health monitoring (SHM) for safety-critical systems. The most notable among them are PT for surface crack detection, ECT for non-ferrous alloys, and AEM for real-time fatigue and crack monitoring.

9.6.1 Dye Penetrant Testing (PT) for Surface Crack Detection

Principle:
PT is a surface inspection method that utilizes capillary action to detect surface-breaking defects in non-porous materials. The process involves applying a visible or fluorescent liquid penetrant onto the surface of a cleaned component. The penetrant seeps into cracks or voids, and after a dwell period, the excess is removed and a developer is applied, which draws out the trapped penetrant, revealing defect indications under visible or UV light. Figure 9.5 illustrates how penetrant testing reveals surface-breaking flaws. The process involves applying a liquid penetrant that seeps into a surface crack, followed by removal of excess penetrant and application of a developer. The developer draws the trapped penetrant out, forming a visible indication of the flaw under visible or UV light. This technique allows easy detection of surface cracks or porosity in non-ferrous and non-magnetic marine materials such as aluminum and stainless steel.

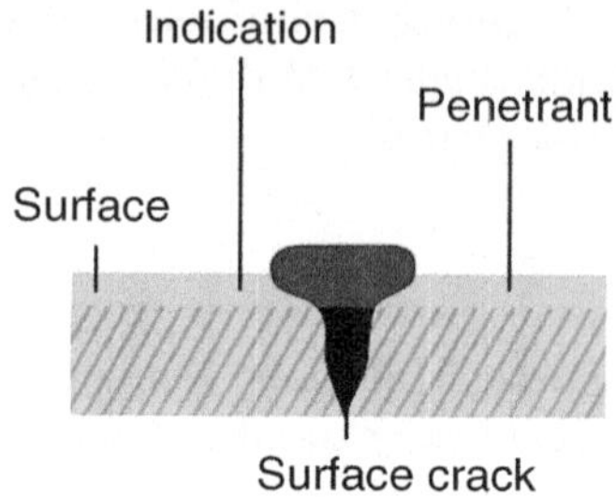

FIGURE 9.5 Principle of dye penetrant testing (PT) for surface crack detection.

Applications in Marine Welding:
PT is widely used in shipyards and offshore fabrication for detecting fine surface flaws such as cracks, porosity, laps, and leaks in non-ferrous or non-magnetic materials where MT cannot be applied. It is especially useful for:

Aluminum and stainless steel components in ship superstructures and hulls.
Weld joints on propeller blades, valves, and turbine housings.
Heat exchangers and piping systems made of copper-nickel or titanium alloys.
Repair verification of re-welded or machined surfaces in dry dock maintenance.

Advantages:

Simple, low-cost, and highly sensitive for small surface flaws.
Applicable to non-magnetic materials.
Portable and suitable for field use.
Provides visual evidence of defect location and size.

Limitations:

Detects only surface-breaking flaws (no subsurface capability).
Requires meticulous cleaning and preparation.
Not suitable for porous or rough surfaces.
Environmental conditions (humidity, salt, and lighting) can affect visibility offshore.
PT remains one of the most practical and effective inspection methods for non-ferromagnetic marine materials, providing essential support during both fabrication and in-service maintenance operations.

9.6.2 Eddy Current Testing (ECT) for Non-Ferrous Alloys

Principle: ECT is an electromagnetic NDT technique used primarily for conductive materials. When an AC passes through a probe coil, it generates an oscillating magnetic field. This field induces eddy currents within the test material, and any discontinuity (such as a crack, corrosion pit, or change in thickness) disturbs the eddy current flow. The resulting impedance change in the probe coil is measured and analyzed to determine defect characteristics. Figure 9.6 illustrates how AC in the excitation coil generates a changing magnetic field that induces eddy currents in the conductive test material. When a surface flaw or discontinuity is present, the flow of eddy currents is disturbed, causing a measurable impedance change in the coil circuit. The ECT instrument detects this variation and displays it for analysis. This principle enables precise identification of cracks, corrosion, and thickness variations in non-ferrous marine materials such as aluminum, copper-nickel, and stainless steel alloys.

Applications in Marine Welding:
ECT is extensively used for non-ferrous marine materials and thin-walled structures where magnetic methods are ineffective. Common applications include:

- **Aluminum Ship Hulls and Superstructures**: Detecting fatigue cracks and corrosion.
- **Heat Exchanger and Condenser Tubes**: Locating pitting, wall thinning, and cracking in copper-nickel or titanium tubing.
- **Aircraft and Marine Propellers**: Assessing surface damage or microcracks.
- **Welded Joints in Non-Magnetic Alloys**: Evaluating aluminum and austenitic stainless steel welds.

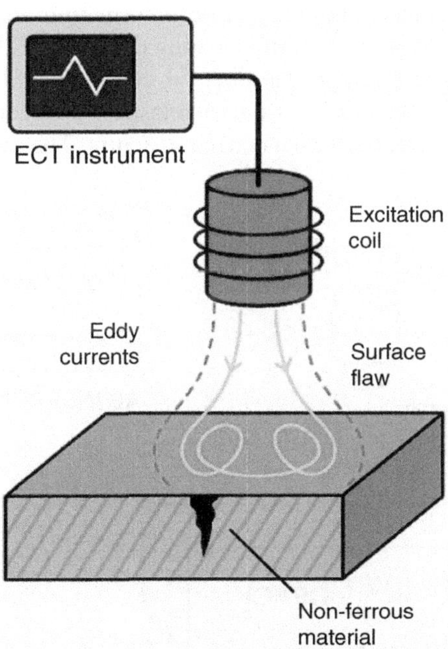

FIGURE 9.6 Principle of eddy current testing (ECT) for non-ferrous alloys.

Advantages:

Non-contact and rapid inspection method.
Highly sensitive to small surface and near-surface defects.
Effective through coatings and paints without surface removal.
No need for couplant (unlike UT).
Can be automated for continuous inspection or scanning in shipyards.

Limitations:

Applicable only to conductive materials.
Sensitivity decreases with increasing material thickness.
Requires surface accessibility.
Interpretation demands experienced operators due to complex signal patterns.

In shipyards and offshore installations, ECT complements UT and PT by providing an efficient, non-contact method for corrosion mapping and crack detection, particularly in aluminum, copper-nickel, and stainless steel structures.

9.6.3 Acoustic Emission Monitoring (AEM) for Structural Health

Principle:
AEM detects the transient elastic waves produced by the rapid release of localized strain energy during crack initiation, plastic deformation, or corrosion processes. Unlike other NDT methods that identify existing defects, AEM focuses on detecting active damage as it occurs, making it ideal for continuous SHM of marine assets. Figure 9.7 illustrates the detection of transient elastic waves generated by crack initiation or material deformation within a marine structure. The diagram shows acoustic

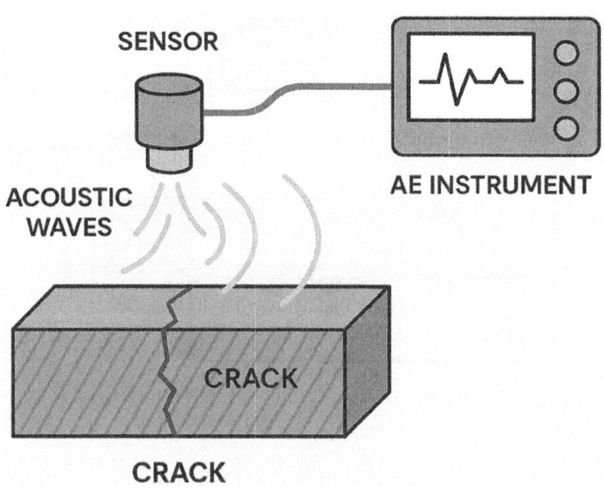

FIGURE 9.7 Principle of acoustic emission monitoring (AEM) for structural health.

waves emitted from a crack source propagating through the structure and being detected by AE sensors, which transmit signals to an AEM instrument for real-time analysis. This principle enables continuous monitoring of fatigue, corrosion, or fracture events, supporting predictive maintenance and structural integrity management in ships and offshore installations.

Applications in Marine Welding:
AEM has gained prominence in monitoring critical offshore and subsea structures where conventional inspection access is limited. It is used for:

Fatigue monitoring of offshore platform legs, risers, and braces subjected to cyclic loading.
Crack propagation detection in pressure vessels and hull structures.
Corrosion activity measurement in ballast tanks and subsea pipelines.
In-service monitoring during load testing, welding, and hydrostatic trials.

Sensors mounted on the structure detect and localize acoustic signals generated by defect growth. Data are transmitted to monitoring units or control centers for real-time evaluation and predictive maintenance.

Advantages:

Real-time detection of crack initiation and progression.
Covers large inspection areas with minimal equipment.
Non-intrusive and suitable for in-service monitoring.
Enables predictive maintenance and failure prevention.

Limitations:

Cannot distinguish between defect types without supplementary data.
Requires low background noise and stable monitoring conditions.
Interpretation depends on signal processing accuracy.
Limited sensitivity in thick or heavily damped structures.

Despite these limitations, AEM represents a modern evolution of NDT transitioning from reactive inspection to proactive, condition-based monitoring. It aligns with the marine industry's shift toward smart asset management and digital twin integration for offshore structures.

9.7 INSPECTION TECHNIQUES FOR WELDED MARINE STRUCTURES

The inspection of welded marine structures is a crucial component of structural integrity management in shipbuilding, offshore engineering, and subsea operations. Marine structures such as ship hulls, decks, offshore platforms, and subsea pipelines are subjected to highly variable stresses, corrosion, and fatigue loading, which can compromise weld quality over time. To ensure long-term performance and safety, NDT methods are systematically employed throughout the fabrication, commissioning, and service life of these assets.

The choice of inspection technique depends on factors such as material type, weld geometry, accessibility, environmental conditions, and criticality of the joint. A combination of conventional and advanced NDT methods—visual, ultrasonic, radiographic, magnetic, and digital—forms a multi-tiered inspection strategy that ensures early flaw detection, reduced downtime, and regulatory compliance.

9.7.1 Hull and Deck Weld Inspection Procedures

Ship hulls and decks form the primary load-bearing structures of a vessel, and their welded joints experience continuous dynamic stresses from wave action, vibration, and thermal cycles. Inspection in these areas is conducted both during fabrication and in-service maintenance phases to identify defects that could lead to structural fatigue or corrosion-related failure.

Inspection Procedures Include:

Visual Inspection (VT):

Initial and routine inspection for weld surface defects such as undercuts, spatter, or incomplete fusion.

Carried out using magnifiers, mirrors, or digital imaging devices with borescopes for confined areas.

Ultrasonic Testing (UT) and Phased Array UT (PAUT):

Used to detect internal flaws and laminar defects in thick hull plates and deck panels.

PAUT provides volumetric mapping and defect sizing, enhancing reliability in large welded structures.

Radiographic Testing (RT):

Applied to critical hull joints and bulkhead welds where full penetration is required.

DR enables faster image capture and reduces radiation exposure in shipyards.

Dye Penetrant Testing (PT):

Utilized on aluminum and stainless steel sections for detecting surface-breaking cracks, especially in superstructure welds.

Routine hull inspections are guided by standards such as the International Association of Classification Societies (IACS) and ABS, ensuring weld integrity under operational stresses and compliance with international maritime safety norms.

9.7.2 Offshore Platform Weld Integrity Assessment

Offshore platforms fixed, floating, or subsea are subject to extreme mechanical loading, wave impact, and corrosive seawater exposure. Welded joints in critical load-bearing regions such as legs, braces, nodes, and risers are continuously monitored to prevent catastrophic failures.

Key Assessment Techniques:

Magnetic Particle Testing (MT): Detects surface cracks and fatigue damage in ferromagnetic steel structures, particularly at tubular joints and node intersections.
Ultrasonic Testing (UT/PAUT): Provides volumetric analysis of welds, detecting internal discontinuities in structural braces and load transfer zones.
Time-of-Flight Diffraction (TOFD): Offers accurate crack depth measurement and defect characterization in high-stress welds.
Acoustic Emission Monitoring (AEM): Enables real-time monitoring of fatigue crack propagation and structural deformation under dynamic loading.
Corrosion Mapping (ECT/IRT): Used for assessing corrosion-induced wall thinning in risers and pipelines.

Offshore inspection programs are typically risk-based, focusing on critical welds identified through structural reliability analysis. Inspections are performed during scheduled shutdowns, maintenance campaigns, or in-service using remotely operated tools. Compliance is ensured through standards such as DNV-ST-F101, API RP 2A, and ISO 17635.

9.7.3 Underwater Weld Inspection Techniques (ROV and Diver-Assisted)

Inspection of underwater welds presents unique challenges due to restricted visibility, high pressure, and environmental variability. These inspections are performed either by divers equipped with portable NDT tools or ROVs equipped with advanced sensor systems.

Diver-Assisted Inspection:
Divers use MT and UT adapted for wet environments to examine offshore platform nodes, riser connections, and subsea flanges. Wet fluorescent magnetic particles and immersion-type UT probes enable inspection under water with real-time communication to surface control stations. Diver-based systems are effective in shallow and mid-depth inspections but are limited by depth and safety considerations.

ROV-Based Inspection:
ROVs integrate PAUT, visual cameras, laser scanning, and acoustic sensors for automated inspection of deep-sea welds up to 3,000 meters. These systems operate remotely, transmitting live data and 3D visualizations to surface

operators. Advanced AI-based defect recognition assists in automated flaw detection and classification, reducing human error. ROV and diver-assisted techniques complement each other—ROVs are preferred for deep and continuous inspection, while divers are used for localized examination and minor repairs. The use of digital twins and remote monitoring enhances the reliability and safety of underwater weld inspection in subsea assets.

9.8 STANDARDS AND CERTIFICATION FOR MARINE WELD NDT

The standardization and certification framework governing NDT in the marine and offshore sectors ensures uniform quality, safety, and reliability in welded structures subjected to harsh ocean environments. Marine welds form the backbone of ship hulls, offshore platforms, subsea pipelines, and pressure systems—all of which must comply with stringent inspection criteria set by classification societies, international standard organizations, and certification authorities.

These standards define inspection methodologies, acceptance criteria, qualification procedures, and documentation requirements. Moreover, they establish mandatory training and certification levels for NDT personnel to ensure technical competence and consistent inspection performance across global shipyards and offshore installations.

9.8.1 Classification Society Requirements (ABS, DNV, LR)

Classification societies play a central role in setting and enforcing NDT requirements for marine welds. Their rules govern inspection procedures during fabrication, commissioning, and in-service inspection to maintain structural integrity and compliance with maritime safety regulations.

American Bureau of Shipping (ABS):

Specifies NDT requirements under ABS Rules for Building and Classing Steel Vessels and Offshore Installations.

Defines NDT method selection (UT, RT, MT, PT) based on weld type, material, and service condition.

Mandates NDT during critical fabrication stages, particularly in hull girder connections, pressure boundaries, and offshore jacket welds.

Requires inspectors to be certified in accordance with the American Society for Non-destructive Testing (ASNT) SNT-TC-1A or ISO 9712 standards.

Det Norske Veritas (DNV):

Provides detailed requirements in DNV-ST-F101 (Submarine Pipeline Systems) and DNV-OS-C401 (Fabrication and Testing of Offshore Structures).

Specifies acceptance levels for weld defects according to ISO 5817 and BS EN 1714 standards.

Promotes the use of advanced NDT techniques such as PAUT, TOFD, and AEM for offshore structural monitoring.
Emphasizes risk-based inspection (RBI) and CBM approaches to optimize inspection intervals.

Lloyd's Register (LR):

Details NDT procedures in the Rules for the Manufacture, Testing, and Certification of Materials and Welding Procedures for Ship Construction.
Requires welding procedure qualification records (WPQR) and NDT reports as part of construction documentation.
Stipulates that NDT operators must hold Level II or III certification as per international standards.
Encourages the integration of digital data management systems for inspection traceability and QA.

Collectively, these classification societies ensure that marine and offshore welded structures meet global standards of safety, reliability, and longevity, while facilitating international vessel registration and certification.

9.8.2 International Standards (ISO, ASME, AWS)

In addition to classification societies, several international organizations establish standardized procedures, acceptance criteria, and quality management frameworks for NDT in marine and offshore applications.

ISO Standards:

ISO 9712: Qualification and certification of NDT personnel by competence levels (Levels I, II, and III).
ISO 17635: General rules for NDT of welds in metallic materials.
ISO 10893 Series: Covers NDT of steel tubes and pipelines.
ISO 5817: Provides acceptance levels for weld imperfections.
ISO 23277/23278: Specific standards for penetrant and MT in welds.
ISO 16810/17640: UT methods and calibration procedures for weld inspection.

American Society of Mechanical Engineers (ASME):

ASME Section V (Boiler and Pressure Vessel Code): Defines NDT methods, procedures, and personnel qualification requirements for pressure-retaining components.
ASME B31.3: Establishes weld inspection guidelines for piping systems used in marine and offshore facilities.
ASME BPVC Section IX: Covers welding procedure and performance qualification, linking NDT results to weld quality certification.

American Welding Society (AWS):

AWS D1.1/D1.2/D1.5 Codes: Set inspection and testing criteria for steel, aluminum, and bridge welds used in marine and offshore structures.
AWS QC1: Establishes requirements for certified welding inspectors (CWIs).
AWS B1.10M/B1.10: Provides guidelines for NDT of welds, including procedural control and acceptance standards.

Together, these standards ensure global consistency in inspection practice, enabling interoperability between international shipyards, offshore fabricators, and certification authorities. They also serve as the foundation for regulatory audits, warranty validation, and failure analysis documentation.

9.8.3 Personnel Qualification and Training

Competent personnel are fundamental to the success of NDT programs in marine engineering. Inadequate training or certification can lead to misinterpretation of flaws, false acceptances, or missed defects, potentially compromising safety and compliance.

Qualification Levels:
Personnel certification is typically based on ISO 9712 or ASNT SNT-TC-1A, which defines three levels of qualification:

Level I: Technicians qualified to perform specific NDT operations under supervision.
Level II: Personnel competent to set up, calibrate, interpret, and evaluate test results independently.
Level III: Experts responsible for developing procedures, training personnel, and managing NDT programs.

Training and Competency Requirements:

Training Hours and Experience: Minimum practical training and on-the-job experience hours are prescribed for each level and method (e.g., UT, RT, MT, PT, ECT).
Certification Renewal: Periodic recertification ensures continued proficiency and awareness of emerging technologies.
Health, Safety, and Environmental (HSE) Awareness: NDT personnel must adhere to safety protocols, especially when handling radiation or performing underwater inspections.
Digital and Automated System Proficiency: Increasing reliance on digital NDT systems requires training in phased array software, data logging, image processing, and AI-assisted analysis.

Prominent Certification Bodies Include:

The American Society for Non-destructive Testing (ASNT)
The British Institute of Non-Destructive Testing (BINDT)
The Indian Society for Non-Destructive Testing (ISNT)
The European Federation for NDT (EFNDT)

In the marine industry, personnel qualification is not just a regulatory formality, but it is a critical assurance mechanism ensuring that inspections meet the exacting standards required for ship classification, offshore safety, and environmental protection.

9.9 FUTURE TRENDS IN MARINE WELD INSPECTION

The future of marine weld inspection is being transformed by the convergence of automation, AI, and digital technologies. As shipbuilding and offshore operations move toward Industry 5.0, inspection processes are evolving from periodic manual testing to autonomous, data-driven, and predictive monitoring systems. Traditional NDT methods while effective are limited by accessibility, operator dependency, and the need for human presence in hazardous environments. The next generation of marine inspection technologies aims to overcome these limitations through robotic automation, AI-based analysis, and digital twin integration, ensuring higher reliability, real-time decision-making, and life cycle optimization of marine structures.

9.9.1 Automated and Robotic NDT Systems

The introduction of automated and robotic systems marks one of the most significant advancements in marine weld inspection. These systems combine traditional NDT methods—such as UT, PAUT, radiography (RT), and ECT with autonomous robotic platforms to perform precise, repeatable inspections without human exposure to hazardous environments.

Key Developments Include:

Crawling Robots for Ship Hulls: Magnetic or suction-based crawlers equipped with ultrasonic probes and cameras can inspect vertical and curved surfaces of ship hulls for corrosion, cracking, and weld degradation.

ROV-Based Subsea Inspection: ROVs integrated with UT, PAUT, or laser sensors conduct high-resolution inspections of subsea welds, risers, and pipelines in deepwater conditions up to 3,000 meters.

Drone-Assisted Visual Inspection: Aerial drones equipped with high-resolution imaging and IRT are increasingly used to inspect upper decks, cranes, and tall offshore structures.

Automated Radiographic and Laser Scanning Systems: These enable continuous inspection during shipbuilding and dry dock operations, improving productivity and traceability.

Automation minimizes human intervention, enhances consistency, and enables operation in environments that are inaccessible, confined, or hazardous. Furthermore, robotic systems can be programmed for periodic inspection routines, ensuring continuous monitoring and data collection throughout the asset's life cycle.

9.9.2 AI-Based Defect Recognition

The integration of AI and ML into marine weld inspection represents a paradigm shift toward intelligent, real-time flaw detection, and classification. AI algorithms analyze data from ultrasonic signals, radiographic images, and AEs to identify defects such as cracks, porosity, inclusions, or corrosion with high accuracy and speed.

AI Applications in Marine NDT Include:

Automated Image Analysis: Deep learning models can interpret complex NDT images (RT, PAUT, IRT) to automatically locate and categorize weld defects, reducing reliance on human interpretation.

Pattern Recognition in Signal Data: ML algorithms trained on large datasets can recognize subtle waveform variations indicative of defect initiation or propagation.

Predictive Defect Analytics: AI-based predictive models analyze trends in inspection data to estimate remaining service life and forecast potential failure zones.

Anomaly Detection in Real Time: Integrated sensors and AI systems detect deviations from baseline conditions, triggering alerts for corrective maintenance.

AI-driven inspection not only enhances accuracy and speed but also supports standardization and objectivity in defect evaluation. The combination of AI with digital data storage enables traceable, auditable inspection records, aligning with the digital transformation goals of modern shipyards and offshore operations.

9.9.3 Integration with Digital Twins for Continuous Monitoring

The concept of the digital twin, a virtual replica of a physical asset that continuously mirrors its condition and performance, is revolutionizing how marine weld inspections are conducted and managed. By integrating NDT data streams, sensor feedback, and simulation models, digital twins enable continuous, real-time monitoring of structural integrity across ships, platforms, and subsea systems.

Key Features and Benefits:

Real-Time Condition Monitoring: Data from ultrasonic, AE, and strain sensors embedded in structures are transmitted to the digital twin for continuous analysis.

Predictive Maintenance: The digital twin uses historical inspection data and AI algorithms to predict defect growth and recommend optimal maintenance schedules.

Life Cycle Data Integration: Combines inspection, welding, material, and environmental data into a single digital model for improved decision-making.

Remote Collaboration: Engineers and inspectors can access real-time asset data from anywhere, enabling coordinated diagnostics across shipyards and offshore control centers.

Visualization and Simulation: The digital twin provides 3D visual representations of weld conditions, allowing simulation of stress effects, crack propagation, and repair outcomes.

In offshore and subsea environments, digital twins coupled with sensor networks and Internet of Things connectivity form the foundation of smart inspection ecosystems, where autonomous robots collect data, AI interprets results, and digital twins visualize the evolving condition of the structure. This integrated approach transforms inspection from a reactive process into a predictive, continuous integrity management system, significantly improving safety, operational efficiency, and asset lifespan.

9.10 SUMMARY

The chapter provides a comprehensive overview of modern inspection methodologies essential for ensuring the structural integrity, safety, and reliability of welded components in shipbuilding, offshore platforms, and subsea structures. It discusses the principles and applications of key NDT techniques, UT, RT, MT, PT, ECT, and AEM, emphasizing their roles in detecting surface and internal defects without compromising material integrity. The chapter highlights inspection practices for hulls, decks, offshore platforms, and underwater welds, supported by real-world case studies demonstrating the prevention of marine failures. It also outlines the standards and certification frameworks established by organizations such as ABS, DNV, LR, ISO, ASME, and AWS, ensuring global consistency in inspection quality and personnel competence. Finally, the chapter explores emerging trends, including robotic and automated NDT systems, AI-based defect recognition, and digital twin integration, which are revolutionizing marine weld inspection through real-time monitoring and predictive maintenance, marking the transition from conventional testing to intelligent, data-driven structural health management in modern marine engineering.

9.11 REVIEW QUESTIONS

1. Define non-destructive testing (NDT) and explain its importance in marine welding applications.
2. What are the main objectives of NDT in ensuring quality assurance for marine welded structures?
3. Discuss the role of NDT in maintaining safety and regulatory compliance in offshore engineering.

4. What environmental challenges affect the performance of NDT techniques in marine environments?
5. Differentiate between surface, subsurface, and volumetric NDT methods with suitable marine examples.
6. Describe the working principle of ultrasonic testing (UT) and how it detects internal flaws in welds.
7. What is the difference between conventional ultrasonic testing (UT) and phased array ultrasonic testing (PAUT)?
8. List and explain three advantages of using PAUT in shipbuilding and offshore fabrication.
9. What are the main limitations of ultrasonic testing (UT) in marine applications?
10. Explain the principle of radiographic testing (RT) and how radiation interacts with material discontinuities.
11. Compare X-ray and gamma-ray radiography in terms of their suitability for shipyard and offshore inspections.
12. Identify key safety considerations when performing RT in confined or populated marine environments.
13. Discuss the advantages of digital radiography (DR) over conventional film-based RT for marine weld inspection.
14. Explain the principle of magnetic flux leakage (MFL) in magnetic particle testing (MT).
15. What types of defects can be effectively detected using MT, and what materials are best suited for this method?
16. List the major advantages and limitations of MT in shipyard and offshore use.
17. Describe the principle and application of dye penetrant testing (PT) for surface crack detection.
18. Why is PT particularly useful for non-ferrous marine materials such as aluminum and stainless steel?
19. Explain the principle of Eddy Current Testing (ECT) and its applications in non-ferrous alloy inspection.
20. What are the advantages of ECT for inspecting aluminum hulls and copper-nickel heat exchanger tubes?
21. Define Acoustic Emission Monitoring (AEM) and explain how it contributes to structural health monitoring in marine systems.
22. Discuss the benefits of real-time monitoring using AEM in offshore platforms and subsea pipelines.
23. Outline the main procedures for hull and deck weld inspection in shipyards.
24. Describe how ROV-based underwater inspection systems enhance deep-sea weld inspection capability.
25. Present a case study example where NDT successfully prevented a major marine structural failure.
26. List three key standards or codes governing marine weld NDT and their primary functions.

27. Explain how classification societies such as ABS, DNV, and Lloyd's Register (LR) define NDT requirements for ship and offshore structures.
28. What are the certification levels defined under ISO 9712 or ASNT SNT-TC-1A, and what do they represent?
29. How are AI and machine learning transforming defect recognition and predictive maintenance in marine NDT?
30. Describe the role of Digital Twins in integrating NDT data for continuous structural monitoring in modern shipyards and offshore platforms.

10 Welding Standards and Regulations in the Marine Industry

10.1 INTRODUCTION

Welding plays a critical role in the construction, repair, and maintenance of ships, offshore platforms, and subsea systems—forming the structural backbone of the global marine industry. The performance, safety, and longevity of these welded structures depend not only on sound engineering design and fabrication practices but also on strict adherence to internationally recognized standards, codes, and regulatory frameworks. These standards ensure that every weld, from a ship's hull joint to an offshore riser connection, meets rigorous criteria for strength, durability, corrosion resistance, and quality assurance under some of the harshest environmental conditions on earth (Olalerea et al. 2023).

The marine environment subjects welded components to a complex combination of stresses including hydrostatic pressure, wave loading, thermal cycling, and corrosive seawater exposure that can accelerate degradation or cause structural failure if weld quality is compromised. Therefore, the implementation of standardized welding procedures, inspection criteria, and qualification systems is essential to maintain safety, reliability, and compliance throughout the life cycle of a vessel or offshore asset. The purpose of welding standards and regulations is to establish uniform guidelines governing material selection, joint design, fabrication techniques, inspection, testing, and personnel certification across the marine industry worldwide.

Leading organizations such as the International Organization for Standardization (ISO), the American Welding Society (AWS), the American Bureau of Shipping (ABS), Det Norske Veritas (DNV), and Lloyd's Register (LR) have developed comprehensive codes and specifications that govern marine welding operations. These frameworks define welding procedure specifications (WPS)/welding procedure qualification requirements (WPQR), welder performance testing, nondestructive examination (NDE) criteria, and acceptance standards for weld imperfections. In addition, international maritime conventions and national regulations such as those established by the International Maritime Organization (IMO) ensure that welding activities align with environmental protection and occupational safety mandates.

In recent years, the evolution of advanced materials, automation, and digital technologies has led to updates in welding standards to accommodate innovations such as hybrid laser-arc welding, friction stir welding (FSW), and additive manufacturing (WAAM). These advancements demand continuous alignment between regulatory

DOI: 10.1201/9781003217350-10

requirements and industrial practices to ensure both innovation and safety coexist within the marine manufacturing ecosystem.

Ultimately, welding standards and regulations form the foundation of quality assurance and certification in marine fabrication. They not only guarantee structural integrity and global interoperability but also protect human lives, property, and the marine environment. This chapter provides a detailed exploration of the regulatory framework, classification society guidelines, international standards, certification procedures, and quality management systems (QMSs) governing welding operations in the marine and offshore industries, highlighting their significance in achieving safe, sustainable, and high-performance maritime engineering.

10.1.1 Importance of Welding Standards in Marine Fabrication

Welding standards serve as the technical backbone of marine fabrication, ensuring consistency, reliability, and safety across all welded joints in ships, offshore structures, and subsea components. Given the critical role of welds in maintaining structural integrity under complex loading conditions, the marine industry depends heavily on well-defined standards to control every aspect of the welding process from material selection and joint preparation to inspection, testing, and certification (Moore 2009).

Standards such as ISO 15614 (Welding Procedure Qualification [WPG]), AWS D1.1 (Structural Welding Code—Steel), and DNV-OS-C401 (Fabrication and Testing of Offshore Structures) establish detailed guidelines for the qualification of welding procedures and personnel, ensuring that welds meet required performance levels under marine operating conditions. By enforcing uniformity across shipyards and offshore fabrication facilities, these standards eliminate ambiguity, minimize human error, and facilitate interoperability between global suppliers, classification societies, and regulatory authorities.

Furthermore, welding standards ensure that materials and processes are optimized to withstand saltwater corrosion, cyclic fatigue, impact loading, and thermal variations. In essence, they act as a common engineering language that bridges design, fabrication, and inspection disciplines guaranteeing that the final structure performs safely throughout its service life.

10.1.2 Role of Regulations in Safety, Quality, and Compliance

In the marine sector, regulations complement standards by providing the legal and operational framework that ensures safety, environmental protection, and quality assurance. Regulatory requirements are enforced by both national maritime authorities and classification societies, who verify that all welding operations comply with established design codes and inspection protocols.

Regulations define the mandatory compliance criteria for vessel construction, offshore installations, and repair activities. They govern areas such as:

- Welding procedure and personnel qualification,
- Inspection and non-destructive testing (NDT) requirements,

- Material traceability and documentation, and
- Periodic surveys and recertification of welded structures.

For example, the IMO mandates safety and environmental compliance through conventions such as Safety of Life at Sea (SOLAS) and Marine Pollution (MARPOL), while classification societies like ABS, DNV, and LR enforce fabrication and inspection requirements during vessel construction and operation.

By integrating these regulations into fabrication processes, shipyards and offshore operators ensure that weld quality aligns with international best practices and that the resulting structures are fit for purpose, environmentally responsible, and legally compliant. In this context, regulations not only safeguard human life and property but also reinforce public and environmental trust in maritime operations.

10.1.3 Global vs. Regional Standardization in the Marine Sector

The marine industry operates across international boundaries, making global standardization essential for ensuring consistent welding quality, safety, and certification across shipyards worldwide. However, regional adaptations often exist to address local industrial capabilities, regulatory environments, and environmental conditions.

At the global level, organizations such as the ISO, the AWS, and the International Institute of Welding (IIW) define unified welding and inspection standards applicable across nations. These global standards promote compatibility, trade efficiency, and mutual recognition of certifications among shipbuilders, suppliers, and inspection agencies. For instance, ISO and AWS standards are often cross-referenced in DNV and ABS rules, ensuring worldwide coherence in fabrication and inspection procedures (Krause et al. 2025).

At the regional level, specific standards may be developed to reflect national policies or operational contexts:

Europe: European Norm (EN) and ISO harmonized standards are widely used in shipbuilding and offshore fabrication under the European Committee for Standardization (CEN).

North America: AWS and ASME codes dominate welding and pressure vessel regulations in US and Canadian shipyards.

Asia-Pacific: Japan's JIS, India's Bureau of Indian Standards (BIS), and China's GB/T codes align with international frameworks but include region-specific requirements for environmental and operational conditions.

While global harmonization fosters uniformity, regional standardization allows flexibility in addressing localized materials, climatic challenges, and operational safety laws. The ongoing collaboration between international and regional standard-setting bodies ensures that marine welding remains both globally consistent and regionally adaptable, maintaining the balance between innovation, safety, and sustainability.

10.2 INTERNATIONAL MARITIME ORGANIZATION (IMO) WELDING STANDARDS

The IMO serves as the primary global authority responsible for establishing and maintaining international regulations governing the safety, environmental protection, and operational integrity of ships and offshore installations. While IMO does not issue welding procedure codes directly, it provides framework conventions, safety requirements, and performance standards that guide the application of welding-related rules in shipbuilding and repair. These standards are integrated into the technical regulations of classification societies and national maritime authorities, forming the foundation for uniform global compliance in marine fabrication (Mandal 2017).

Through its conventions, codes, and guidelines, the IMO ensures that welding and fabrication activities contribute to structural safety, pollution prevention, and long-term operational reliability, especially for vessels and offshore units operating under harsh maritime conditions.

10.2.1 Overview of IMO's Role in Maritime Safety

The IMO, established in 1948 and operational since 1959, functions as a specialized agency of the United Nations dedicated to ensuring safe, secure, and environmentally sound shipping. Its mandate encompasses the formulation, adoption, and enforcement of international conventions and technical codes governing all aspects of maritime operations, including construction, stability, fire safety, and pollution control.

In the context of welding, IMO's role focuses on ensuring that fabrication and repair standards used in ship and offshore structure construction meet minimum safety and quality criteria to prevent structural failures and environmental hazards. While IMO does not define specific welding parameters, it mandates compliance with recognized international standards (e.g., ISO, AWS, ASME) that are referenced within its conventions.

Key IMO objectives related to welding include the following:

Ensuring structural integrity of hulls, decks, tanks, and pressure boundaries.
Preventing loss of life, cargo, or environmental pollution due to weld failure.
Promoting the use of qualified personnel, certified procedures, and approved materials.
Encouraging standardized inspection and documentation practices across the global fleet.

IMO's framework ensures that the quality of welding directly supports the broader goal of maritime safety and environmental stewardship across all shipbuilding and offshore fabrication sectors.

10.2.2 Relevant IMO Conventions and Guidelines for Welding

Several IMO conventions and associated codes contain provisions that indirectly or directly govern welding operations, inspection, and maintenance in marine structures. These conventions establish mandatory safety and quality requirements that member states incorporate into their national regulations and ship classification systems.

Key IMO Conventions and Codes Relevant to Welding:

SOLAS (Safety of Life at Sea) Convention:

Regulates the construction and structural integrity of ships.
References approved welding standards and procedures for hull, deck, and bulkhead fabrication.
Requires periodic surveys of critical welds and structural components to ensure ongoing compliance.

MARPOL (International Convention for the Prevention of Pollution from Ships):

Addresses environmental protection by ensuring the structural integrity of fuel tanks, pipelines, and cargo containment systems.
Mandates that welding and repair processes used in such systems minimize the risk of leakage or contamination.

IMO Resolution MSC.47(66): International Code for the Construction and Equipment of Ships Carrying Liquefied Gases in Bulk (IGC Code):

Specifies material and welding procedure qualifications for cryogenic containment systems.
Ensures compatibility between welding methods and low-temperature operational environments.

International Code for the Construction and Equipment of Ships Carrying Dangerous Chemicals in Bulk (IBC Code):

Defines requirements for welded structures exposed to corrosive or reactive chemicals.
Emphasizes the use of corrosion-resistant materials and welding techniques.

International Code on Intact Stability (IS Code) and International Code for the Construction and Equipment of Mobile Offshore Drilling Units (MODU Code):

Establish standards for structural welding in offshore platforms and drilling units to ensure stability and fatigue resistance.

IMO Guidelines for Ship Recycling (Hong Kong Convention, 2009):

Promotes safe and environmentally responsible dismantling of welded ship structures.
Encourages the documentation of welding materials and processes used during fabrication.

These conventions collectively ensure that welding activities in ship construction, modification, and maintenance adhere to, reducing the likelihood of catastrophic failures or environmental incidents.

10.2.3 Integration of IMO Standards into Shipbuilding Practices

IMO standards are not applied in isolation; they are implemented through national maritime authorities, classification societies, and industrial codes of practice that translate IMO conventions into actionable engineering and inspection requirements.

Integration Mechanisms Include:

Adoption into National Legislation:

Member states incorporate IMO conventions into domestic maritime regulations, requiring all vessels registered under their flag to comply with applicable welding and inspection standards.

Incorporation into Classification Society Rules:

Societies such as **ABS, DNV, LR, Bureau Veritas (BV)**, and **RINA** interpret and enforce IMO requirements through their construction and survey rules.

These rules specify welding procedure qualification (WPS/WPQR), welder certification, and NDT inspection frequency in alignment with IMO objectives.

Design and Fabrication Documentation:

Shipyards must maintain approved WPS, welder qualification records (WQR), and NDT reports to demonstrate compliance with IMO-based regulations.

Survey and Audit Systems:

Periodic classification surveys and audits verify the continued compliance of ship and offshore structures with IMO standards, ensuring weld integrity throughout their operational life.

The integration of IMO welding standards ensures a uniform level of safety and quality across the global maritime fleet. It bridges the gap between regulatory oversight **and** industrial practice, promoting a culture of proactive risk management, continuous quality assurance, and environmental responsibility in shipbuilding and offshore fabrication.

10.3 ABS WELDING STANDARDS

The **ABS** is one of the world's leading classification societies responsible for establishing and enforcing technical standards for the design, construction, and maintenance of ships and offshore structures. Welding plays a central role in the ABS rule framework, as it directly influences the strength, safety, and durability of marine structures operating under extreme environmental and mechanical stresses.

ABS welding standards define comprehensive requirements covering materials, welding procedures, personnel qualification, consumable approval, inspection, and testing. These rules ensure that all welded joints conform to recognized mechanical properties, corrosion resistance, and fatigue life standards, thereby reducing the likelihood of structural failure in service. The ABS approach emphasizes traceability, quality assurance, and certification, forming a key element of the global marine quality control system.

10.3.1 ABS Rules for Materials and Welding Procedures

The core requirements for welding under ABS classification are outlined in the publication **Rules for Materials and Welding—Part 2**, which governs all welding-related activities in the fabrication of ships, offshore units, and subsea components.

Key Provisions Include:

Material Qualification:

All base materials used for welding must be certified by ABS and comply with recognized standards such as **ASTM**, **ASME**, or **ISO**.

Each heat of steel, aluminum, or alloy must undergo chemical and mechanical testing to verify conformity with grade specifications.

Welding Procedure Qualification (WPS/WPQR):

Welding procedures must be pre-qualified through testing and documented in accordance with **AWS D1.1**, **ASME Section IX**, or **ISO 15614**.

The **Welding Procedure Qualification Record (WPQR)** must include parameters such as current, voltage, heat input, preheat/interpass temperatures, and post-weld treatment details.

Procedure tests include tensile, bend, impact, and hardness tests to validate the mechanical performance of welds under simulated marine conditions.

Structural Application Requirements:

Specific rules govern hull, deck, bulkhead, and pressure vessel welds, considering material thickness, joint geometry, and service loading.

Weld symbols, joint designs, and acceptance levels must comply with ABS-approved drawings and fabrication standards.

Weld Inspection and Testing:

NDT methods such as **ultrasonic testing (UT)**, **radiography (RT)**, **magnetic particle testing (MT)**, and **penetrant testing (PT)** must be performed as per ABS surveyor-approved procedures.

Critical joints in high-stress zones (e.g., longitudinal bulkheads, tank boundaries, riser supports) are subject to enhanced inspection criteria.

Through these detailed requirements, ABS ensures that welding processes maintain uniform quality and reliability, regardless of shipyard location or fabrication method.

10.3.2 Approval Processes for Welders and Welding Consumables

To maintain high standards of weld quality, ABS mandates the approval and certification of both welding personnel and welding consumables (such as electrodes, wires, and fluxes) used in fabrication.

Welder Qualification:

All welders and welding operators must pass qualification tests in accordance with **ABS Rules Part 2,** Section 3, and standards such as **AWS D1.1** or **ASME Section IX.**

Qualification tests simulate production conditions, including welding position, joint type, and base metal thickness.

Successful candidates receive **ABS Welder Qualification Certificates**, valid for a specified period, subject to continuity of practice.

Periodic requalification is required if a welder has not performed welding within a defined time frame or if process changes occur.

Welding Consumable Approval:

All consumables must be **ABS type approved**, ensuring consistent mechanical and metallurgical performance.

Consumable approval tests include:

Chemical composition analysis,

Mechanical property verification (tensile, impact, and yield strength),

Weldment toughness and diffusible hydrogen testing, and

Performance assessment under simulated marine conditions (salt spray, temperature cycles).

Approved consumables are listed in the **ABS Recognized Manufacturer Database**, providing traceability for shipyards and suppliers.

Procedure and Production Audits:

ABS surveyors conduct on-site audits during procedure qualification, production welding, and inspection stages.

Any deviation from approved parameters requires formal re-approval to maintain class certification.

This rigorous approval and monitoring process ensures consistent quality control from raw materials to final weld inspection, reducing variability and enhancing safety across global shipbuilding operations.

10.3.3 Case Examples of ABS Compliance in Shipyards

10.3.3.1 Case 1: Hull Construction of a Crude Oil Tanker (South Korea)

During the fabrication of a 300,000 DWT crude carrier, ABS required full procedure qualification for gas metal arc welding (GMAW) of high-tensile steel plates (EH36 grade). Welds underwent ultrasonic and radiographic testing, achieving 100%

acceptance under ABS class notation. The project demonstrated how ABS standards ensure consistent weld quality and fatigue resistance in large hull structures subjected to cyclic ocean loading.

10.3.3.2 Case 2: Offshore Platform Fabrication (Gulf of Mexico)

An offshore fixed platform constructed under ABS MODU Rules required compliance with AWS D1.1 and ASME IX for all structural and pressure boundary welds. ABS surveyors verified WPS approval, NDT compliance, and WQR, resulting in successful certification. The project highlighted ABS's role in standardizing offshore welding quality and minimizing operational risk.

10.3.3.3 Case 3: Repair of Liquefied Natural Gas (LNG) Containment System (Japan)

ABS approved the repair welding procedures for a cryogenic containment system in an LNG carrier, referencing the IGC Code and ABS Rules Part 5C. Special welding consumables with low-temperature fracture toughness were used, and post-weld heat treatment (PWHT) was verified through inspection. This case illustrated ABS's ability to adapt its standards to specialized materials and service conditions.

10.3.3.4 Case 4: Additive Manufacturing Qualification (Europe)

In line with recent innovations, ABS collaborated with shipyards using WAAM for large structural components. ABS validated the process by qualifying deposition parameters and verifying the mechanical properties of the printed sections, demonstrating its commitment to integrating emerging fabrication technologies into established standards.

10.4 DET NORSKE VERITAS (DNV) WELDING STANDARDS

DNV, now operating as **DNV—Classification Society**, is one of the foremost authorities in establishing technical standards, certification rules, and quality assurance frameworks for the global marine and offshore industries. DNV's welding standards are internationally recognized for their comprehensive coverage of shipbuilding, offshore structures, subsea systems, and renewable marine installations.

The DNV standards focus on ensuring that welded joints meet stringent requirements for strength, fatigue resistance, fracture toughness, and corrosion durability, key properties essential for long-term operation in demanding oceanic environments. The DNV rule framework integrates welding into a broader system of design validation, material selection, fabrication control, and inspection, ensuring consistency from concept to commissioning.

10.4.1 DNV Rules for Ship and Offshore Structure Welding

DNV defines its welding and fabrication standards primarily through the following key documents:

DNV-ST-N001: Integrity management of marine structures.

DNV-OS-C401: Fabrication and testing of offshore structures.
DNV-ST-F101: Submarine pipeline systems.
DNV-RP-C203: Fatigue design of offshore steel structures.
DNV-RP-B401: Cathodic protection design for welded steel.

These standards establish the baseline requirements for material preparation, welding processes, consumable selection, qualification, and inspection in both shipbuilding and offshore fabrication.

Key Provisions Include:

Material and Joint Preparation:

Base materials must comply with **DNV-certified grades** (e.g., NV-A, NV-DH36, NV-EH40). Joints are to be prepared according to DNV-specified bevel angles, root gaps, and surface cleanliness requirements.

Welding Process Control:

All welding processes, **SMAW, GMAW, FCAW, SAW, GTAW, and hybrid methods,** must be pre-approved with process parameter ranges defined in WPS. Preheat and interpass temperatures, filler materials, and shielding gases are closely regulated to ensure mechanical integrity.

Structural Application-Specific Guidance:

Ship Hulls: Emphasis on fatigue resistance and crack-arresting design.
Offshore Platforms: Enhanced toughness and ductility for low-temperature, high-stress conditions.
Subsea Pipelines: Strict requirements for weld metal yield strength matching and corrosion resistance.

Inspection and Acceptance Criteria:

Weld inspection follows **DNV-ST-N101** and **ISO 5817** acceptance levels. Critical joints in tension and fatigue-sensitive regions undergo 100% volumetric NDT. Visual, magnetic, ultrasonic, and radiographic testing are mandatory at different fabrication stages.

Through these rules, DNV establishes a rigorous fabrication and inspection framework ensuring that welded marine structures maintain structural integrity throughout their service life, even under cyclic, thermal, and corrosive loading.

10.4.2 Welding Procedure Qualification under DNV

DNV mandates a structured approach to **WPQ** to validate that proposed welding methods produce joints with adequate mechanical and metallurgical properties for their intended service.

Key Steps in DNV Procedure Qualification:

Preparation of Welding Procedure Specification (WPS): The WPS defines all welding variables, including process type, base metal, filler wire classification, shielding gas composition, heat input, and position. It is reviewed and pre-approved by DNV surveyors before testing.

Welding Procedure Qualification Record (WPQR): A qualification test weld is produced under representative conditions. The test joint undergoes visual inspection, radiography, macro-examination, tensile and bend testing, impact toughness, and hardness profiling. For offshore steels, Charpy V-notch testing is performed at service temperatures (e.g., –40°C).

Approval and Documentation: The WPQR results are submitted to DNV for evaluation. If compliant, a WPQ Certificate (WPQC) is issued, valid for the range of essential variables tested.

Welder and Operator Qualification: Individual welders must demonstrate competency through approved test joints following ISO 9606 or ASME Section IX standards. DNV requires requalification if process parameters or material thickness exceed qualified limits.

Revalidation: For offshore or subsea projects, DNV may require periodic requalification due to environmental severity or extended production campaigns.

This procedure ensures traceability and standardization of welding methods, minimizing variability in production and guaranteeing that all welds perform predictably under real-world marine conditions.

10.4.3 DNV's Approach to Quality Assurance and Inspection

DNV's quality assurance philosophy integrates welding into a risk-based inspection (RBI) and risk-based verification (RBV) framework designed to prevent structural failure and ensure life cycle integrity. Its approach combines pre-fabrication audits, in-process surveillance, and post-fabrication testing.

Core Elements of DNV Quality Assurance:

Fabrication Quality Plan (FQP): Each project must include an FQP outlining quality control checkpoints, NDT stages, and documentation protocols. DNV surveyors verify compliance through on-site audits and review of test records.

Inspection and Testing Regime: Visual inspection precedes all NDT to identify surface discontinuities. UT, phased array UT (PAUT), and time-of-flight diffraction (TOFD) are preferred for volumetric examination. Radiography is applied for welds in inaccessible or complex geometries.

Defect Acceptance and Repair: Defect evaluation follows **DNV-ST-N101** acceptance levels, often referencing **ISO 5817 (Quality Level B or C).** Defects exceeding limits must be removed and repaired using approved procedures, followed by reinspection.

Documentation and Traceability: DNV emphasizes full documentation of material certificates, WPS/WPQR approvals, welder qualifications, and NDT results. All records must be retained for classification audits and future structural assessments.

Digital and Predictive Quality Management: Recent DNV initiatives integrate digital weld tracking systems, automated inspection data analytics, and artificial intelligence (AI)-based defect recognition into quality assurance workflows. These digital tools enable predictive maintenance and real-time integrity monitoring, aligning with DNV's focus on smart and sustainable shipbuilding.

Through its structured, data-driven approach, DNV ensures that every welded joint satisfies both technical integrity and safety compliance, supporting the long-term operability and environmental reliability of marine assets.

10.5 LLOYD'S REGISTER (LR) WELDING STANDARDS

LR, one of the oldest and most respected classification societies, plays a pivotal role in setting technical standards and certification systems for welding in shipbuilding, offshore structures, and marine repair industries. Its welding standards emphasize safety, structural reliability, material integrity, and environmental sustainability, ensuring that marine welded structures maintain performance throughout their operational life cycle.

LR's approach to welding integrates design, fabrication, inspection, and certification within a unified quality framework. Through its Rules and Regulations for the Classification of Ships and Offshore Units, LR establishes detailed guidelines covering materials, welding procedures, consumable approval, welder qualification, and inspection documentation. These rules align with international codes such as **ISO**, **ASME**, and **AWS**, ensuring global consistency and recognition across shipyards and fabrication facilities.

10.5.1 LR Requirements for Marine Welding Procedures and Materials

LR requires that all welding procedures, materials, and consumables used in marine fabrication comply with its Rules for the Manufacture, Testing, and Certification of Materials and Rules for Welding and Fabrication of Ships and Offshore Installations.

Key Requirements Include:

Material Approval and Certification: Base materials must be **LR-approved grades,** typically conforming to recognized standards such as **BS EN, ISO, ASTM**, or **ASME.** Each batch of material must carry a **Material Test Certificate (MTC)** verifying chemical composition, mechanical properties, and traceability.

Welding Procedure Specification (WPS) and Qualification (WPQR): All welding operations must be performed under a **qualified and approved Welding Procedure Specification (WPS)**. The **WPQR** must demonstrate the mechanical integrity of the weld through destructive and non-destructive testing. Tests include tensile, bend, impact (Charpy V-notch), and hardness tests, as well as macro/micro examinations for metallurgical evaluation. Welding procedures must be approved by LR surveyors prior to production use.

Consumable Approval: Welding consumables (electrodes, wires, fluxes, shielding gases) must be **type approved by LR** to ensure consistent quality and compatibility with base metals. Each consumable must undergo batch testing to verify weld metal strength, ductility, and toughness.

Welding Environment Control: LR specifies strict control over preheat and interpass temperatures, humidity, and PWHT to prevent cracking and hydrogen embrittlement. Environmental conditions are monitored during welding operations, particularly in shipyards with variable climatic exposure.

Through these detailed requirements, LR ensures that every weld produced under its classification meets the highest standards of strength, ductility, and corrosion resistance, suitable for long-term service in marine and offshore environments.

10.5.2 Certification and Welder Qualification Processes

LR enforces a comprehensive certification system to ensure that only qualified personnel and approved processes are used in marine fabrication.

Welder and Welding Operator Qualification: Welders must qualify through testing in accordance with **ISO 9606, ASME Section IX,** or equivalent LR standards. Qualification tests simulate actual working conditions, including welding position, joint type, and material grade. Successful candidates receive **LR Welder Qualification Certificates**, valid for a specified range of materials, processes, and thicknesses. Requalification is required if significant changes occur in welding parameters, processes, or material groups. Continuous employment and verification of skills are mandatory to maintain certification validity.

Welding Procedure and Fabricator Approval: Fabrication facilities are required to obtain **LR welding approval,** which includes audits of personnel competence, equipment calibration, and quality systems. Each WPS and WPQR must be **approved by LR surveyors**, with full traceability of test results. For critical structures (e.g., offshore platforms, LNG containment systems), **LR may witness qualification testing** to verify compliance.

Approval of Welding Consumables and Manufacturers: Welding consumables are subjected to **type approval** and **batch certification** by LR. Manufacturers of welding consumables, electrodes, and filler materials must maintain a **QMS** compliant with **ISO 9001**. Approved consumables and manufacturers are listed in the **LR Type Approval Database,** ensuring transparency and traceability.

Through its certification framework, LR guarantees that personnel, processes, and materials used in marine welding are controlled, traceable, and compliant with international standards—maintaining consistency and integrity across the global supply chain.

10.5.3 LR Inspection and Audit Practices

LR employs a structured and proactive approach to inspection, verification, and auditing, ensuring that welding operations consistently meet required quality and safety standards.

Core Elements of LR Inspection Framework:

Pre-Fabrication Inspection: Review of approved WPS/WPQR documents and verification of material traceability. Inspection of equipment calibration (e.g., ovens, wire feeders, preheat devices).

In-Process Inspection: Continuous monitoring of welding parameters such as voltage, current, travel speed, and heat input. Verification of environmental conditions, electrode storage, and interpass temperature control. LR surveyors perform random sampling and on-the-spot audits during production.

Post-Weld Inspection and Testing: Completed welds are visually examined and subjected to **NDT**, including **RT, ultrasonic testing (UT), MT,** and **dye penetrant testing (PT)** in accordance with **ISO 17635** and LR-specific guidelines. Critical welds (e.g., hull girder connections, riser joints, pressure boundaries) undergo 100% NDT inspection. Results are documented and reviewed by LR surveyors for compliance and traceability.

Audit and Surveillance Activities: LR conducts periodic audits of shipyards and fabrication yards to assess QMSs, personnel competence, and adherence to approved procedures. Deviations or non-conformities (NCs) trigger **corrective action reports (CARs)** and may require requalification or reinspection. Digital inspection records are maintained for audit traceability and future life cycle integrity assessments.

Life cycle Integrity and Reinspection: LR mandates periodic in-service inspection and maintenance audits to assess weld degradation, fatigue, or corrosion over time. Advanced methods such as **PAUT** and **acoustic emission monitoring (AEM)** are recommended for detecting early-stage damage.

This systematic approach to inspection and auditing ensures that marine welded structures remain compliant, safe, and durable throughout their service life. LR's auditing philosophy emphasizes prevention over correction, encouraging continuous improvement and the integration of digital inspection technologies for predictive maintenance.

10.6 SHIPBUILDING AND OFFSHORE STRUCTURAL WELDING CODES

Shipbuilding and offshore fabrication involve complex welded structures that must withstand dynamic mechanical loads, cyclic fatigue, corrosion, and extreme environmental conditions. To ensure the structural safety and reliability of such constructions, international welding codes provide standardized rules and acceptance criteria governing materials, joint design, welding procedures, inspection, and testing.

Among the most widely applied frameworks are the AWS codes, ISO and EN standards, and offshore-specific codes developed by organizations such as the

American Petroleum Institute (API) and Norwegian Petroleum Industry Standards (NORSOK). These standards serve as the engineering foundation for global interoperability, ensuring that shipyards and offshore fabrication facilities maintain consistent weld quality regardless of geographic location or project type.

10.6.1 Overview of AWS D1.1, AWS D3.6M, and Related Codes

AWS has established a family of codes that are widely adopted in the marine, offshore, and shipbuilding industries, particularly in regions following US or ASME-based design practices.

Key AWS Codes Used in Marine Applications:

AWS D1.1—Structural Welding Code: Steel: The most widely used code for structural steel fabrication in shipbuilding and offshore platforms. It covers design, prequalification, qualification testing, fabrication, and inspection of steel weldments. It specifies essential variables for procedure and welder qualification, NDT methods, and acceptance criteria. Commonly referenced for hull, deck, bulkhead, and superstructure welds.

AWS D1.2—Structural Welding Code: Aluminum: It governs welding of aluminum alloys in ship superstructures, small craft, and lightweight marine components and addresses preheat requirements, heat-affected zone (HAZ) control, and mechanical testing for corrosion resistance.

AWS D3.6M—Underwater Welding Code: Specifically developed for wet and dry underwater welding used in subsea repair and maintenance. It defines procedures for hyperbaric welding, consumable selection, and testing requirements for welds made under pressure. Categorizes weld quality into Classes A, B, and C, based on service criticality and environmental exposure.

AWS D14.3—Welding of Earthmoving, Construction, and Offshore Equipment: Applied in the fabrication of cranes, lifting systems, and offshore handling structures.

AWS QC1—Standard for AWS Certified Welding Inspectors (CWI): Defines training, examination, and certification requirements for welding inspection professionals.

The AWS code series ensures that welding operations in shipyards meet mechanical performance, fatigue, and corrosion resistance standards required for marine service. Its global acceptance also facilitates cooperation between US-based shipbuilders, offshore contractors, and international classification societies.

10.6.2 ISO and EN Standards Relevant to Marine Welding

The **ISO** and the **CEN** provide harmonized international standards widely recognized across European, Asian, and global shipyards. These standards ensure compatibility, safety, and traceability in marine fabrication.

Key ISO and EN Standards Include:

ISO 15614 Series: Specifies procedure qualification for welding metallic materials; ISO 15614-1 applies to steels and ISO 15614-2 to aluminum alloys.

ISO 9606 Series: Defines welder qualification testing across multiple processes (arc, gas, resistance, etc.).

ISO 5817: Sets acceptance levels for weld imperfections (Quality Levels B, C, and D).

ISO 17635: Provides general rules for NDT of welded joints.

ISO 3834: Establishes quality requirements for fusion welding of metallic materials.

EN 1011 Series: Offers detailed welding recommendations for steels, aluminum, and nickel alloys used in shipbuilding.

EN ISO 14731: Defines responsibilities and competencies of welding coordinators.

These ISO and EN standards complement classification society rules (DNV, ABS, LR) by providing uniform testing, qualification, and inspection protocols. They are especially important in European and Asian shipyards**,** where hybrid fabrication standards are used to align with both regional and international requirements.

10.6.3 Offshore Structural Welding Requirements (API, NORSOK)

Offshore structures such as fixed platforms, floating production units, and subsea installations operate under high stresses, pressure fluctuations, and corrosive seawater conditions. Specialized codes developed by **API** and **NORSOK** govern welding and fabrication in these environments.

Key Offshore Welding Codes:

API RP 2A—Recommended Practice for Planning, Designing, and Constructing Fixed Offshore Platforms: Provides detailed welding requirements for offshore structural members; specifies procedure qualification testing, fatigue resistance, and fracture toughness; includes inspection and repair protocols under ABS and DNV oversight.

API 1104—Welding of Pipelines and Related Facilities: Used for subsea and topside pipeline welding, including risers and flowlines; defines qualification of procedures and personnel, defect acceptance criteria, and repair procedures.

NORSOK M-101—Structural Steel Fabrication: Norwegian offshore standard specifying welding, inspection, and documentation for offshore structures; emphasizes fatigue resistance, preheat control, and use of low-hydrogen consumables; aligns closely with DNV and ISO standards for compatibility in the North Sea region.

NORSOK M-601—Welding and Inspection of Piping: Defines procedure qualification, welder testing, and NDT requirements for high-integrity piping systems used in offshore production facilities.

These codes prioritize integrity under cyclic loading and extreme marine conditions, ensuring that offshore structures maintain safety and operability throughout their design life. API and NORSOK are frequently used in combination with classification society rules and ISO standards to achieve multi-code compliance.

10.6.4 Harmonization of Multiple Codes in Global Shipyards

Modern shipyards and offshore fabrication facilities operate in a globally integrated environment, where projects often involve multinational stakeholders, cross-border fabrication, and diverse classification requirements. Consequently, harmonization of multiple welding codes and standards has become essential to ensure uniform quality control and project certification.

Key Harmonization Practices Include:

Dual-Standard Qualification: Shipyards often qualify welding procedures under **both AWS and ISO standards** to satisfy clients and classification societies with differing code preferences.

Cross-Recognition Agreements: Classification societies such as **ABS, DNV, LR, and BV** recognize certain ISO, ASME, or AWS-based qualifications to minimize redundancy and streamline project approval.

Unified Documentation and Traceability: Digital welding management systems (WMS) are used to integrate data from multiple standards linking WPS/WPQR, welder qualifications, and NDT records into a single auditable platform.

Hybrid Design and Inspection Strategies: Shipyards constructing hybrid vessels (e.g., LNG carriers, FPSOs) often use **AWS D1.1 for structural elements, ISO 15614 for material testing**, and **API/NORSOK for offshore modules**, ensuring comprehensive compliance.

Global Competence and Training Alignment: Personnel training and certification under **ISO 9606** and **AWS QC1** are harmonized through mutual recognition initiatives, promoting workforce mobility and standardization.

This harmonized approach enables shipyards to manage complex international projects efficiently, reducing administrative overhead while maintaining safety, consistency, and quality assurance across all fabrication stages.

10.7 COMPLIANCE AND QUALITY CONTROL IN MARINE WELDING

Compliance and quality control form the foundation of reliability and safety assurance in marine and offshore welding. Given the demanding operational conditions of ships and offshore structures, exposure to cyclic loads, hydrostatic pressure, saltwater corrosion, and temperature fluctuations, maintaining rigorous control over every aspect of the welding process is essential.

Marine welding compliance is governed by a structured framework involving documentation, inspection, testing, and third-party verification based on international standards (ISO, AWS, ASME) and classification society rules (ABS, DNV, LR, BV). This framework ensures that welding activities adhere to approved procedures, traceable materials, and qualified personnel, all of which contribute to the overall structural integrity, safety, and longevity of marine assets.

10.7.1 Documentation and Traceability Requirements

In marine fabrication, documentation and traceability are critical to verifying that all welding operations conform to approved procedures and meet prescribed quality standards. The documentation process enables complete traceability from base materials and consumables to welders, inspection results, and final certification.

Essential Documentation Components Include:

Welding Procedure Specifications (WPS) and Qualification Records (WPQR): Define approved parameters such as current, voltage, travel speed, preheat, and interpass temperature; each WPS must be qualified through WPQR testing witnessed by a classification surveyor.

Welder and Operator Qualification Records: Detail the credentials and test results of personnel authorized to perform welding; records must indicate the process, position, material group, and qualification expiry date.

Material Certificates and Consumable Traceability: Each batch of base metal and filler material must have an **MTC** confirming compliance with chemical and mechanical specifications; consumables (electrodes, wires, fluxes) must be from approved manufacturers and stored according to procedure to prevent contamination.

Inspection and NDT Reports: Include results of **visual, ultrasonic, radiographic, magnetic particle, and penetrant tests;** must specify acceptance criteria per ISO 5817 or AWS D1.1 quality levels.

Weld Maps and Identification Records: Every weld is assigned a unique identification number linked to its WPS, welder ID, inspection reports, and repair history; digital weld tracking systems (WMSs) are increasingly used to automate this process.

Quality Control Plan (QCP) and Inspection Test Plan (ITP): Outline stages of fabrication where inspections and tests occur and withhold points for classification surveyor approval.

Such meticulous documentation ensures traceability, transparency, and accountability throughout the project life cycle, allowing shipyards and regulators to reconstruct weld histories even decades after construction.

10.7.2 Role of Third-Party Inspection and Classification Societies

Third-party inspection agencies and classification societies play a crucial role in independent verification and quality assurance of marine welding activities. Their involvement ensures that fabrication practices meet international safety, design, and environmental protection standards.

Primary Functions of Third-Party Inspectors and Classification Societies:

Procedure and Design Approval: Review and approve WPS and Procedure Qualification Records (WPQR) prior to production; Verify compliance with applicable standards such as **DNV-OS-C401, AWS D1.1,** and **ISO 15614.**

Material and Consumable Certification: Approve base materials and consumables through type testing and manufacturer audits; Issue **Type Approval Certificates** confirming compliance with LR, ABS, or DNV requirements.

In-Process Inspection and Witnessing: Attend welding qualification tests, mechanical testing, and NDT examinations; perform surveillance audits during fabrication to verify that parameters remain within approved limits.

Non-Destructive Testing (NDT) Review: Evaluate NDT reports and ensure that defect indications meet acceptance criteria under **ISO 5817 Level B/C** or equivalent; confirm reinspection after defect rectification.

Certification and Classification: Upon successful inspection and documentation review, the classification society issues **Certificates of Compliance (CoC) or Fabrication Approval Certificates.** These certificates confirm that the welded structure meets all applicable regulatory and safety requirements.

In-Service and Periodic Surveys: Classification societies conduct regular surveys during a vessel's service life to assess weld condition and fatigue performance; findings may trigger maintenance or repair actions to maintain class certification.

By acting as impartial authorities, organizations like **LR, ABS, BV,** and **DNV** ensure that marine welding practices achieve the highest levels of safety, reliability, and environmental compliance.

10.7.3 Audit and Certification Processes in Marine Projects

Quality audits and certification form the final layer of compliance verification in marine welding projects. These processes ensure that shipyards, contractors, and fabricators maintain continuous adherence to approved procedures, qualified personnel, and international quality management standards.

Key Components of the Audit and Certification Process:

Initial Certification Audit: Conducted by classification societies or accredited third-party auditors to verify compliance with standards such as **ISO 3834 (Quality Requirements for Welding) and ISO 9001 (QMSs)**; focuses on the shipyard's organization, documentation control, welder qualification, and inspection systems.

Surveillance and Renewal Audits: Periodic audits verify ongoing compliance and address changes in materials, processes, or workforce; NCs are documented, and corrective actions are required within defined time limits.

Process and Equipment Verification: Auditors inspect welding equipment calibration, storage conditions for consumables, and maintenance of NDT instruments; verification ensures that process variables (current, voltage, heat input) remain within qualified limits.

Fabrication and Project Certification: Upon completion of the project, documentation and inspection data are compiled into a **Final Data Book** reviewed by classification surveyors; Certificates of Conformity or **Fabrication Approval Certificates** are issued, confirming that all welded components meet specified standards and are eligible for class entry.

Digital Compliance and Data Integration: Modern shipyards implement **digital audit platforms** that integrate inspection, NDT, and material traceability records into centralized databases. This enhances transparency, reduces manual errors, and allows real-time tracking of compliance across global project teams.

The audit and certification process thus ensures not only that **standards are met during construction,** but also that **continuous improvement and accountability** are embedded within the organization's quality management culture.

10.8 FUTURE TRENDS IN WELDING STANDARDS

The evolution of marine and offshore welding standards is being shaped by rapid advances in digitalization, AI, automation, and sustainability initiatives. As shipyards transition toward Industry 4.0 and Smart Manufacturing, welding standards are increasingly adapting to support real-time quality monitoring, predictive defect detection, and environmental responsibility.

Future regulatory frameworks aim to integrate digital traceability, machine learning (ML)-based inspection systems, and low-emission manufacturing processes into the quality assurance cycle. These developments not only enhance safety and productivity but also align with global objectives for carbon reduction, life cycle integrity, and resource efficiency in shipbuilding and offshore engineering.

10.8.1 Digitalization and Real-Time Welding Data Compliance

The digital transformation of marine welding processes marks a major shift from traditional, document-based quality control to data-driven, automated compliance systems.

Key Aspects of Digital Welding Compliance Include:

Welding Data Acquisition and Monitoring Systems (WDMS): Modern shipyards are implementing sensors and data loggers integrated into welding power sources to continuously capture current, voltage, travel speed, arc stability, and heat input in real time. Data are stored in cloud-based platforms linked to WPS, enabling automatic verification against approved parameters.

Digital Welding Management Systems (WMS): These platforms centralize documentation such as **WPS/WPQR, welder qualifications, material certificates, and inspection reports**. Automated alerts notify supervisors of deviations or NCs during welding operations. Systems like **DNV's**

Veracity and **LR's Digital Compliance Hub** provide audit-ready traceability through digital signatures and blockchain-backed verification.

Integration with Classification Society Databases: Real-time data sharing with classification bodies (ABS, DNV, LR) streamlines audits and reduces manual verification effort. Digital compliance allows surveyors to remotely approve welding procedures, reducing inspection times and improving efficiency.

Life cycle Data Retention: Digital records enable **end-to-end weld life cycle tracking** from fabrication to in-service inspection supporting maintenance, repair, and recycling decisions decades later. The adoption of digital welding standards ensures transparency, efficiency, and accountability, transforming compliance from a reactive process into a continuous, real-time assurance system for global marine operations.

10.8.2 AI-Driven Quality Monitoring and Standardization

AI and ML are revolutionizing the monitoring, inspection, and standardization of marine welding operations. By analyzing vast amounts of process and inspection data, AI enhances both predictive maintenance and real-time quality control.

Applications of AI in Welding Standardization:

Predictive Quality Analytics: AI models trained on historical welding data can predict defect formation based on process fluctuations (e.g., arc instability, heat input variation). These systems enable adaptive control, adjusting parameters in real time to maintain compliance with WPS requirements.

Automated Defect Recognition: Advanced computer vision algorithms integrated with radiography, ultrasonic, and acoustic emission systems automatically detect and classify weld defects with high accuracy. AI-assisted NDT aligns with ISO and AWS inspection acceptance standards, reducing human subjectivity in evaluation.

Machine Learning-Based Standardization: ML tools can harmonize data from multiple code frameworks (AWS, ISO, API, NORSOK) to generate unified compliance reports. This supports global projects requiring multi-standard certification, reducing redundancy and error.

AI-Enabled Training and Qualification: Virtual reality (VR) and AI-based simulators are increasingly used to train welders, ensuring they adhere to standardized procedures while optimizing arc control and deposition rates.

AI integration within welding standards promotes a proactive, intelligent, and adaptive compliance ecosystem, where systems continuously learn from data to enhance both productivity and safety in shipbuilding and offshore fabrication.

10.8.3 Sustainability and Environmental Compliance in Marine Welding

As the maritime industry moves toward carbon-neutral and sustainable manufacturing, future welding standards are expanding to include environmental performance metrics alongside traditional mechanical and quality parameters.

Key Sustainability Trends Influencing Welding Standards:

Low-Emission Welding Processes: Transition from high-energy processes (e.g., SMAW, SAW) to energy-efficient methods like **laser hybrid welding, FSW, and WAAM**; standards are being updated to evaluate and certify **energy consumption and greenhouse gas emissions** associated with welding operations.

Eco-Friendly Consumables and Materials: Development of **low-fume electrodes, green fluxes**, and **recyclable filler materials** that reduce environmental impact during fabrication and disposal; classification societies are incorporating sustainability audits into consumable approval systems.

Waste and Emission Control: Implementation of fume extraction standards and emission measurement protocols as part of workplace safety and environmental certification; marine fabrication yards are adopting **ISO 14001 Environmental Management Systems** integrated with welding operations.

Life Cycle Assessment (LCA) Integration: Future welding standards emphasize full cradle-to-grave analysis, covering material sourcing, fabrication, operation, and recycling stages.

This ensures that the environmental footprint of welded structures aligns with global initiatives like the **IMO's GHG Reduction Strategy and UN Sustainable Development Goals (SDGs).**

By embedding sustainability into technical codes and inspection standards, the marine industry is steering toward responsible manufacturing and repair practices, ensuring that future shipbuilding not only meets safety and strength criteria but also contributes to a cleaner, more sustainable maritime ecosystem.

10.9 SUMMARY

The chapter provides a comprehensive overview of the global frameworks, classification rules, and evolving technologies that govern welding quality, safety, and compliance in shipbuilding and offshore fabrication. It explains how international organizations such as the IMO, ABS, DNV, LR, ISO, AWS, and NORSOK establish unified procedures for welding qualification, inspection, material selection, and certification to ensure the structural integrity and environmental safety of marine assets. The chapter highlights the essential role of standards and regulations in harmonizing global practices, minimizing fabrication defects, and maintaining life cycle traceability across diverse shipyards. It also emphasizes the increasing integration of digitalization, AI-driven monitoring, and sustainability standards, which are transforming conventional quality assurance into a real-time, data-driven, and environmentally responsible system. By combining regulatory rigor with modern innovation, these evolving standards ensure that marine welding remains safe, efficient, and sustainable, supporting the long-term reliability and global competitiveness of the maritime industry.

10.10 REVIEW QUESTIONS

1. What is the primary purpose of welding standards and regulations in the marine industry?
2. Explain how welding standards contribute to ensuring safety and structural integrity in shipbuilding.
3. Discuss the difference between a *standard* and a *regulation* in the context of marine welding.
4. Why is adherence to international welding codes crucial for global shipyard interoperability?
5. How do environmental and operational conditions in the marine sector influence the development of welding standards?
6. Describe the role of the International Maritime Organization (IMO) in establishing global safety and fabrication guidelines.
7. Which IMO conventions are most relevant to welding operations in shipbuilding and offshore structures?
8. How does the SOLAS Convention ensure structural integrity in welded marine components?
9. Explain how IMO guidelines are integrated into national and classification society rules.
10. What role does the IGC Code play in defining welding requirements for LNG and gas carrier systems?
11. Outline the main objectives of the ABS welding standards.
12. What key parameters are evaluated during the Welding Procedure Qualification Record (WPQR) process under ABS rules?
13. Discuss the DNV approach to risk-based quality assurance in offshore welding.
14. What types of testing and documentation are required by Lloyd's Register (LR) before a welding procedure is approved?
15. Compare how ABS and DNV manage welder qualification and consumable approval processes.
16. Differentiate between AWS D1.1, AWS D3.6M, and ISO 15614 in terms of scope and marine applicability.
17. What are the key ISO and EN standards governing welding quality and inspection in shipyards?
18. Explain the purpose of NORSOK M-101 and API RP 2A in offshore structural welding.
19. How do AWS and ISO standards complement each other in international marine fabrication projects?
20. What challenges arise in harmonizing multiple welding codes (AWS, ISO, API, NORSOK) across global shipyards?
21. What is the significance of documentation and traceability in marine welding quality assurance?
22. List the essential records included in a Weld Data Book or Fabrication Dossier.

23. Explain the role of classification societies as third-party inspectors in marine fabrication projects.
24. What are the key stages of inspection and certification performed by ABS, DNV, or LR surveyors?
25. Describe the audit process used to verify compliance with ISO 3834 and ISO 9001 in marine welding.
26. How does digitalization transform compliance and traceability in marine welding operations?
27. Explain how AI-driven monitoring systems improve weld defect detection and process control.
28. Describe how real-time data logging enhances welding standardization and audit readiness.
29. What role does sustainability play in the future evolution of marine welding standards?
30. How are modern welding standards adapting to include environmental compliance, low-emission processes, and life cycle assessments?

11 Corrosion and Fatigue in Marine Welds

11.1 INTRODUCTION

Welded joints serve as critical structural links in ships, offshore platforms, and subsea systems, where they are constantly exposed to some of the most aggressive environmental and mechanical conditions on earth. The combined effects of saltwater corrosion, cyclic loading, temperature variation, and hydrodynamic pressure create a highly demanding service environment that challenges even the most advanced welding technologies and materials. Among the key failure mechanisms affecting the long-term reliability of welded marine structures, corrosion and fatigue remain the most significant and interdependent threats.

In marine engineering, corrosion leads to material degradation, loss of cross-sectional area, and initiation of surface flaws, while fatigue results from repeated cyclic stresses that cause microcrack formation and progressive crack propagation. When these two phenomena act together a process known as corrosion fatigue their combined effect is far more severe than their individual impact, often resulting in premature structural failure. Such failures are not only costly to repair but also pose risks to safety, the environment, and vessel operability.

The susceptibility of marine welds to corrosion and fatigue is influenced by multiple factors, including welding process parameters, residual stresses, heat-affected zone (HAZ) microstructure, weld geometry, and environmental exposure. Improper control of these variables can lead to defects such as porosity, undercut, inclusions, and uneven fusion, which act as stress concentrators and accelerate crack initiation. Moreover, the electrochemical nature of the marine environment promotes localized attack modes such as pitting, crevice corrosion, and galvanic corrosion, particularly at weld interfaces where metallurgical heterogeneity exists.

To combat these challenges, the marine industry relies on a combination of design optimization, advanced materials, surface protection systems, and continuous inspection protocols. Standards and guidelines from organizations such as Det Norske Veritas (DNV), American Bureau of Shipping (ABS), and ISO provide frameworks for evaluating fatigue life, corrosion allowance, and weld quality in ships and offshore structures. In parallel, modern research focuses on corrosion-fatigue modeling, residual stress mitigation, cathodic protection (CP), and the application of high-performance alloys and coatings to extend service life and minimize maintenance costs.

This chapter explores the mechanisms, influencing factors, and mitigation strategies related to corrosion and fatigue in marine welds. It discusses the electrochemical and mechanical fundamentals, presents failure analysis case studies, and outlines

DOI: 10.1201/9781003217350-11

inspection and prevention techniques that enhance the durability of welded marine structures. By integrating insights from materials science, corrosion engineering, and structural mechanics, the chapter aims to provide a holistic understanding of how corrosion and fatigue interact and how advanced welding design and monitoring practices can ensure safe, reliable, and long-lasting performance in marine and offshore applications.

11.1.1 Importance of Corrosion and Fatigue Considerations in Marine Welding

In the marine and offshore industries, welded joints are among the most critical and vulnerable components of structural systems. They serve as load-transmitting elements in hulls, decks, risers, and subsea assemblies, where corrosion and fatigue collectively determine the service life and reliability of the entire structure. Understanding and controlling these phenomena are therefore essential to prevent catastrophic failures, unplanned downtime, and costly repairs.

Corrosion and fatigue are not isolated effects but synergistic degradation mechanisms. Corrosion weakens the weld by reducing its cross-sectional area and introducing surface pits, which act as stress concentrators under cyclic loading. Conversely, fatigue-induced microcracks expose fresh metal surfaces that accelerate localized corrosion. This interaction known as corrosion fatigue is particularly detrimental in welded marine components subjected to wave-induced vibrations, fluctuating hydrostatic pressures, and dynamic bending stresses.

Effective marine welding design must therefore incorporate corrosion-fatigue considerations from the earliest stages of material selection and joint configuration. Proper heat input control, weld geometry optimization, residual stress management, and post-weld treatment can greatly enhance resistance to these coupled mechanisms. Moreover, international standards such as DNV-RP-C203 (Fatigue Design of Offshore Steel Structures) and ISO 12944 (Corrosion Protection of Steel Structures) provide guidance on design criteria and protective measures to minimize degradation. By addressing corrosion and fatigue collectively, shipyards and offshore fabricators ensure long-term safety, structural integrity, and regulatory compliance across diverse operating conditions.

11.1.2 Influence of Marine Environment on Weld Durability

The marine environment presents one of the most aggressive natural conditions for metallic materials and welded joints. Its combination of high salinity, humidity, dissolved oxygen, temperature gradients, and mechanical forces creates a unique set of challenges that accelerate both corrosion and fatigue processes.

Seawater acts as a highly conductive electrolyte, promoting electrochemical reactions between dissimilar metals and between the weld metal (WM), HAZ, and base material. These microstructural differences establish galvanic cells, where the anodic regions corrode preferentially. Moreover, turbulent seawater flow, splash zones, and alternate wetting and drying cycles intensify corrosion, especially in the presence of oxygen and chlorides.

In addition to electrochemical effects, mechanical factors such as wave loading, vibration, and thermal cycling induce fluctuating stresses that lead to fatigue crack initiation. Structures in offshore environments, such as jackets, risers, and subsea pipelines, are subjected to millions of stress cycles over their service life. The combination of saltwater exposure and cyclic loading causes microcrack formation in surface imperfections and weld toes, leading to progressive crack propagation and eventual failure.

Temperature and oxygen variations with water depth further influence degradation modes. For instance, warm tropical waters accelerate corrosion rates, while deep-sea, low-oxygen conditions promote hydrogen embrittlement and stress corrosion cracking (SCC). The marine environment's complex chemistry and dynamic mechanical forces thus make it one of the most severe testbeds for welded structures, requiring continuous attention to design robustness, protective coatings, and CP systems to ensure durability.

11.1.3 Overview of Common Degradation Mechanisms in Marine Structures

Marine welded structures are exposed to multiple degradation mechanisms, often acting simultaneously and interacting in complex ways. The most significant of these include:

General (Uniform) Corrosion:

A uniform attack across exposed metal surfaces, typically in unprotected carbon steel welds.
Leads to material thinning and gradual loss of strength.

Pitting and Crevice Corrosion:

Localized corrosion initiated at surface defects or confined areas (e.g., under deposits, seals, or weld undercuts).
Common in stainless steel welds exposed to stagnant seawater or chloride-rich zones.

Galvanic Corrosion:

Occurs when dissimilar metals (e.g., steel and aluminum) are electrically connected in seawater.
Weld zones, with varying compositions and microstructures, can form natural galvanic couples.

Stress Corrosion Cracking (SCC):

A result of the combined action of tensile stress and a corrosive environment.
Particularly critical in HAZs of high-strength steels and aluminum alloys.

Fatigue and Corrosion Fatigue:

Progressive crack initiation and propagation under cyclic stress, accelerated by corrosive attack.

Responsible for the majority of weld-related failures in marine and offshore structures.

Hydrogen-Induced Cracking (HIC) and Hydrogen Embrittlement:

Caused by the absorption of atomic hydrogen during welding or CP.

Leads to brittle fracture, especially in high-strength steels.

Microbiologically Influenced Corrosion (MIC):

Corrosion initiated or accelerated by marine microorganisms (e.g., sulfate-reducing bacteria).

Often occurs in submerged or sediment-covered weld zones.

Each of these mechanisms contributes differently depending on the material composition, weld geometry, residual stress, and service environment. Therefore, understanding their interactions is essential for developing effective corrosion prevention, fatigue design, and inspection strategies. Modern approaches increasingly employ predictive modeling, surface engineering, and advanced non-destructive testing (NDT) techniques to identify degradation at early stages and ensure structural reliability throughout the marine asset's life cycle.

11.2 TYPES OF CORROSION IN MARINE WELDS

Corrosion in marine welds manifests in multiple forms, influenced by the electrochemical nature of seawater, microstructural heterogeneity of weld zones, and mechanical stresses imposed during service. The variety of corrosion mechanisms in welded structures is attributed to differences in metallurgical composition, residual stress distribution, and environmental exposure across the WM, HAZ, and base metal (BM).

Understanding the major corrosion types, general, galvanic, pitting, crevice, and SCC, is essential for designing and maintaining long-term durable marine components. Each mechanism requires specific preventive strategies, ranging from material selection and protective coatings to CP and stress control.

11.2.1 General Corrosion in Welded Joints

General corrosion, also known as uniform corrosion, is the most common and widespread form of degradation in marine welds. It occurs when the entire exposed metal surface is attacked relatively evenly by electrochemical reactions with seawater.

In carbon and low-alloy steel welds, general corrosion typically results in uniform thinning, leading to a gradual reduction in cross-sectional area and mechanical strength. Although predictable in nature, it can still pose serious problems if corrosion allowances are not properly considered during design.

The rate of general corrosion depends on several factors:

Seawater salinity and temperature (higher chloride levels accelerate reaction kinetics).
Oxygen concentration (more oxygen promotes anodic dissolution).
Surface roughness and cleanliness (residual slag or spatter can create localized corrosion sites).
Weld thermal cycles that alter the microstructure of the HAZ, making it more anodic than the BM.

Protective coatings, CP systems, and regular maintenance schedules are widely used to mitigate uniform corrosion. When properly managed, this type of corrosion can be accounted for in structural design and life expectancy calculations. Figure 11.1 presents a schematic representation of general corrosion, also referred to as uniform corrosion, as it commonly occurs in welded marine structures exposed to seawater environments. This figure illustrates the uniform attack of the metal surface across three distinct regions: the WM, HAZ, and BM. Each of these areas undergoes a similar rate of corrosion due to the electrochemical reactions taking place between the metallic surface and the chloride-containing marine electrolyte.

The weld bead in the center of the illustration shows the deposited filler metal, flanked by the HAZ, where microstructural transformations have occurred as a result of the welding thermal cycle. The BM regions represent the unaffected parent material adjacent to the weld. A corrosion layer forms continuously over these surfaces, depicting uniform metal loss across the exposed area. Arrows indicate the direction of corrosion progression, emphasizing that the attack occurs evenly and predictably over the surface rather than at specific localized points.

This figure highlights that, although general corrosion is less catastrophic than localized forms such as pitting or crevice corrosion, it can result in significant thickness reduction and strength loss over extended exposure periods. Such gradual deterioration compromises the load-bearing capacity of marine structures, particularly in hull plating, offshore risers, and underwater frameworks. In practical shipbuilding and offshore fabrication, designers typically account for this type of degradation by

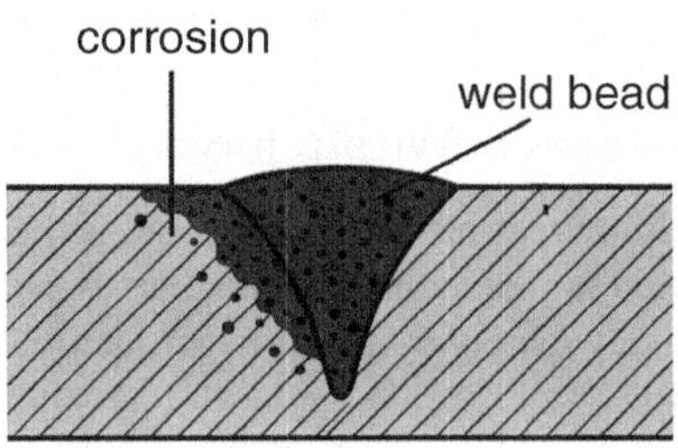

FIGURE 11.1 General corrosion in welded joints.

incorporating corrosion allowances, applying protective coatings, or implementing CP systems to ensure long-term durability.

11.2.2 Galvanic Corrosion between Dissimilar Metals

Galvanic corrosion occurs when two dissimilar metals are electrically connected in the presence of an electrolyte such as seawater. In a galvanic couple, the metal with the lower electrochemical potential (anode) corrodes preferentially, while the more noble metal (cathode) remains protected.

In marine welds, galvanic corrosion is often triggered by differences between the WM, filler metal, and base material, each possessing distinct electrochemical properties. For instance, stainless steel welded to carbon steel or aluminum joined to steel components creates strong galvanic potentials.

Key Contributing Factors:

Potential difference between materials (greater difference = higher corrosion rate).
Area ratio of anode to cathode (small anode coupled with a large cathode accelerates corrosion).
Seawater conductivity and oxygen content.
Inconsistent surface protection, where coating damage on the anodic metal exposes it to accelerated attack.

Examples in Practice:

Aluminum deck fittings bolted to steel hulls.
Stainless steel piping welded onto carbon steel structures.
Brass or bronze propellers connected to steel shafts.

Mitigation Strategies:

Use of **insulating gaskets or sleeves** to break electrical continuity.
Application of **sacrificial anodes (e.g., zinc, magnesium)** for CP.
Material pairing based on galvanic series proximity.
Uniform coating systems applied to both metals to minimize potential difference.

Without proper management, galvanic corrosion can lead to rapid localized material loss, particularly at the weld interfaces and bolted joints. Figure 11.2 illustrates the principle of galvanic corrosion that occurs when two dissimilar metals, such as aluminum and steel, are electrically connected and exposed to an electrolyte like seawater. The diagram shows an aluminum alloy plate welded to a steel substrate through a weld bead, with seawater acting as the conductive medium.

In this setup, the aluminum alloy, being less noble (more active), functions as the anode, while the steel acts as the cathode. The flow of electrons from the anodic

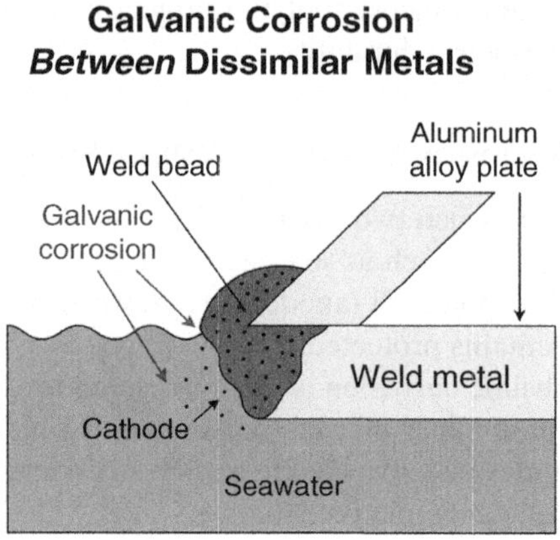

FIGURE 11.2 Galvanic corrosion between dissimilar metals.

aluminum to the cathodic steel through the metallic connection drives the corrosion process, resulting in metal loss at the anodic region near the weld interface. The labeled arrows indicate the direction of electron flow and corrosion attack, emphasizing the anodic dissolution occurring at the aluminum surface adjacent to the WM.

This figure demonstrates how potential differences between metals can lead to accelerated localized corrosion in the more active material, especially in marine and offshore environments where conductivity and chloride content are high. Preventive measures include electrical insulation between dissimilar metals, the use of sacrificial anodes, and uniform coating systems to minimize potential differences.

11.2.3 Crevice and Pitting Corrosion in Weld Zones

Crevice and pitting corrosion represent localized forms of electrochemical attack that are particularly dangerous because they cause deep, penetrating damage while affecting only small surface areas. These forms are most prevalent in stainless steels, aluminum alloys, and nickel-based materials used in marine environments.

Crevice Corrosion:
This occurs in narrow, shielded areas where oxygen diffusion is restricted such as under gaskets, fasteners, or weld overlaps. In these regions, oxygen depletion causes a shift in electrochemical potential, forming anodic zones that corrode rapidly. In welds, crevice corrosion often initiates in overlapping joints, slag inclusions, or undercut zones.

Pitting Corrosion:
Pitting develops when protective oxide films break down locally due to chloride ion attack. Once initiated, pits propagate rapidly due to the formation of a localized acidic environment within the cavity. Pitting is common along weld beads and HAZ, where microstructural in homogeneities and inclusions act as initiation points.

Critical Influences:

Chloride concentration and temperature (pitting increases in warm, saline water).
Residual tensile stress and microstructural segregation in HAZ.
Surface contamination (halides, oxides, or metallic debris).

Preventive Measures:

Post-weld surface cleaning and passivation treatments to restore protective oxide layers.
Optimized weld parameters to minimize undercuts and slag inclusions.
Use of molybdenum-bearing stainless steels (e.g., 316L) or super duplex alloys for higher pitting resistance.
Application of high-performance epoxy or polyurethane coatings.

Pitting and crevice corrosion can lead to rapid perforation and structural failure, often without visible warning, making regular inspection and surface maintenance essential. Figure 11.3 illustrates the occurrence of crevice and pitting corrosion in the welded zones of marine structures exposed to seawater. The diagram shows a welded steel plate, highlighting the weld bead, HAZ, and BM each subject to localized electrochemical attack under marine conditions. These corrosion forms are particularly dangerous because they cause deep penetration in confined regions without extensive surface damage, often leading to undetected material failure.

The figure depicts crevice corrosion occurring beneath overlapping joints, gaskets, or deposits near the weld interface. In these narrow, oxygen-depleted zones, seawater becomes stagnant, creating an acidic and chloride-rich microenvironment that accelerates anodic metal dissolution. Arrows in the figure show the direction of ion movement and localized corrosion progression under the crevice.

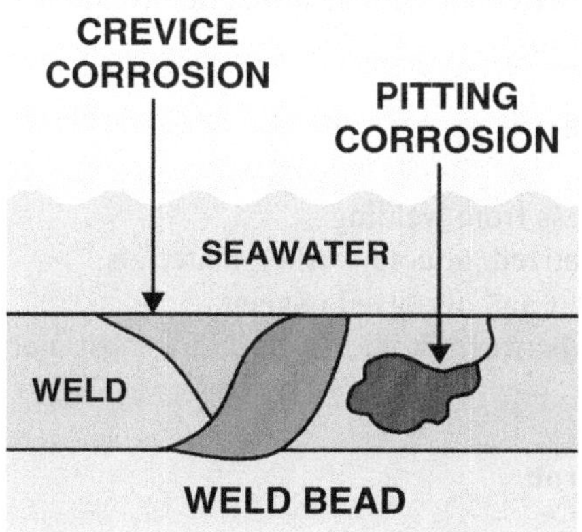

FIGURE 11.3 Crevice and pitting corrosion in weld zones.

Adjacent to these areas, pitting corrosion is represented as small, isolated cavities forming on the weld surface and HAZ. These pits initiate when the passive oxide film on the metal surface breaks down due to chloride ion attack, creating self-sustaining corrosion cells. Once formed, pits grow rapidly downward, leading to potential perforation even in thick welded components.

This figure emphasizes that microstructural heterogeneity, surface roughness, and residual stress in weld zones enhance susceptibility to localized corrosion. To mitigate such damage, post-weld cleaning, passivation, optimized welding parameters, and high-alloy or molybdenum-containing materials are recommended. Thus, Figure 11.3 effectively visualizes the dual threat of crevice and pitting corrosion as key contributors to the premature degradation of marine welded joints.

11.2.4 Stress Corrosion Cracking (SCC) in Marine Applications

SCC is a form of brittle, environmentally assisted cracking that occurs due to the combined action of tensile stress and a corrosive environment. It is one of the most insidious and dangerous degradation mechanisms in marine welds, as it can lead to sudden, catastrophic failure even under normal service loads.

SCC typically initiates in high-strength steels, aluminum alloys, and certain stainless steels, particularly in the HAZ, where tensile residual stresses and microstructural changes are most pronounced. The chloride-rich marine environment provides ideal conditions for SCC, especially at elevated temperatures.

Mechanism:

Localized anodic dissolution begins at microscopic flaws, inclusions, or weld toe regions.
Crack nucleation occurs under tensile stress, often perpendicular to the applied load.
Crack propagation advances through intergranular or transgranular modes depending on alloy type and environment.
Sudden fracture may occur with minimal deformation, often without visible corrosion evidence.

Contributing Factors:

Residual tensile stress from welding.
High-strength, sensitized, or cold-worked materials.
Presence of chlorides and dissolved oxygen.
Temperature range between 30°C and 80°C for most marine steels and stainless alloys.

Prevention and Control:

Post-weld heat treatment (PWHT) to relieve residual stresses.
Use of **low-sulfur, low-carbon, and stabilized stainless steels**.

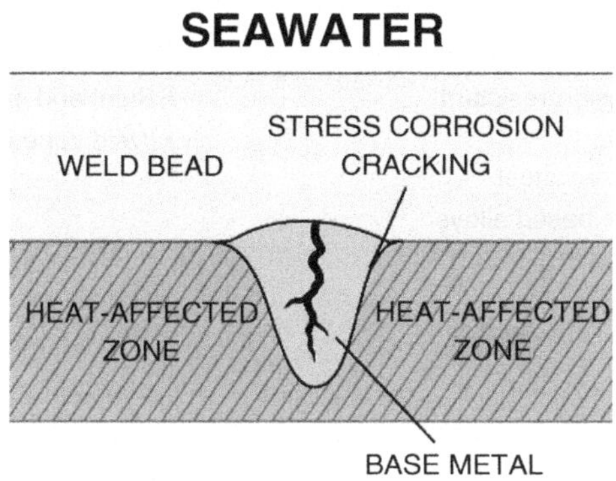

FIGURE 11.4 Stress corrosion cracking (SCC) in marine applications.

CP and **corrosion inhibitors** in seawater systems.
Design modification to minimize stress concentrators (e.g., smooth transitions at weld toes).

SCC is particularly concerning in pressure hulls, piping systems, and offshore risers, where failure can compromise both safety and environmental containment. Preventing SCC requires integrated control of metallurgical, mechanical, and environmental factors throughout the fabrication and service stages. Figure 11.4 depicts the phenomenon of SCC in a welded steel structure exposed to a seawater environment. The illustration shows the interaction between tensile stress and a chloride-rich corrosive medium, leading to crack initiation and propagation primarily within the HAZ. The figure identifies the key components of the welded joint, the weld bead, HAZ, and BM with the SCC crack path extending downward from the weld surface into the substrate.

The diagram visually represents how chloride ions in seawater penetrate microstructural flaws or surface imperfections at the weld interface. Under sustained tensile stress, these regions act as anodic sites, where localized corrosion begins, initiating fine cracks that propagate over time. Arrows indicate the direction of stress and crack propagation, emphasizing that SCC develops perpendicular to the applied load and can occur even in materials that are otherwise resistant to uniform corrosion.

This figure highlights that SCC is a synergistic failure mechanism, combining electrochemical and mechanical effects. It most often affects high-strength steels, stainless steels, and aluminum alloys used in ship hulls, pipelines, and offshore risers. Preventive measures include PWHT to relieve residual stresses, the use of low-sensitivity alloys, and CP in submerged conditions.

11.3 CORROSION PREVENTION TECHNIQUES IN MARINE WELDED STRUCTURES

Corrosion prevention in marine welded structures is an essential aspect of shipbuilding, offshore fabrication, and subsea system design. Because welded joints act as

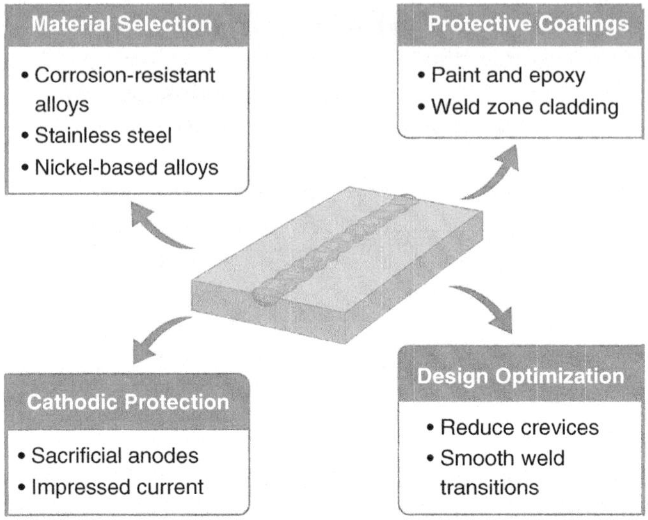

FIGURE 11.5 Corrosion prevention techniques in marine welded structures.

natural weak points due to microstructural variations, residual stress, and geometric discontinuities, targeted prevention strategies are required to enhance their long-term performance. Effective corrosion control combines material engineering, protective systems, electrochemical control, and design optimization to resist the aggressive marine environment. This section outlines key approaches from material selection to post-weld finishing that collectively improve the durability and safety of welded components exposed to seawater and cyclic loading conditions. Figure 11.5 illustrates the integrated strategies used to enhance the durability and longevity of welded components in ships, offshore platforms, and subsea systems. The diagram summarizes five key approaches: material selection, protective coatings and cladding, CP, weld design modification, and post-weld treatments that collectively mitigate the effects of the harsh marine environment. Corrosion-resistant materials such as stainless steel, duplex alloys, and aluminum are chosen for their inherent resistance to chloride attack, while coatings and claddings provide an external barrier against seawater exposure. CP, using sacrificial anodes or impressed current systems, electrochemically prevents oxidation of structural steel. Proper weld design minimizes crevices and stress concentration, and post-weld treatments such as stress relief, passivation, and polishing restore protective surface layers. Together, these coordinated techniques form a comprehensive defense system that ensures long-term structural integrity, reduced maintenance, and improved safety of marine welded structures operating in corrosive seawater conditions.

11.3.1 Material Selection for Corrosion Resistance

The foundation of corrosion control lies in choosing materials with inherent resistance to the electrochemical processes that dominate in marine environments. The material selection process must consider both mechanical performance **and** environmental compatibility, ensuring that the WM, filler, and base material are electrochemically balanced.

Common corrosion-resistant materials for marine welds include:

Austenitic Stainless Steels (e.g., 316L, 317L): Excellent resistance to pitting and crevice corrosion due to high chromium and molybdenum content.

Duplex and Super Duplex Stainless Steels (e.g., UNS S32205, S32750): Combine high strength with exceptional chloride SCC resistance.

Nickel-based Alloys (e.g., Inconel 625, Monel 400): Provide superior resistance to seawater corrosion, especially in heat exchangers and offshore risers.

Marine-Grade Aluminum Alloys (5xxx and 6xxx series): Offer light weight, good corrosion resistance, and are widely used in ship superstructures.

Corrosion-Resistant Claddings: Used to overlay carbon steel with a corrosion-resistant alloy (CRA) in high-risk areas.

Matching filler materials with compatible corrosion potentials and mechanical properties is critical to prevent galvanic corrosion between the WM and BM. Additionally, low-carbon or stabilized grades minimize sensitization, reducing the risk of intergranular corrosion.

11.3.2 Protective Coatings and Cladding in Weld Zones

Protective coatings and claddings act as physical barriers between the metal surface and the marine environment, effectively minimizing the interaction of water, oxygen, and chloride ions with the underlying metal.

Protective Coatings:

Epoxy and polyurethane coatings are the most widely used for hulls, decks, and ballast tanks, offering high adhesion and chemical resistance.

Zinc-rich primers provide sacrificial protection to steel surfaces, particularly near weld joints.

Thermally sprayed aluminum (TSA) coatings are effective for long-term corrosion protection in offshore platforms and splash zones.

For underwater components, **ceramic-epoxy hybrid coatings** offer high abrasion and corrosion resistance.

Cladding Techniques:

Weld overlay cladding uses CRAs like Inconel or stainless steel to protect carbon steel substrates.

Explosive bonding or roll bonding is used to produce bimetallic plates for ship hulls and offshore pressure vessels.

Proper coating application after surface cleaning and post-weld grinding ensures uniform protection. Coating repair procedures must also be integrated into maintenance routines to prevent localized corrosion at damaged areas.

11.3.3 Cathodic Protection (Sacrificial Anode and Impressed Current)

CP is a widely adopted electrochemical technique that prevents corrosion by converting the protected metal surface into a **cathode** of an electrochemical cell. It is highly effective for submerged and buried marine structures, including hulls, subsea pipelines, and offshore platforms.

Sacrificial Anode Cathodic Protection (SACP):

Uses more active metals such as **zinc, aluminum, or magnesium** as sacrificial anodes.

These anodes corrode preferentially, supplying electrons to the protected steel structure and preventing its oxidation.

Common in ship hulls and ballast tanks, where periodic replacement of anodes is part of maintenance.

Impressed Current Cathodic Protection (ICCP):

Employs an external **DC power source** to supply current to inert anodes (titanium, mixed-metal oxide).

Provides adjustable protection levels suitable for large offshore structures.

Requires continuous monitoring to avoid overprotection, which may cause **hydrogen embrittlement** in high-strength steels.

Both CP systems are often combined with coatings for enhanced protection, a strategy known as **duplex protection,** providing redundancy and extended service life in harsh marine conditions.

11.3.4 Weld Design Modifications to Minimize Corrosion Risk

The geometric and structural design of a weld plays a crucial role in determining its corrosion susceptibility. Poor weld profiles, sharp corners, and misalignments can trap moisture and promote crevice formation, leading to accelerated degradation.

Design Guidelines to Reduce Corrosion:

Use **smooth transitions** between weld beads and BM to reduce stress concentration.

Avoid **overlapping joints** and **sharp angles**, which can entrap seawater and debris.

Minimize **weld undercuts and surface irregularities** through optimized process control.

Ensure **complete fusion and penetration** to eliminate crevices at the weld root.

Apply **seal welding or intermittent welding** judiciously to prevent moisture ingress.

Additionally, maintaining a uniform corrosion potential between welded components by using compatible materials and coatings helps prevent galvanic effects. Proper design not only improves corrosion resistance but also enhances fatigue performance and inspectability.

11.3.5 Post-Weld Treatments and Surface Finishing

After welding, a range of **post-weld treatments** is employed to restore corrosion resistance and mechanical integrity compromised during fabrication. These processes address **residual stress, surface oxidation, and microstructural heterogeneity**, which otherwise promote corrosion initiation.

Key Post-Weld Treatments Include:

Post-Weld Heat Treatment (PWHT): Relieves residual stresses that contribute to SCC.
Pickling and Passivation: Removes oxide scales and restores the natural protective chromium oxide layer on stainless steels.
Mechanical Polishing and Grinding: Smoothens weld surfaces, reducing pitting initiation sites.
Shot Peening: Induces beneficial compressive surface stresses, enhancing fatigue and corrosion resistance.
Application of Corrosion Inhibitors or Sealants: Provides additional barrier protection in crevices and hard-to-reach areas.

In subsea and offshore components, combining **surface finishing with protective coatings** ensures long-term stability and minimizes maintenance requirements. These finishing operations, when integrated into the fabrication workflow, significantly extend the **service life of welded joints** operating in chloride-rich marine environments.

11.4 FATIGUE IN MARINE WELDS

Fatigue is a critical mode of structural degradation in marine welded joints, caused by repeated cyclic loading that induces progressive crack initiation and growth over time. In the marine environment, the combination of mechanical stress fluctuations, corrosion, and environmental exposure accelerates fatigue failure, making it one of the leading causes of structural deterioration in ships, offshore platforms, and subsea components. Unlike static failure, fatigue occurs gradually and often without visible warning, resulting in unexpected failures that compromise safety and serviceability. Understanding the fatigue mechanisms, influencing factors, and stress concentration effects in welds is essential for designing reliable and durable marine structures. Figure 11.6 illustrates the progressive fatigue failure mechanism in a welded joint subjected to cyclic loading. The diagram shows the weld bead, HAZ, and BM, with arrows depicting the sequence of crack initiation, crack propagation, and eventual fracture.

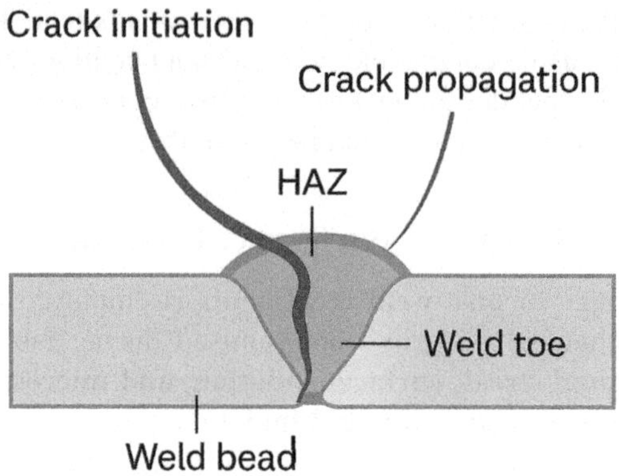

FIGURE 11.6 Mechanism of fatigue failure in welded joints.

The fatigue process typically begins at the weld toe, where geometric discontinuities and residual tensile stresses create localized stress concentration. Under repeated cyclic stresses from wave loading or vibration, microcracks form and extend along the HAZ—the region most affected by thermal cycling and microstructural hardening. Over time, these cracks propagate incrementally with each load cycle, eventually leading to the final fracture once the remaining cross section can no longer support the applied stresses.

This figure emphasizes that fatigue failure in marine welds is governed by both mechanical and metallurgical factors, including weld geometry, surface finish, and residual stress distribution. Understanding the fatigue crack path helps engineers implement effective mitigation techniques such as toe grinding, peening, and optimized weld design, all aimed at extending fatigue life in marine and offshore welded structures.

11.4.1 Mechanism of Fatigue Failure in Welded Joints

The fatigue failure mechanism in welded joints occurs through a three-stage process: crack initiation, crack propagation, and final fracture.

Crack Initiation: Microcracks develop at stress concentration points, such as weld toes, roots, undercuts, and surface imperfections, due to cyclic tensile stresses. Residual stresses from welding, microstructural discontinuities, and surface roughness intensify local strain, accelerating initiation.

Crack Propagation: Once initiated, cracks extend incrementally with each loading cycle. Propagation typically occurs through the HAZ, where microstructural gradients and hardness variations promote brittle behavior. The crack growth rate depends on stress range, frequency, and environmental conditions particularly seawater exposure, which leads to corrosion fatigue and hydrogen-assisted crack growth.

Final Fracture: When the remaining cross section can no longer sustain the applied load, sudden brittle or ductile fracture occurs. The failure surface typically exhibits beach marks or striations, characteristic of cyclic loading.

In marine applications, fatigue is aggravated by wave-induced loading, vibration, and dynamic stresses in structural members such as hull plates, risers, and deck connections. To mitigate fatigue, engineers employ smooth joint geometries, controlled welding parameters, and post-weld treatments (e.g., grinding or peening) to reduce stress concentrations and extend fatigue life.

11.4.2 Factors Affecting Fatigue Life in Marine Structures

Fatigue life, the number of cycles a welded joint can endure before failure, is influenced by multiple mechanical, metallurgical, environmental, and design-related factors.

Key Factors Include:

Weld Geometry and Surface Condition: Irregularities such as undercuts, porosity, and misalignment create stress raisers, reducing fatigue strength. A smooth, well-profiled weld surface minimizes crack initiation sites.

Residual Stress and Microstructure: Welding introduces **tensile residual stresses** that reduce fatigue resistance. Microstructural inhomogeneities in the HAZ, including grain coarsening and phase transformation, contribute to crack initiation.

Loading Conditions: Fluctuating tensile or bending loads from waves, current, or vibration are primary fatigue drivers. The **stress ratio (*R* = minimum/maximum stress) and frequency of cycles** affect fatigue behavior.

Environmental Effects: Seawater accelerates fatigue damage through **corrosion fatigue,** where anodic dissolution at crack tips increases propagation rates. Temperature, salinity, and oxygen concentration further influence fatigue life.

Material Type and Strength Level: Higher strength steels exhibit shorter fatigue lives under similar stress ranges due to their reduced ductility. Austenitic stainless steels and duplex alloys provide better performance under cyclic marine conditions.

Weld Quality and Inspection: Defects such as lack of fusion, inclusions, and cracks drastically reduce fatigue life. Rigorous **NDT methods** (ultrasonic, radiographic, or magnetic testing) ensure weld integrity before deployment.

Design and Maintenance: Proper joint configuration, load distribution, and corrosion protection systems are essential for fatigue control. Periodic inspections help detect early crack formation, allowing timely repair.

In practice, fatigue design curves (*S-N* curves) and standards such as **DNV-RP-C203** or **BS 7608** are used to predict fatigue life and classify welded joints according to their stress categories and geometries.

11.4.3 Stress Concentration Effects at Weld Toes and Roots

Stress concentration is the primary trigger for fatigue crack initiation in welded marine joints. The weld toe (junction of weld bead and BM) and weld root (internal joint surface) are especially critical due to geometric discontinuities and local hardening effects.

At the weld toe, the abrupt change in geometry and microstructure causes local stress amplification under cyclic loads. This region experiences high tensile stresses, particularly in fillet and butt joints subjected to bending or axial loading. Imperfections such as undercuts, overlap, or poor transition profiles further increase the stress concentration factor (SCF), often exceeding the nominal applied stress by several times.

The weld root is equally vulnerable, especially in partial-penetration or double-sided welds. Internal defects like lack of fusion, slag inclusions, or porosity serve as internal crack initiation sites. Since the root is often inaccessible for inspection or finishing, cracks can propagate undetected until failure occurs.

To mitigate stress concentration effects, the following practices are commonly employed:

Toe grinding or inert gas welding (TIG) dressing to smooth the weld transition and reduce local SCF.

Shot peening to introduce compressive residual stresses, improving fatigue resistance.

Improved joint design, such as larger toe radii or full-penetration welds, to reduce geometric discontinuities.

Accurate weld alignment and dimensional control during fabrication.

By reducing localized stresses, these measures significantly extend the fatigue life of welded components, particularly in high-cycle load regions such as offshore platforms, ship hull girders, and subsea pipelines.

11.5 FATIGUE LIFE ASSESSMENT OF WELDED JOINTS IN SHIPS AND OFFSHORE PLATFORMS

The fatigue life assessment of welded joints is a critical aspect of the structural design and maintenance of ships, offshore platforms, risers, and subsea components. These structures are continuously subjected to cyclic loading from waves, currents, wind, and operational stresses, which lead to gradual crack initiation and propagation in weld zones. Assessing fatigue life allows engineers to predict when a welded component will fail and to design for safe service life through appropriate material selection, welding procedures, and inspection intervals.

Fatigue assessment methods in the marine industry rely primarily on two approaches: the *S-N* curve (stress-life) method, suitable for design and classification, and the fracture mechanics (crack growth) approach, used for detailed analysis of damaged or existing structures. Both methods are supported by guidelines from classification societies and international standards such as DNV-RP-C203, ABS Guidance Notes, and BS 7608, ensuring consistency and safety in structural design and maintenance.

11.5.1 *S-N* Curve Approach for Fatigue Evaluation

The *S-N* curve approach (stress versus number of cycles) is the most commonly used method for estimating fatigue life in marine welded joints. It is based on empirical data obtained from laboratory testing, where specimens are subjected to cyclic loads until failure. The results are plotted on a logarithmic scale, relating the applied stress range (*S*) to the corresponding number of cycles to failure (*N*).

Fundamental Relationship:

$$S = A.\ N^{-m}$$

where *A* and *m* are material constants derived from experimental data.

In the marine context, *S-N* curves are categorized into classes (A to F2, per DNV-RP-C203 and BS 7608) that reflect the weld geometry, quality, and expected fatigue performance. Smooth, high-quality welds have higher fatigue strength categories (Class B or C), while rough, as-welded joints fall into lower categories (Classes D to F2).

Key Factors in *S–N* Approach:

Stress Range (Δσ): The difference between maximum and minimum stress in a cycle, often induced by wave loading or vibration.

Stress Concentration: Local geometric discontinuities at the weld toe or root amplify cyclic stresses.

Environmental Correction Factors: Seawater and corrosion effects reduce fatigue life, typically accounted for by applying **reduction factors (Ke).**

Mean Stress and Load Ratio (*R*): The ratio of minimum to maximum load affects crack initiation behavior.

The *S–N* method is highly effective for design-phase evaluations and classification society verification, offering a straightforward means to estimate fatigue design life for welded marine structures.

11.5.2 Fracture Mechanics Approach to Fatigue Analysis

The fracture mechanics approach provides a more analytical framework for assessing fatigue behavior, focusing on the crack propagation phase rather than total life. It is particularly useful for structures already in service or containing detectable flaws.

This method applies crack growth laws, such as Paris' law, which relates crack growth rate (*da/dN*) to the stress intensity factor range (ΔK):

$$\frac{da}{dN} = (K)^m$$

where *C* and *m* are material constants determined experimentally, and ΔK depends on stress, crack size, and geometry.

Stages in Fracture Mechanics Assessment:

Crack Detection and Characterization: Identifying initial crack size and location using **NDT techniques** (e.g., ultrasonic or radiographic testing).

Crack Growth Analysis: Using Paris' law or its variants to estimate how fast cracks will grow under cyclic loading.

Critical Crack Size Determination: Calculating the crack size at which catastrophic fracture occurs using **fracture toughness.**

Remaining Life Prediction: Estimating time or number of cycles until failure based on current crack size and operational stresses.

Fracture mechanics enables damage tolerance design, allowing for planned inspection intervals and controlled repair before failure. In offshore applications, this method is essential for the integrity management of welded risers, jacket nodes, and Floating Production Storage and Offloading (FPSO) hulls subjected to complex cyclic loads.

11.5.3 Influence of Welding Defects on Fatigue Performance

Welding defects are among the most significant factors reducing fatigue life in marine structures. Even minor imperfections can act as stress concentrators and crack initiation sites, accelerating failure under cyclic loading.

Common Welding Defects and Their Impact:

Undercuts and Overlaps: Sharp geometric transitions at weld toes concentrate stress, leading to early crack initiation.

Porosity and Inclusions: Entrapped gas or slag weakens the WM and interrupts crack propagation paths.

Lack of Fusion or Penetration: Creates discontinuities where fatigue cracks can initiate internally, often undetectable by visual inspection.

Residual Tensile Stresses: From rapid cooling, these stresses increase effective cyclic stress levels.

Mitigation Strategies:

Optimized Welding Parameters: Proper control of heat input and filler selection.

Post-Weld Treatments: Techniques like **toe grinding, TIG dressing, and shot peening** to smooth stress concentrators and introduce compressive surface stresses.

Rigorous Inspection: Employing **ultrasonic testing (UT)** and **magnetic particle testing (MT)** to detect subsurface flaws before deployment.

In fatigue-critical structures such as ship hulls and offshore jackets, even small defects can reduce fatigue life by **50% or more**, underscoring the need for precision welding and thorough quality control.

11.6 TESTING AND MONITORING OF CORROSION AND FATIGUE

The testing and monitoring of corrosion and fatigue behavior in marine welded structures are crucial for ensuring the structural reliability, safety, and longevity of

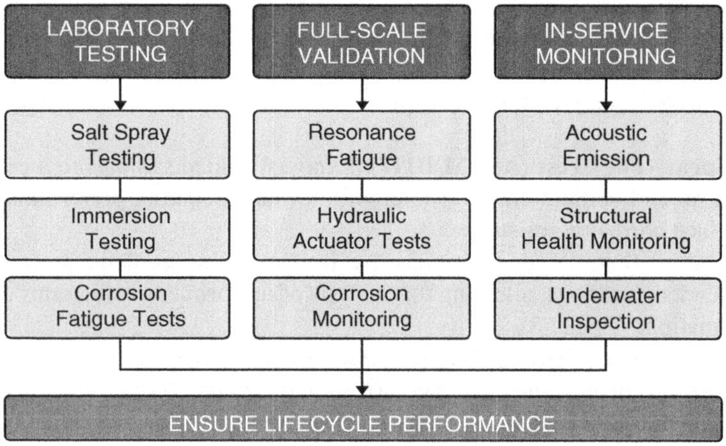

FIGURE 11.7 Testing and monitoring framework for corrosion and fatigue in marine welds.

ships and offshore platforms. Since marine welds are subjected to harsh environmental conditions, including saltwater immersion, cyclic loading, and fluctuating temperatures, comprehensive laboratory testing, full-scale evaluations, and continuous in-service monitoring are necessary to predict performance and detect early signs of degradation. These techniques help engineers validate design models, improve material selection, and establish maintenance strategies aligned with classification society standards and industry best practices. Figure 11.7 illustrates a comprehensive framework for evaluating and maintaining the integrity of marine welded structures through three integrated stages: laboratory testing, full-scale validation, and in-service monitoring.

The laboratory testing stage focuses on controlled experimental methods such as salt spray testing, immersion testing, and corrosion-fatigue tests to analyze material resistance under simulated seawater conditions. The full-scale validation stage includes resonance fatigue and hydraulic actuator tests that replicate real operational stresses, ensuring structural endurance before deployment. In the in-service monitoring phase, techniques like acoustic emission (AE) monitoring, structural health monitoring (SHM), and underwater inspection provide real-time feedback on corrosion progression and fatigue crack development.

Together, these interconnected processes form a closed-loop system that enables continuous assessment and feedback, guiding design optimization, maintenance planning, and operational decision-making. Ultimately, this framework ensures the life cycle performance, safety, and reliability of welded marine components in demanding offshore and shipboard environments.

11.6.1 Laboratory Corrosion Testing for Marine Welds

Laboratory corrosion testing plays a vital role in assessing the resistance of welded joints to various corrosion mechanisms such as pitting, crevice, galvanic, and SCC.

Controlled laboratory environments allow for the simulation of seawater exposure and electrochemical conditions that marine structures experience in service.

Common Laboratory Corrosion Tests:

Salt Spray (Fog) Test (ASTM B117): Exposes welded samples to a continuous mist of sodium chloride solution to evaluate coating performance and surface corrosion resistance.

Immersion Test: Welded coupons are immersed in synthetic seawater for extended durations, allowing for analysis of uniform corrosion rates and pit formation.

Electrochemical Testing (Potentiodynamic Polarization, EIS): Measures corrosion current density and polarization behavior to determine the **corrosion potential (E_corr)** and **pitting resistance** of WMs and HAZ.

Cyclic Corrosion Testing (ISO 14993): Combines wet–dry cycles to better replicate offshore and splash-zone environments, assessing coating degradation and weld surface attack.

Stress Corrosion Cracking (SCC) Tests (ASTM G36, G39): Evaluates crack initiation under simultaneous mechanical loading and corrosive conditions, especially for stainless steels and high-strength alloys.

Post-test characterization involves scanning electron microscopy (SEM) and energy-dispersive spectroscopy (EDS) to analyze microstructural corrosion morphology, crack propagation paths, and passive film breakdown. These laboratory insights form the basis for selecting coatings, claddings, and heat treatments that improve the long-term performance of marine welds.

11.6.2 Full-Scale Fatigue Testing of Welded Components

While laboratory tests provide valuable micro-level data, full-scale fatigue testing validates the performance of welded components under realistic loading conditions representative of ship and offshore service environments. Such testing ensures that design assumptions based on *S–N* curves **and** fracture mechanics hold true for large structural assemblies.

Key Approaches in Full-Scale Fatigue Testing:

Resonance Fatigue Testing: Utilizes high-frequency cyclic loading to simulate millions of stress cycles efficiently. Commonly applied to **ship hull panels, deck structures**, and **offshore joints.**

Hydraulic Actuator Testing: Replicates **wave-induced bending, torsion, and vibration** in large specimens, enabling fatigue crack growth observation at critical weld locations.

Corrosion-Fatigue Testing: Combines cyclic mechanical loading with **saltwater exposure** to reproduce the synergistic effects of corrosion and fatigue.

This method is essential for evaluating components operating in splash zones or submerged conditions.

Large-Scale Mock-Up Testing: Used in shipyards and offshore fabrication facilities to test **welded stiffeners, nodes, and risers** under representative loading and environmental conditions before certification.

Data from full-scale tests are analyzed to generate fatigue life curves, validate finite-element models (FEMs), and improve weld geometry and repair methods. These tests are integral to design validation, helping ensure compliance with standards like **DNV-RP-C203, ISO 19902,** and **American Welding Society (AWS) D1.1** for offshore and marine structures.

11.6.3 In-Service Monitoring Techniques for Offshore and Ship Structures

Even after deployment, welded marine structures must be continuously monitored to detect early signs of corrosion and fatigue. Modern in-service monitoring systems integrate sensors, NDT, and digital tools to track structural health in real time, enhancing safety and reducing maintenance costs.

Key Monitoring Techniques:

Corrosion Monitoring Sensors:

Linear Polarization Resistance (LPR) Probes: Measure corrosion rate directly in seawater environments.

Galvanic and Reference Electrodes: Track potential variations to evaluate CP performance.

Thickness Gauges (Ultrasonic): Detect uniform and localized metal loss in hull plating and subsea pipelines.

Fatigue Crack Detection and Growth Monitoring:

AE Sensors: Detect stress waves emitted by crack initiation and propagation.

Strain Gauges and Fiber Bragg Grating (FBG) Sensors: Record stress cycles and fatigue accumulation.

Phased Array Ultrasonic Testing (PAUT) and Time of Flight Diffraction (TOFD): Used periodically for detailed crack imaging and size estimation.

Structural Health Monitoring (SHM) Systems: Integrated SHM networks use **Internet of Things (IoT)** and **digital twin** technologies to provide continuous feedback on structural conditions. Real-time data analytics help predict remaining fatigue life and schedule maintenance before failure.

Underwater Robotic Inspection: Remotely operated vehicles (ROVs) equipped with high-resolution cameras, ultrasonic transducers, and corrosion probes enable inspection of **subsea welds and offshore risers** without dry docking or diver intervention.

By combining traditional NDT with modern digital monitoring, shipyards and offshore operators can transition from reactive to predictive maintenance, ensuring long-term safety, efficiency, and compliance with class society standards such as **ABS, DNV**, and **Lloyd's Register (LR).**

11.7 STANDARDS AND GUIDELINES FOR CORROSION AND FATIGUE DESIGN

The design and maintenance of welded marine structures are governed by a robust framework of international standards and classification society requirements that ensure structural integrity, safety, and durability under corrosion and fatigue loading. These guidelines establish the minimum design criteria, testing methods, and inspection requirements that must be met to obtain certification for ships, offshore platforms, and subsea installations.

By integrating corrosion resistance design principles and fatigue life assessment procedures, such standards provide engineers with uniform methods for material selection, welding procedure qualification, inspection scheduling, and maintenance planning. This section summarizes the key regulatory bodies and technical codes relevant to the corrosion and fatigue performance of marine welds.

11.7.1 Classification Society Requirements (ABS, DNV, LR)

Classification societies play a critical role in setting mandatory design and inspection standards for the marine and offshore industries. The most prominent organizations, namely, ABS, DNV, and LR, have developed comprehensive rules addressing corrosion protection, fatigue design, and weld integrity.

11.7.1.1 American Bureau of Shipping (ABS)

The ABS Rules for Building and Classing Steel Vessels and Offshore Installations provide specific requirements for welded joint design, corrosion allowance, and fatigue life prediction.

Fatigue Design Criteria: ABS uses ***S–N* curve–based fatigue assessment**, accounting for SCFs, residual stresses, and environmental effects.
Corrosion Protection: ABS mandates the use of protective coatings, CP systems, and CRAs in critical zones like ballast tanks and splash zones.
Inspection and Maintenance: Periodic NDT and coating surveys are required to maintain classification status.

11.7.1.2 Det Norske Veritas (DNV)

DNV standards, particularly DNV-RP-C203 (*Fatigue Design of Offshore Steel Structures*) and DNV-ST-F101 (*Submarine Pipeline Systems*), are widely used references for fatigue and corrosion assessment in offshore engineering.

Fatigue Classification System: DNV categorizes welds into **fatigue design classes (FAT)** based on geometry, weld quality, and surface condition.
Corrosion Allowance and Protection: Specifies quantitative corrosion allowances and combined protection systems (coating + CP).

Probabilistic Fatigue Design: Incorporates load spectra and stochastic sea-state modeling for accurate life prediction.

11.7.1.3 Lloyd's Register (LR)

LR Rules for the Manufacture, Testing, and Certification of Materials (Part 2) and Ship Rules Part 4 provide extensive coverage of weld quality and fatigue design.

Welding Procedure Approval (WPS): All procedures must be qualified under LR supervision, ensuring compliance with material and joint design standards.

Fatigue and Corrosion Guidelines: Emphasize **inspection intervals, weld repair methods**, and **structural redundancy** in critical load-carrying joints.

Corrosion Management: LR promotes **design-integrated corrosion protection**, combining coatings, CP, and material optimization for life cycle safety.

Together, ABS, DNV, and LR establish the foundation of global marine welding practices, harmonizing design methodologies for fatigue and corrosion control across shipyards and offshore installations.

11.7.2 ISO, AWS, and API Standards for Marine Weld Durability

Beyond classification societies, international standards from organizations such as ISO (International Organization for Standardization), AWS, and API (American Petroleum Institute) provide the technical framework for weld quality, testing, and performance under corrosive and cyclic conditions.

11.7.2.1 ISO Standards

ISO standards offer globally harmonized methods for assessing corrosion, fatigue, and material performance in marine applications:

ISO 19902: *Fixed Steel Offshore Structures* covers fatigue design, load modeling, and corrosion protection requirements.

ISO 15156: *Materials for H_2S-Containing Environments in Oil and Gas Production* specifies material resistance to SCC.

ISO 9227 / ISO 14993: Define laboratory corrosion test methods (salt spray, cyclic corrosion) for marine coatings and welds.

ISO 5817: Specifies weld quality levels for imperfections affecting fatigue life (Classes B, C, D).

11.7.2.2 AWS Standards

The AWS standards define welding quality and procedure qualification for marine and offshore fabrication:

AWS D1.1: *Structural Welding Code Steel* provides rules for design, workmanship, and inspection, including allowable discontinuities.

AWS D3.6M: *Underwater Welding Code* details wet and dry welding methods, electrode types, and corrosion mitigation for subsea repair.

AWS B4.0 and B5.1: Cover welder qualification and performance criteria to ensure consistency in weld integrity.

These standards ensure welds meet specific mechanical and corrosion performance criteria, essential for marine durability and certification.

11.7.2.3 API Standards

The **API** focuses on offshore structures and pipeline systems subjected to combined fatigue and corrosion loading:

API RP 2A-WSD: *Recommended Practice for Planning, Designing, and Constructing Fixed Offshore Platforms* includes fatigue life evaluation and corrosion design allowance.

API 1104: *Welding of Pipelines and Related Facilities* specifies welding procedures, inspection, and testing of joints exposed to seawater and cyclic stresses.

API RP 17N: *Subsea Production System Reliability and Integrity Management* integrates fatigue life monitoring with corrosion prevention strategies.

API's guidance is particularly influential in offshore oil and gas projects, where structural fatigue and corrosion fatigue directly affect safety and operational reliability.

11.8 FUTURE TRENDS IN CORROSION AND FATIGUE MITIGATION

The future of corrosion and fatigue mitigation in marine welded structures lies in the integration of advanced materials, intelligent monitoring systems, and digital technologies. As ships, offshore platforms, and subsea systems operate in increasingly demanding environments, traditional inspection-based maintenance is being replaced by data-driven, predictive, and adaptive approaches. Emerging solutions such as nanostructured coatings, artificial intelligence (AI)-based analytics, and digital twins are redefining how engineers predict degradation, extend service life, and ensure safety. This section highlights key innovations that will shape the next generation of corrosion- and fatigue-resistant marine structures.

11.8.1 Advanced Coatings and Nanomaterials for Weld Protection

Recent advancements in materials science have led to the development of next-generation protective coatings and nanomaterial-based surface treatments that provide superior resistance to corrosion, erosion, and fatigue damage in welded joints.

11.8.1.1 Nanostructured Coatings and Composite Films

Nanoceramic coatings (e.g., Al_2O_3, TiO_2, SiO_2) offer high hardness and thermal stability, effectively shielding welds from saltwater-induced corrosion and abrasion.

Graphene and carbon nanotube (CNT)-reinforced coatings improve barrier performance by reducing porosity and increasing electrical conductivity, minimizing galvanic corrosion.

Sol-gel hybrid coatings form dense, self-healing layers with enhanced adhesion to WM and HAZs.

11.8.1.2 Smart and Self-Healing Coatings

Coatings embedded with microcapsules containing corrosion inhibitors can automatically release protective agents when surface damage occurs.

Stimuli-responsive coatings, activated by pH or potential change, provide adaptive corrosion resistance under dynamic seawater conditions.

11.8.1.3 Additive Manufacturing and Surface Engineering

Advanced deposition techniques such as cold spraying, laser cladding, and high-velocity oxygen fuel (HVOF) enable localized reinforcement of weld areas prone to corrosion or fatigue. These coatings not only enhance corrosion resistance but also reduce residual stress and improve fatigue strength through refined microstructures.

In the near future, hybrid coating systems combining nanomaterials, ceramics, and polymers will dominate marine applications, providing multifunctional protection that integrates corrosion resistance, wear reduction, and self-monitoring capabilities.

11.8.2 AI-Based Predictive Maintenance for Fatigue and Corrosion

The adoption of AI and machine learning (ML) is transforming the maintenance paradigm for marine welded structures from reactive to predictive. By analyzing large volumes of data from sensors, inspection records, and environmental inputs, AI can identify early degradation patterns and predict remaining life with remarkable accuracy.

11.8.2.1 Data-Driven Condition Monitoring

AI models trained on strain gauge, ultrasonic, and AE data can detect microcrack initiation before visual signs appear.

Algorithms evaluate corrosion growth rates, pitting evolution, and weld stress behavior under variable sea states, providing early-warning indicators.

11.8.2.2 Predictive Maintenance and Decision Support

ML regression models (e.g., neural networks, random forests) predict fatigue life based on cyclic loading history and corrosion exposure.

AI-assisted inspection planning optimizes the frequency and location of NDT activities, reducing operational downtime.

Real-time anomaly detection algorithms alert operators to unexpected deterioration trends, allowing targeted intervention.

11.8.2.3 Integration with Digital Platforms

AI systems are increasingly embedded within digital twin architectures (discussed in Section 8.3), creating feedback loops where operational data continuously refine predictive models. This integration supports condition-based maintenance (CBM) strategies aligned with classification society guidelines such as DNV-RP-A204 (Digital Assurance).

The result is a paradigm shift from periodic, human-dependent inspections to autonomous, continuous, and intelligent monitoring, greatly enhancing structural reliability and reducing life cycle costs.

11.8.3 Integration of Digital Twins for Weld Life Prediction

The digital twin concept represents the next frontier in corrosion and fatigue management. A digital twin is a real-time virtual replica of a physical asset that mirrors its structural condition, operational environment, and degradation state through continuous data exchange.

11.8.3.1 Digital Representation and Real-Time Simulation

In marine welding applications, the digital twin incorporates geometric, material, and operational data from the computer-aided design (CAD) model, welding process simulations, and in-service sensors. It continuously updates based on live inputs such as:

- Strain, temperature, and vibration data.
- Corrosion potential and oxygen concentration levels.
- Environmental parameters (salinity, wave load, temperature cycles).

  Using FEM and ML algorithms, the digital twin simulates weld degradation over time, predicting crack initiation and corrosion rates under specific operational conditions.

11.8.3.2 Predictive and Prescriptive Capabilities

- Enables real-time fatigue life estimation and corrosion progression mapping.
- Identifies critical hotspots for inspection and maintenance prioritization.
- Supports life cycle management by predicting when weld sections will reach end-of-life criteria.

11.8.3.3 Integration with Smart Shipyards and Offshore Operations

Digital twins are being integrated into Industry 4.0 shipyards, connecting welding robots, inspection drones, and IoT sensors into a unified platform. This creates a closed-loop system where design, fabrication, and operational feedback continuously improve each other. In offshore platforms, digital twins interface with SHM systems to provide 24/7 weld performance visualization.

In the future, digital twins combined with AI and blockchain-based traceability will enable fully automated, self-learning corrosion and fatigue management systems, revolutionizing the life cycle maintenance of marine welded structures.

11.9 SUMMARY

The chapter provides a comprehensive understanding of the degradation mechanisms that affect the durability and safety of welded joints in ships, offshore platforms, and subsea structures. It explains how corrosion and cyclic loading interact to cause progressive material deterioration, emphasizing key types such as general, galvanic, pitting, crevice, and SCC. The chapter discusses the metallurgical, mechanical, and environmental factors influencing weld performance and outlines preventive strategies, including material selection, protective coatings, CP, optimized weld design, and post-weld treatments. It also examines fatigue mechanisms, stress concentration effects, and life assessment methods based on *S–N* curves and fracture mechanics, supported by real-world case studies. Further sections highlight testing and monitoring frameworks, relevant standards and classification society requirements, and emerging innovations such as nanocoating, AI-driven predictive maintenance, and digital twin integration. Collectively, the chapter underscores a shift from conventional inspection-based practices to smart, predictive, and sustainable corrosion-fatigue management, ensuring the long-term integrity, reliability, and efficiency of marine welded structures in harsh oceanic environments.

11.10 REVIEW QUESTIONS

1. Explain why corrosion and fatigue are considered the most critical degradation mechanisms in marine welded structures.
2. Describe how the marine environment influences the corrosion and fatigue performance of welded joints.
3. What are the key differences between corrosion fatigue and pure mechanical fatigue in marine applications?
4. Discuss the electrochemical reactions that drive corrosion in seawater-exposed welds.
5. How do residual stresses from welding contribute to fatigue crack initiation and corrosion susceptibility?
6. Differentiate between general corrosion, galvanic corrosion, and localized corrosion in welded joints.
7. What factors govern the rate of general corrosion in marine welds, and how can it be controlled?
8. Explain the mechanism of galvanic corrosion between dissimilar metals and provide examples in ship structures.
9. Describe how crevice and pitting corrosion develop in weld zones and why they are particularly dangerous.
10. What conditions promote stress corrosion cracking (SCC) in marine environments, and how can it be prevented?
11. Discuss the criteria for selecting corrosion-resistant materials for marine welded structures.
12. Compare the roles of protective coatings and cladding in corrosion prevention.
13. Explain the working principles of sacrificial anode and impressed current cathodic protection systems.
14. How can weld design modifications minimize corrosion risks in ship and offshore structures?

15. Describe the purpose of post-weld treatments such as passivation and shot peening in corrosion mitigation.
16. Outline the stages of fatigue failure in welded joints and explain where cracks typically initiate.
17. Identify the main factors influencing the fatigue life of welded marine structures.
18. What role does stress concentration at the weld toe and root play in fatigue failure?
19. Explain the difference between high-cycle and low-cycle fatigue in welded components.
20. How do surface roughness and weld geometry affect fatigue crack initiation and propagation?
21. Describe the *S–N* curve approach for fatigue life estimation and its relevance to marine welds.
22. Explain how the fracture mechanics approach is used for assessing existing cracks in welded joints.
23. Discuss the effects of welding defects such as undercuts, porosity, and lack of fusion on fatigue performance.
24. How can full-scale fatigue testing help validate fatigue design methods for offshore structures?
25. Summarize one real-world case study of fatigue failure in a marine or offshore welded structure.
26. What laboratory corrosion tests are commonly used to evaluate the performance of welded marine materials?
27. Explain the importance of full-scale fatigue testing in predicting service life under real conditions.
28. List and describe key in-service monitoring techniques used for corrosion and fatigue detection in ships and offshore platforms.
29. How does the integration of sensor networks and digital twins improve the accuracy of fatigue life prediction?
30. Discuss how emerging technologies such as nanomaterial coatings, AI-based predictive maintenance, and digital twin systems are transforming corrosion and fatigue management in marine welded structures.

12 Sustainability and Environmental Considerations in Marine Welding

12.1 INTRODUCTION

The evolution of modern shipbuilding and offshore engineering increasingly emphasizes sustainability, environmental responsibility, and energy efficiency as central design and fabrication objectives. Marine welding, being a core process in the construction and maintenance of vessels, offshore platforms, and subsea structures, plays a pivotal role in determining the environmental footprint of the maritime industry. Traditionally, welding processes have been associated with high energy consumption, greenhouse gas emissions, and material wastage, particularly through inefficient heat input, fume generation, and consumable use. However, the global transition toward green manufacturing and sustainable maritime operations is transforming welding practices through cleaner technologies, digital optimization, and life cycle-based design principles (Di Bella et al. 2022).

This chapter explores how sustainable welding technologies, eco-friendly materials, and energy-efficient practices can minimize environmental impact while maintaining structural integrity and safety. It examines the integration of low-emission welding processes, recyclable alloys, and renewable energy sources, alongside innovations such as artificial intelligence (AI)-driven process optimization and additive manufacturing, which reduce waste and improve precision. Furthermore, the chapter discusses regulatory frameworks and international standards such as those by the International Maritime Organization (IMO), ISO, and classification societies that guide the industry toward compliance with global sustainability goals.

By addressing both technical and ecological dimensions, this chapter underscores that sustainability in marine welding is not limited to environmental protection but also encompasses resource efficiency, operational safety, and life cycle performance optimization. Through the adoption of intelligent automation, digital monitoring, and circular economy principles, the marine sector is steadily moving toward a future where welding processes contribute to a cleaner, safer, and more sustainable maritime ecosystem.

12.1.1 Importance of Sustainability in the Marine Welding Sector

Sustainability in the marine welding sector has become a cornerstone of modern shipbuilding and offshore fabrication, driven by the global commitment to reducing

DOI: 10.1201/9781003217350-12

carbon emissions, resource depletion, and ecological degradation. Welding is one of the most energy-intensive processes in marine manufacturing, consuming significant quantities of electricity, shielding gases, filler metals, and consumables. As the maritime industry transitions toward green and low-carbon operations, sustainable welding practices are essential to achieving energy efficiency, cost optimization, and compliance with environmental standards.

Adopting sustainable approaches such as low-emission welding technologies, recyclable materials, renewable energy integration, and digital process control not only minimizes environmental impact but also improves operational performance and weld quality. Moreover, sustainability enhances occupational health and safety by reducing toxic fume emissions and waste generation. In a broader context, sustainable marine welding contributes to the United Nations Sustainable Development Goals (SDGs), particularly those related to responsible production, clean energy, and climate action. Thus, sustainability in marine welding is not merely an environmental obligation but a strategic imperative that ensures economic viability, global competitiveness, and environmental stewardship for the maritime sector.

12.1.2 Environmental Impacts of Conventional Shipyard Welding

Conventional shipyard welding techniques such as shielded metal arc welding (SMAW), gas metal arc welding (GMAW), and flux-cored arc welding (FCAW) have long been associated with substantial environmental burdens. These processes often generate high carbon emissions, metallic particulates, toxic fumes, and slag waste, which collectively affect air, water, and soil quality in shipyards. The extensive use of fossil-fuel-generated electricity and shielding gases such as carbon dioxide (CO_2) and argon contributes significantly to the carbon footprint of shipbuilding operations.

Moreover, improper management of welding residues, solvents, and consumable waste leads to contamination of nearby marine ecosystems. Energy inefficiencies arising from excessive heat input and material overuse further compound environmental degradation. Additionally, traditional shipyard operations often lack effective fume extraction systems and waste recycling facilities, resulting in health risks to workers and long-term ecological impacts.

In sum, conventional welding methods, though effective for fabrication, are inherently resource-intensive and pollutive, underscoring the urgent need for the marine industry to adopt energy-efficient technologies, eco-friendly materials, and closed-loop waste management systems. Sustainable welding innovations such as laser-arc hybrid systems, low-hydrogen consumables, and AI-based energy optimization are pivotal in mitigating these impacts and advancing toward greener shipyard operations.

12.1.3 Global Regulations and Maritime Environmental Goals (IMO, EU Green Deal)

The international maritime community, led by the IMO, has established a comprehensive framework for reducing environmental impact and promoting sustainability in all aspects of ship design, construction, and operation including welding and fabrication processes. The IMO Initial GHG Strategy (2018) aims to reduce total greenhouse gas emissions from international shipping by 50% by 2050 compared to 2008 levels, driving shipyards to adopt energy-efficient manufacturing and repair technologies. Additionally, the MARPOL (International Convention for the Prevention of Pollution from Ships) Convention (Annex VI) regulates emissions of sulfur oxides, nitrogen oxides, and particulate matter, indirectly influencing the choice of welding consumables and fume control measures.

The European Green Deal, launched by the European Union, further strengthens maritime environmental objectives by promoting carbon-neutral shipbuilding, clean energy integration, and circular economy principles in manufacturing. It encourages the use of renewable electricity in shipyards, low-emission fabrication technologies, and recycling of metallic waste. Other regional initiatives, such as ISO 14001 Environmental Management Systems and Det Norske Veritas' (DNV's) Green Shipyard Standards, provide operational guidelines for achieving measurable sustainability targets.

Collectively, these global and regional frameworks signify a paradigm shift in the maritime industry, where environmental performance is now a critical measure of technical excellence. Compliance with IMO and EU sustainability mandates compels shipyards to embrace innovative welding technologies, data-driven energy management, and sustainable design philosophies, aligning the marine sector with global climate goals and future-ready manufacturing practices.

12.2 REDUCING CARBON FOOTPRINT IN SHIPYARD WELDING

Reducing the carbon footprint of shipyard welding operations is a vital component of sustainable shipbuilding and offshore fabrication. Traditional welding methods consume large amounts of electrical energy, rely on fossil-fuel-generated power, and emit CO_2, nitrogen oxides (NO_x), and particulate matter. With the maritime sector aligning toward net-zero emission goals, there is a growing emphasis on adopting energy-efficient, low-emission, and resource-optimized welding technologies. This section outlines practical strategies for minimizing carbon output through process improvements, renewable energy integration, and comprehensive life cycle assessment (LCA). Figure 12.1 presents a comprehensive overview of the primary strategies employed to reduce the carbon footprint in shipyard welding operations. At the center of the diagram lies the goal "Sustainable Marine Welding Operations" surrounded by five interconnected approaches that collectively contribute to low-emission and energy-efficient shipbuilding practices.

The first segment, energy-efficient welding processes, highlights advanced methods such as GMAW, FCAW, and hybrid laser-arc welding (HLAW), which improve arc efficiency, reduce heat input, and minimize energy waste. The second, use of renewable energy, emphasizes integrating solar, wind, and hydrogen-based systems to power welding operations sustainably. The third area, low-emission consumables and shielding gases, promotes the use of eco-friendly electrodes, recycled filler metals, and inert gas blends to cut down CO_2 and particulate emissions. The fourth component, process optimization with AI and automation, demonstrates how digital monitoring, adaptive control, and smart sequencing reduce unnecessary energy use

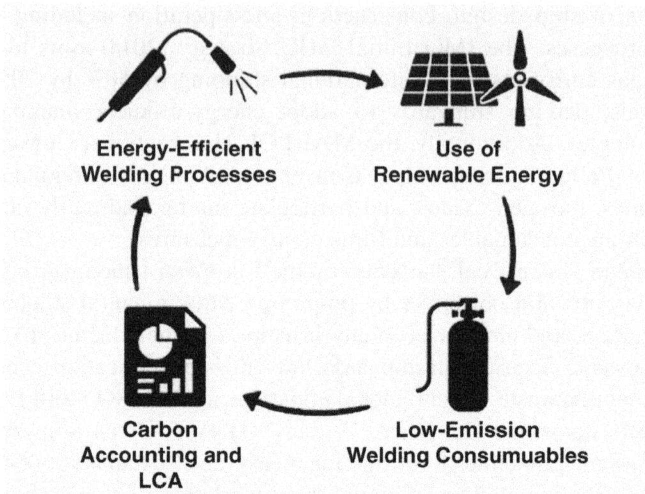

FIGURE 12.1 Strategies for reducing carbon footprint in shipyard welding.

while enhancing weld precision. Finally, carbon accounting and LCA provide the analytical framework to quantify and track total emissions across material sourcing, fabrication, and operation stages.

Together, these interlinked strategies form a closed-loop sustainability model, enabling shipyards to transition from conventional high-emission welding practices to environmentally responsible, data-driven, and energy-optimized fabrication processes aligned with global maritime decarbonization goals.

12.2.1 Energy-Efficient Welding Processes (GMAW, FCAW, HLAW)

Energy efficiency in welding directly influences both operational cost and environmental sustainability. Among conventional techniques, GMAW and FCAW have evolved with process controls that optimize arc stability, deposition rate, and heat input, thereby reducing energy wastage.

GMAW (MIG welding) achieves high metal transfer efficiency, minimizing spatter and post-weld finishing requirements. When automated, it provides up to 25%–30% energy savings compared to manual SMAW operations.

FCAW offers superior deposition rates and reduced rework in thick-section ship structures, lowering overall process time and energy demand.

HLAW, a combination of laser and arc energy sources, significantly enhances penetration depth and travel speed while using less heat input. This results in up to 50% lower energy consumption and a reduced carbon footprint per meter of weld.

By transitioning from traditional SMAW to modern GMAW, FCAW, and HLAW systems, shipyards can achieve substantial reductions in power consumption, filler waste, and greenhouse gas emissions while maintaining or improving weld quality.

12.2.2 Use of Renewable Energy in Welding Operations

Integrating renewable energy sources into welding operations represents a major step toward decarbonized shipbuilding. Renewable power such as solar photovoltaics (PV), wind energy, and hydrogen-based power systems can be harnessed to operate welding machines, automation systems, and auxiliary shipyard infrastructure.

Solar-powered welding units are increasingly feasible in coastal shipyards, where photovoltaic panels can supply sustainable electricity for daily operations.

Wind-assisted microgrids can provide stable, off-grid power for welding operations in remote offshore fabrication yards.

The use of green hydrogen as an energy carrier is an emerging solution, enabling the generation of electricity with zero-carbon emissions for both welding and preheating applications.

By partially or fully replacing grid electricity derived from fossil fuels, shipyards can drastically reduce Scope 2 emissions and enhance compliance with IMO and EU Green Deal energy targets. Renewable integration also improves energy resilience, reducing dependency on conventional power supply chains.

12.2.3 Low-Emission Welding Consumables and Shielding Gases

Consumables and shielding gases significantly influence the carbon intensity and pollutant output of welding operations. Transitioning to low-emission materials reduces both direct process emissions and indirect environmental impacts associated with consumable production.

Low-hydrogen electrodes and flux-cored wires minimize hydrogen generation and fume emissions, improving both weld integrity and air quality.

Recycled filler metals and eco-designed fluxes lower embodied carbon by reducing the need for virgin metal extraction.

Substituting CO_2-rich shielding gases with inert or mixed gases such as argon–helium blends or argon–hydrogen combinations can reduce process-related carbon emissions by up to 20%.

The adoption of biodegradable flux binders and water-based anti-spatter agents further reduces volatile organic compound (VOC) emissions and toxic waste.

These improvements, when combined with optimized gas flow control and recovery systems, ensure cleaner welding environments while maintaining metallurgical performance.

12.2.4 Process Optimization to Reduce Energy Consumption

Energy optimization in marine welding is achieved by integrating digital control, automation, and process analytics. Smart shipyards increasingly utilize real-time sensors, AI-driven algorithms, and weld parameter monitoring systems to minimize energy waste during fabrication.

Adaptive process control automatically adjusts voltage, current, and wire feed rate to maintain optimal arc efficiency.

Weld sequencing and robotic path optimization reduce idle time and minimize unnecessary heat input.

AI and machine learning models analyze energy profiles, identifying inefficiencies and recommending process improvements.

Thermal management systems, including controlled preheating and cooling, ensure consistent weld quality while conserving power.

By employing these digital optimization strategies, shipyards can achieve up to 30%–40% reductions in energy use and corresponding decreases in CO_2 emissions, aligning manufacturing operations with ISO 50001 Energy Management System standards.

12.2.5 Carbon Accounting and Life Cycle Assessment (LCA) for Marine Welding

A key element of sustainable manufacturing is the implementation of carbon accounting and LCA methodologies. These tools provide a quantitative basis for evaluating the total environmental impact of welding operations from raw material extraction to end-of-life recycling.

Carbon accounting tracks emissions across all stages of welding, including electricity usage, consumable production, transportation, and waste management.

LCA evaluates the environmental footprint of welding processes over their complete lifespan, identifying stages with the highest emission intensity.

Parameters such as embodied energy, CO_2 equivalent emissions, and recycling potential are measured to establish a sustainability baseline.

Software tools like SimaPro, GaBi, and OpenLCA are used in shipyards for data analysis, supporting ISO 14040 and ISO 14044 environmental management standards.

By applying LCA and carbon accounting, shipyards can make informed decisions on process selection, material sourcing, and design modifications, enabling continuous improvement toward low-carbon fabrication ecosystems.

12.3 RECYCLING AND ECO-FRIENDLY WELDING TECHNOLOGIES

The transition toward sustainable shipbuilding and offshore fabrication requires not only reducing emissions but also adopting circular economy principles, where materials, consumables, and energy are efficiently reused or recycled. Marine welding operations produce significant volumes of metal scrap, slag, spent fluxes, and consumable waste, all of which can be recovered and reintegrated into the production cycle. Concurrently, innovations in solid-state and hybrid welding processes, along with the development of eco-friendly consumables, are transforming how shipyards approach waste management, resource conservation, and environmental performance. Figure 12.2 illustrates the circular resource flow model in sustainable shipyard welding operations, highlighting the integration of recycling, renewable materials, and hybrid technologies to minimize waste and environmental impact. The flow begins with scrap metal recovery, where leftover materials from shipbuilding and repair are recycled and reintroduced into the fabrication cycle. Recovered flux and reusable consumables are processed through filtration and reconditioning systems to be reused in subsequent welding operations. The

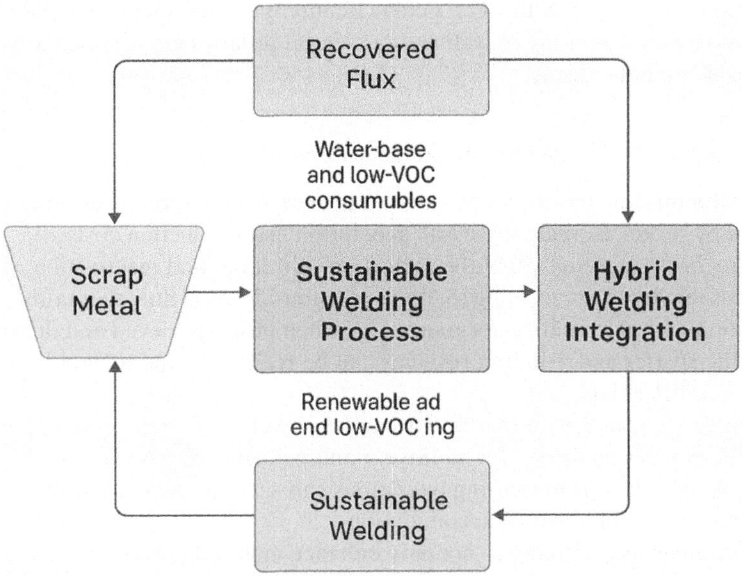

FIGURE 12.2 Eco-friendly welding technologies and circular resource flow in shipyards.

sustainable welding process incorporates water-based and low-VOC consumables, reducing toxic emissions while ensuring high weld integrity. These processes connect to hybrid welding integration, including advanced techniques like HLAW, friction stir welding (FSW), and wire arc additive manufacturing (WAAM), which enhance energy efficiency and material utilization. Finally, all outputs feed back into the system, establishing a closed-loop sustainable cycle where waste is minimized, resources are conserved, and environmental performance is optimized. This model embodies the principles of the circular economy, positioning modern shipyards toward greener, cleaner, and more efficient marine manufacturing.

12.3.1 Recycling of Scrap Metal from Shipbuilding and Repair

Shipbuilding and repair processes generate substantial metallic waste, including off-cuts, rejected welds, and decommissioned structural components. Recycling these materials not only minimizes landfill waste but also conserves natural resources and reduces the energy demand associated with primary metal production.

Recycled metals, especially steel and aluminum alloys, can be remelted and reprocessed into filler materials, hull plates, or structural components. Advanced sorting systems using magnetic separation, optical sorting, and spectroscopic analysis ensure the segregation of ferrous and non-ferrous materials to maintain metallurgical integrity.

Moreover, integrating on-site recycling facilities within shipyards allows for the immediate reuse of scrap metal, lowering transportation-related emissions. Each tonne of recycled steel saves approximately 1.5 tonnes of CO_2 and 60%–70% of energy compared to virgin production.

Such initiatives align with EU Circular Economy Action Plans and IMO sustainability frameworks, making recycling an essential pillar of environmentally responsible marine manufacturing.

12.3.2 Reuse of Weld Consumables and By-Products

Beyond structural materials, shipyards can also recover and reuse welding consumables and by-products, reducing waste generation and production costs.

Flux recovery systems allow the collection, filtration, and reactivation of unused flux from submerged arc welding (SAW) operations. Reused flux maintains chemical composition with minimal performance loss when properly sieved and dried.

Metallic spatter and grinding residues can be reclaimed and melted into secondary alloys or filler rods.

Electrode stubs and wire trimmings can be recycled to produce flux-cored wires or metallic powder feedstock for additive manufacturing (WAAM).

Slag and flux dust from welding operations can serve as cement additives or abrasive media, preventing landfill accumulation.

Such circular-use approaches not only enhance material efficiency but also reflect a broader movement toward zero-waste shipyard operations, where every by-product is rechanneled into a productive resource.

12.3.3 Water-Based and Low-VOC Welding Consumables

Conventional welding often involves solvents, coatings, and anti-spatter agents rich in VOCs, contributing to air pollution and worker health risks. The adoption of water-based and low-VOC consumables represents a major advancement in eco-friendly welding technology.

Water-based anti-spatter sprays and flux binders eliminate hazardous solvent emissions while maintaining weld quality.

Low-VOC primers and coatings applied before or after welding improve surface protection without releasing harmful fumes.

Bio-based fluxes and binders, derived from renewable sources such as lignin or starch, provide biodegradable alternatives to petroleum-based compounds.

By reducing airborne pollutants and hazardous waste, these consumables contribute significantly to ISO 14001 Environmental Management System compliance and worker safety enhancement. They also align with tightening regulatory limits under the European VOC Directive (2004/42/EC) and similar global emission controls.

Figure 12.3 illustrates the concept and mechanism of water-based and low-VOC welding consumables, which are central to sustainable and environmentally safe marine welding operations. The diagram depicts a welding setup where the welding wire is coated with a water-based flux rather than solvent-based formulations, thereby significantly reducing the emission of VOCs during welding. The shielding gas protects the molten metal from atmospheric contamination, while the arc provides the necessary heat for fusion with minimal fume generation.

The use of eco-friendly consumables such as water-based fluxes, low-VOC coatings, and biodegradable anti-spatter sprays helps minimize air pollution, improve worker safety, and ensure compliance with global environmental regulations such as

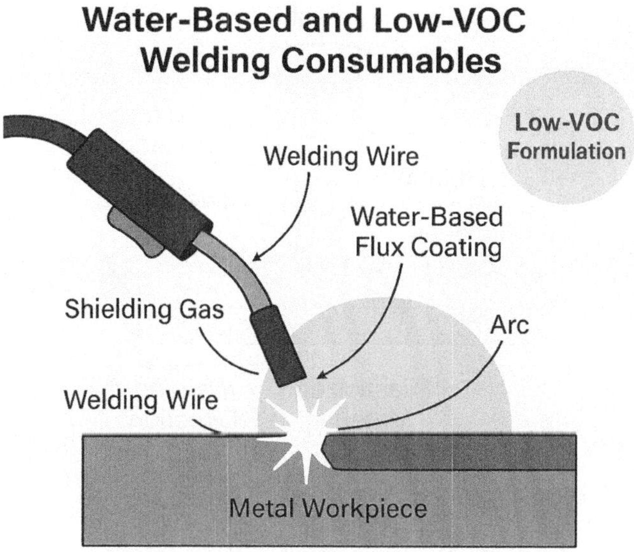

FIGURE 12.3 Water-based and low-VOC welding consumables.

ISO 14001 and the EU VOC Directive. Compared to conventional consumables, these alternatives produce cleaner weld zones, enhance air quality in shipyards, and contribute to the overall reduction of the carbon and pollution footprint in marine fabrication processes.

12.3.4 Solid-State Welding for Reduced Environmental Impact (FSW, WAAM)

Solid-state welding processes, which join materials without melting, are inherently more energy-efficient and environmentally sustainable than traditional fusion welding. Two technologies FSW and WAAM exemplify this shift toward green fabrication.

FSW operates below the melting point, using mechanical friction and plastic deformation to form high-integrity joints with minimal heat input. This eliminates filler material, shielding gas, and fume emissions, reducing total energy consumption by up to 60%.

WAAM, though based on arc welding, minimizes material waste by depositing near-net-shape structures layer by layer. This drastically reduces scrap generation and the need for heavy machining.

Both FSW and WAAM enable the fabrication of lightweight, high-strength components such as ship hull panels, offshore brackets, and structural frames, aligning with circular manufacturing principles and the low-carbon vision of the marine sector.

Figure 12.4 illustrates the principle and working mechanism of the FSW process, a solid-state joining technique widely adopted in marine and offshore fabrication for its superior energy efficiency and weld integrity. In this process, a rotating cylindrical tool equipped with a pin (probe) and shoulder is plunged into the interface of two adjoining workpieces. The friction between the tool and the base material generates heat, softening (but not melting) the metal to form a plasticized region.

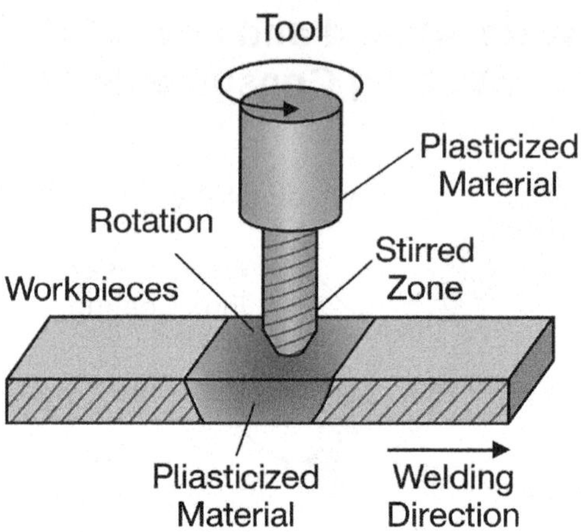

FIGURE 12.4 Friction stir welding (FSW).

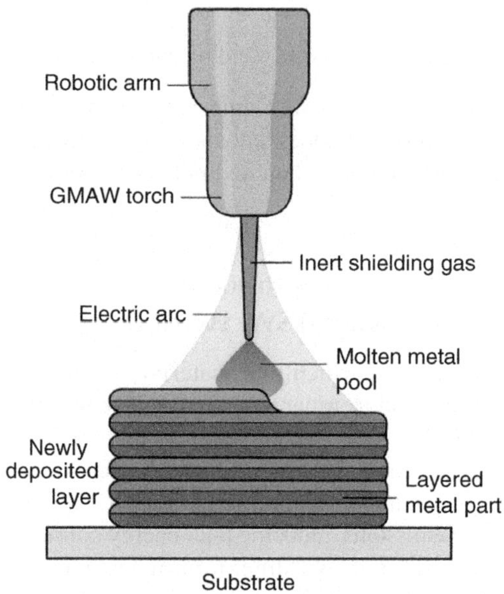

FIGURE 12.5 Wire arc additive manufacturing (WAAM).

As the tool traverses along the joint line, it stirs and forges the plasticized material, creating a high-quality joint known as the stirred zone or weld nugget. The surrounding area, called the heat-affected zone (HAZ), experiences minimal thermal distortion compared to conventional fusion welding. The rotation direction and welding travel direction are shown to indicate material flow and joint formation.

This technique eliminates the need for filler materials, shielding gases, or flux, producing defect-free, low-emission joints with excellent mechanical properties.

FSW is especially advantageous for aluminum and marine alloys, offering a sustainable, low-energy alternative ideal for ship panels, hulls, and structural applications.

Figure 12.5 illustrates the WAAM process, an advanced additive technique that uses a GMAW torch to deposit metal layer by layer from a continuous wire feedstock. The diagram shows the main components of the robotic arm, GMAW torch, electric arc, molten metal pool, and layered deposit formed on the substrate. The inert shielding gas protects the molten metal from oxidation and contamination, ensuring high-quality layer deposition.

During the WAAM process, the electric arc melts the wire, which is precisely fed into the weld pool. As the robotic torch moves along the substrate, molten metal solidifies to form each layer, progressively building the component. This layer-by-layer fabrication method significantly reduces material waste and energy consumption compared to traditional subtractive manufacturing. WAAM is especially valuable in marine and offshore engineering, enabling the production and repair of large metallic structures such as propeller blades, brackets, and hull components with high material efficiency and structural integrity.

12.3.5 Hybrid and Automated Welding for Material Efficiency

The implementation of hybrid and automated welding technologies further enhances material utilization and process sustainability. By combining precision control with energy efficiency, hybrid systems ensure optimal deposition rates and minimal waste generation.

HLAW merges laser and arc sources to deliver deeper penetration, narrower HAZ, and less filler usage.

Automated robotic welding reduces human error, rework, and consumable waste through consistent process parameters and precise arc placement.

AI-assisted weld path optimization ensures uniform heat input and controlled filler distribution, reducing over-welding and excess material usage.

These innovations contribute to the lean manufacturing paradigm of modern shipyards, promoting productivity and sustainability simultaneously. By integrating automation and hybridization, shipyards can achieve material savings up to 20%–25%, reduce energy costs, and lower overall carbon intensity.

Figure 12.6 illustrates the HLAW process, which combines a focused laser beam and a GMAW torch to achieve superior weld penetration, speed, and quality. In this process, the laser beam provides deep, narrow keyhole penetration, while the electric arc enhances filler metal deposition and stabilizes the molten pool. The shielding gas protects the weld area from oxidation, ensuring high metallurgical purity.

The laser and arc heat sources operate simultaneously, creating a synergistic thermal effect that results in faster welding speeds, reduced distortion, and improved mechanical properties compared to conventional arc or laser welding alone. The molten pool solidifies to form a high-integrity joint with minimal porosity and superior fusion. This hybrid technique is particularly suited for ship hull panels, offshore structures, and thick-section marine components, offering significant energy efficiency and process stability in harsh marine fabrication environments.

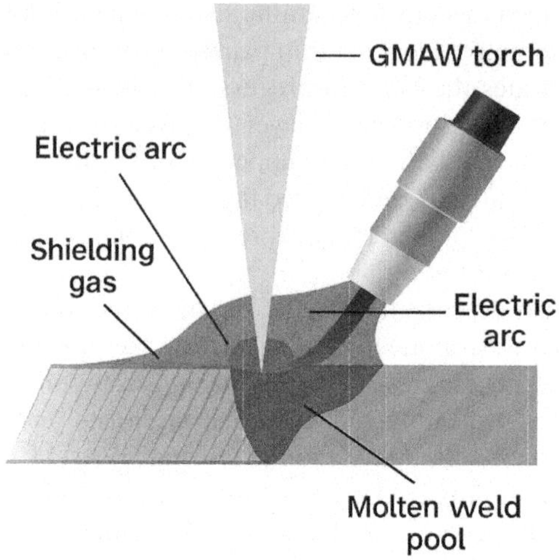

FIGURE 12.6 Hybrid laser-arc welding (HLAW).

12.4 WASTE MANAGEMENT IN MARINE WELDING OPERATIONS

Effective waste management is a critical pillar of sustainability in marine welding operations. Shipyards and offshore fabrication facilities generate a wide variety of solid, liquid, and gaseous wastes, including slag, metallic dust, spent flux, welding fumes, solvents, and scrap materials. If not properly managed, these by-products can pose serious risks to human health, marine ecosystems, and regulatory compliance. The goal of waste management in this context is not only safe disposal but also minimization, recovery, and recycling of welding residues in accordance with international environmental standards. This section outlines the major strategies for handling, treatment, and regulatory compliance in marine welding waste management. Figure 12.7 illustrates the systematic framework for waste management in marine welding operations, encompassing three major stages: waste generation, treatment and recycling, and regulatory compliance. The process begins with marine welding operations, where by-products such as slag, fumes, and metal dust are generated. These materials are then collected and processed through respective treatment pathways.

Slag reuse involves reprocessing and repurposing slag as aggregates in cement or surface preparation media, feeding into broader recycling programs. Fume extraction systems equipped with filters and scrubbers capture airborne contaminants, contributing to hazardous waste minimization by preventing toxic emissions. Dust recycling mechanisms, including wet collectors and magnetic separators, recover metallic particulates that can be reintegrated into manufacturing processes.

The final stage ensures all residual waste streams are managed according to MARPOL compliance and ISO 14001 Environmental Management Standards, guaranteeing safe disposal and traceability. Collectively, this framework forms a closed-loop waste management system, promoting environmental stewardship, workplace safety, and sustainable shipyard operations.

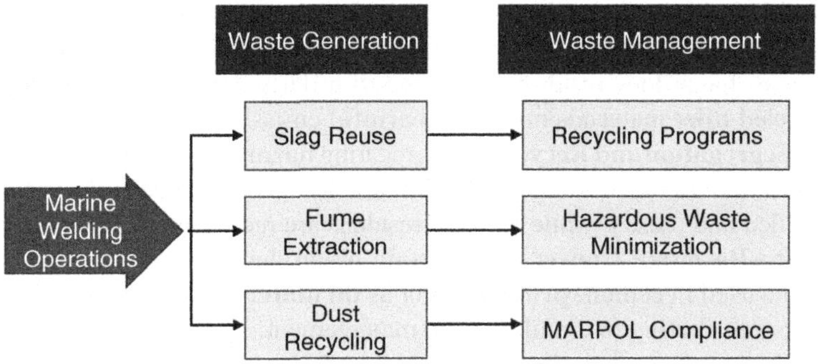

FIGURE 12.7 Waste management framework in marine welding operations.

12.4.1 Handling and Disposal of Slag, Fumes, and Metal Dust

Welding by-products such as slag, fumes, and metal dust represent the primary waste streams in shipyard operations. Proper handling and disposal of these materials are vital for maintaining workplace safety and preventing environmental contamination.

Slag Management: Slag generated from **SMAW**, **FCAW**, and **SAW** contains oxides, silicates, and flux residues. Instead of discarding, it can be **collected, crushed, and reused** as aggregate in **cement or road construction** or as abrasive material for surface cleaning.

Fume Extraction and Filtration: Welding fumes, composed of metal oxides (Fe, Mn, Ni, Cr) and gases such as ozone and nitrogen oxides, are hazardous to both health and the environment. **Local exhaust ventilation (LEV)** systems, **fume extraction torches**, and **HEPA filters** are essential to capture and neutralize airborne particulates before release.

Metal Dust Control: Fine metallic dust from grinding and cutting must be contained through **wet collection systems** or **magnetic dust recovery units**. Collected dust can often be reprocessed into sintered products or recycled as alloying material.

Implementing these measures ensures **compliance with occupational safety norms (e.g., OSHA 1910.252)** and minimizes the release of pollutants into the marine environment. Additionally, **real-time air quality monitoring systems** can help track emissions and verify the effectiveness of filtration technologies.

12.4.2 Hazardous Waste Minimization Techniques

Hazardous waste minimization focuses on reducing the quantity and toxicity of waste produced during welding and associated processes. This is achieved through process optimization, material substitution, and advanced waste treatment technologies.

Process Optimization: Utilizing **automated welding systems** and **adaptive control algorithms** ensures precise material deposition, reducing overwelding and consumable waste. Advanced processes like **FSW** and **HLAW** eliminate the need for flux and produce negligible fume emissions.

Material Substitution: Replacing **lead-based, chromate-containing**, or **chlorinated fluxes** with **eco-friendly, water-based alternatives** greatly reduces hazardous residue generation. Similarly, low-fume electrodes and recycled filler materials minimize harmful emissions.

Waste Segregation and Recycling: Segregating **hazardous and non-hazardous waste streams** allows targeted treatment. For instance, **spent solvents** can be distilled and reused, while metallic residues are recovered for smelting.

Waste-to-Resource Conversion: Certain by-products, such as **welding slag**, can be used in **cement production** or as **fill material** in shipyard infrastructure, contributing to circular waste management.

Through a combination of technological innovation and procedural control, shipyards can achieve significant reductions in both hazardous waste output and overall environmental footprint.

12.4.3 Compliance with Marine Pollution Prevention Standards (MARPOL)

Compliance with international environmental regulations, particularly the MARPOL Convention, is essential for sustainable welding operations in shipyards and offshore fabrication sites. Annex V of MARPOL specifically addresses the control of garbage and solid waste disposal from ships and shipyards, prohibiting the discharge of hazardous materials, metal residues, and industrial waste into the marine environment.

Regulatory Alignment: Welding waste such as slag, metal dust, and spent abrasives must be **collected, documented, and disposed of through licensed waste management contractors**. Shipyards are required to maintain **Waste Reception Facilities (WRFs)** that comply with MARPOL and **ISO 14001 Environmental Management** standards.

Emission and Water Pollution Control: Fume treatment units and stormwater containment systems prevent airborne and liquid contaminants from entering surrounding waters. Compliance with **Annex VI (Air Pollution Control)** ensures the reduction of volatile emissions and particulate discharge.

Documentation and Auditing: Regular environmental audits, **Waste Management Plans (WMPs)**, and **Material Safety Data Sheets (MSDS)** are essential for verification and certification under MARPOL and regional authorities like the **European Maritime Safety Agency (EMSA)**.

Meeting MARPOL and related standards not only safeguards marine ecosystems but also enhances a shipyard's **environmental reputation** and **eligibility for green certification programs** under **DNV, ABS, or Lloyd's Register (LR).**

Meeting MARPOL and related standards not only safeguards marine ecosystems but also enhances a shipyard's **environmental reputation** and **eligibility for green certification programs** under **DNV, ABS, or LR**.

12.5 HEALTH AND SAFETY IN SUSTAINABLE WELDING PRACTICES

Health and safety are inseparable from sustainability in marine welding. Sustainable shipyard operations not only aim to reduce environmental impact but also strive to protect workers from hazardous emissions, heat exposure, and chemical risks. Traditional welding processes emit fumes, ultraviolet radiation, and metal particulates that can adversely affect human health, particularly in confined or poorly ventilated marine environments. By implementing green welding technologies, eco-friendly personal protective equipment (PPE), and comprehensive worker training, shipyards can achieve safer, cleaner, and more sustainable fabrication conditions while maintaining productivity and weld quality.

12.5.1 Reducing Weld Fume Exposure with Green Technologies

Reducing exposure to welding fumes is a major sustainability and occupational safety priority. Traditional welding generates metal oxide fumes, carbon monoxide, and VOCs, which pose respiratory and neurological risks. Green technologies minimize these hazards through process innovation, automation, and cleaner consumables.

Low-Emission Welding Processes: Techniques such as FSW, cold metal transfer (CMT), and HLAW drastically reduce fume formation by operating at lower heat inputs or in solid-state conditions.

FSW, in particular, produces nearly zero fumes since no melting or filler material is involved.

CMT reduces spatter and oxidation, decreasing airborne particulate matter by up to 80%.

Air Filtration and Extraction Systems: Advanced LEV systems and fume extraction torches capture contaminants at the source. Combined with HEPA filtration units and activated carbon filters, these systems ensure that shipyards maintain air quality compliance under OSHA and ISO 45001 standards.

Eco-Consumables:

Use of low-hydrogen electrodes, water-based fluxes, and solvent-free coatings further reduces VOCs and metallic fume emissions, ensuring cleaner breathing environments for welders.

Through these combined approaches, marine industries can lower fume exposure while reducing their carbon footprint, improving both human and environmental health.

12.5.2 Eco-Friendly Personal Protective Equipment (PPE)

Sustainable welding extends beyond process improvements to include eco-conscious PPE design and usage. Conventional protective gear often relies on synthetic materials and disposable components that contribute to waste and pollution. Eco-friendly PPE focuses on renewable materials, reusability, and energy-efficient designs while maintaining compliance with international safety standards.

Sustainable Materials: Flame-resistant garments made from organic cotton, recycled aramid fibers, and bamboo composites offer biodegradability and reduced manufacturing emissions.

Energy-Efficient Helmets and Respirators: Solar-powered auto-darkening welding helmets and rechargeable powered air-purifying respirators (PAPRs) minimize battery waste and energy consumption.

Recyclable Gloves and Footwear: Gloves fabricated from recycled nitrile and natural latex balance comfort with environmental performance. Similarly, eco-certified boots made from renewable rubber compounds reduce synthetic polymer use.

Eco-friendly PPE initiatives align with **ISO 14001 Environmental Management** and **EN 11611/11612 PPE Standards**, promoting both personal safety and sustainable production practices in shipyards.

12.5.3 Worker Training for Sustainable Welding Practices

Technology alone cannot ensure sustainability; human expertise and awareness are equally critical. Effective worker training programs focus on improving environmental consciousness, safety awareness, and operational efficiency among welders, supervisors, and inspection personnel.

Environmental and Safety Awareness: Welders are trained to understand the environmental impact of consumables, safe handling of hazardous materials, and proper waste disposal according to **MARPOL Annex V** and local shipyard regulations.

Energy and Resource Optimization: Training emphasizes parameter optimization (voltage, current, gas flow) to reduce power consumption, material waste, and over-welding.

Digital and Automation Skills: With increasing integration of robotic welding systems and AI-based monitoring, workers learn to interpret energy data, manage predictive maintenance, and operate automated systems efficiently.

Emergency Preparedness and Health Monitoring: Programs include guidelines for managing fume exposure, heat stress, and confined-space welding, ensuring quick response to potential hazards.

By embedding sustainability principles into training, shipyards cultivate a culture of environmental responsibility and occupational safety, ensuring every worker becomes an active participant in achieving green manufacturing goals.

12.6 SUMMARY

The chapter highlights the essential transition of the shipbuilding and offshore fabrication industries toward cleaner, safer, and resource-efficient practices. It emphasizes how sustainable welding integrates energy-efficient processes, renewable power sources, and low-emission consumables to significantly reduce carbon footprints and environmental pollution. The chapter discusses recycling of scrap metals, reuse of flux and by-products, and adoption of eco-friendly, water-based, and low-VOC materials that align with circular economy principles. Advanced technologies such as FSW, HLAW, and WAAM are presented as key enablers of energy and material efficiency in marine fabrication. Furthermore, the chapter addresses waste management, worker safety, and green training initiatives to ensure sustainable shipyard operations in compliance with IMO, MARPOL, and EU Green Deal regulations. Collectively, it establishes that sustainability in marine welding is not only an environmental obligation but also a strategic framework for achieving long-term efficiency, occupational safety, and global regulatory compliance, paving the way for a cleaner and smarter maritime industry.

12.7 REVIEW QUESTIONS

1. What is meant by sustainability in the context of marine welding, and why is it important for modern shipbuilding?
2. Describe how conventional welding processes impact the environment in shipyards.
3. What are the main sources of carbon emissions during shipyard welding operations?
4. Explain how energy-efficient welding processes such as GMAW, FCAW, and HLAW contribute to emission reduction.
5. How can renewable energy be integrated into welding operations in shipyards?
6. Discuss the environmental benefits of using low-emission consumables and shielding gases.
7. What is the significance of life cycle assessment (LCA) in evaluating the sustainability of marine welding?
8. Describe the circular resource flow in sustainable shipyard operations.
9. How does metal scrap recycling reduce environmental impact and energy consumption in marine fabrication?
10. Explain the importance of reusing welding consumables and by-products in sustainable shipbuilding.
11. What are water-based and low-VOC consumables, and how do they improve air quality in shipyards?
12. Compare the environmental performance of solid-state welding (FSW) with traditional arc welding.
13. How does wire arc additive manufacturing (WAAM) support material efficiency in marine applications?
14. Discuss the role of hybrid welding technologies in improving sustainability and weld quality.
15. Outline key steps involved in the waste management framework for marine welding operations.
16. What are the primary types of waste generated during shipyard welding, and how can they be treated sustainably?
17. Explain the concept of hazardous waste minimization in welding operations.
18. How do MARPOL regulations influence waste disposal and pollution control in marine welding?
19. Identify methods used to capture and treat weld fumes and metallic dust in shipyards.
20. What are the advantages of implementing eco-friendly personal protective equipment (PPE) in marine welding?
21. How do sustainable PPE materials differ from conventional protective gear?
22. Describe how worker training and awareness programs contribute to sustainable shipyard practices.

23. Explain how digitalization and automation can improve energy efficiency and process control in welding.
24. What are the key global frameworks guiding environmental compliance in marine welding (e.g., IMO, EU Green Deal, ISO 14001)?
25. How can AI-based monitoring systems help in reducing the carbon footprint of welding operations?
26. Discuss the role of fume extraction and air filtration systems in ensuring occupational health and sustainability.
27. Why is it important to integrate safety and environmental management in sustainable shipyard operations?
28. What are the major challenges faced by shipyards in adopting sustainable welding technologies?
29. How does sustainable welding support the long-term economic and operational goals of the maritime industry?
30. Summarize how sustainability in marine welding contributes to the global decarburization goals of the maritime sector.

13 Future Trends in Marine Welding Engineering

13.1 INTRODUCTION

The field of marine welding engineering is entering a transformative era, driven by rapid advancements in automation, digital technologies, artificial intelligence (AI), and sustainable manufacturing practices. As the global maritime industry evolves toward cleaner, smarter, and more efficient operations, welding, the backbone of shipbuilding and offshore fabrication is being reshaped by innovation. Traditional manual and semi-automatic welding techniques are giving way to robotic, hybrid, and intelligent systems that offer unprecedented levels of precision, quality, and productivity.

Future marine welding will be characterized by the integration of digital technologies, including AI, machine learning (ML), digital twin systems, and Industry 4.0 frameworks, which will enable real-time process monitoring, defect prediction, and adaptive control. These technologies will significantly enhance weld integrity, reduce rework, and ensure data-driven decision-making across the life cycle of ship construction and maintenance. Simultaneously, sustainability imperatives, including carbon reduction, energy efficiency, and green shipyard practices, are accelerating the development of low-emission welding processes, recyclable materials, and eco-friendly consumables (Kah and Martikainen 2012).

Emerging trends such as additive manufacturing (wire arc additive manufacturing [WAAM]), hybrid laser-arc welding (HLAW), and autonomous underwater welding systems are redefining the capabilities of marine fabrication. Furthermore, AI-assisted welding robots, sensor-based quality assurance, and augmented reality (AR) training systems are making shipyards more intelligent, efficient, and safe.

This chapter explores these evolving trends, emphasizing how the fusion of advanced manufacturing, digital intelligence, and environmental responsibility is shaping the future of marine welding. It underscores that the next generation of welding engineering will focus not only on structural performance but also on sustainability, automation, and global competitiveness, heralding the arrival of the "Smart Shipyard" era in maritime fabrication.

13.1.1 Evolution of Welding Practices in the Marine Industry

The evolution of welding practices in the marine industry reflects the technological progression of shipbuilding itself from manual craftsmanship to fully automated and intelligent fabrication systems. Early 20th-century shipbuilding relied heavily on riveting and mechanical fastening, which, while reliable, were labor-intensive and

DOI: 10.1201/9781003217350-13

structurally limited. The introduction of electric arc welding during the mid-1900s revolutionized the field, enabling stronger, lighter, and more watertight joints that replaced riveted structures in both naval and commercial vessels.

By the late 20th century, submerged arc welding (SAW), gas metal arc welding (GMAW), and flux-cored arc welding (FCAW) became standard in large-scale hull fabrication, dramatically improving productivity and quality. The advent of automation and robotics in the 1990s further enhanced efficiency and consistency, leading to the development of computer-controlled welding systems for ship panels, pipelines, and offshore platforms.

Today, welding in the marine sector is moving into the digital and intelligent manufacturing era, where processes are driven by AI-based control, real-time sensing, and digital twin simulation. Emerging techniques such as HLAW and WAAM are reshaping fabrication paradigms, offering high deposition rates, superior weld quality, and material savings. This ongoing evolution marks a transition from manual craftsmanship to data-driven, sustainable, and automated marine welding aligned with Industry 4.0 principles (Anand and Khajuria 2015).

Figure 13.1 illustrates the historical evolution of welding technologies used in the marine industry, tracing the journey from manual riveting in early shipbuilding to AI-driven automated welding systems in the modern era. The timeline highlights major milestones beginning with riveting in the early 1900s, which provided structural reliability but demanded heavy manual labor. This was followed by the introduction of arc welding in the mid-20th century, enabling faster fabrication and stronger joints. The late 20th century saw the rise of robotic and digital welding, integrating programmable control and precision in shipyards. In the 21st century, AI-assisted and automated welding systems have transformed marine fabrication through intelligent control, real-time monitoring, and predictive analytics. This evolution signifies the marine industry's continuous advancement toward efficiency, quality, and sustainability in shipbuilding and offshore engineering.

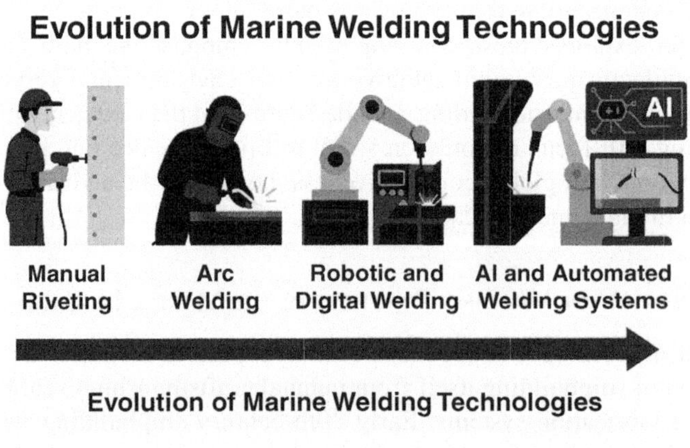

FIGURE 13.1 Evolution of marine welding technologies.

13.1.2 Drivers for Innovation in Marine Welding

Innovation in marine welding is being propelled by a combination of technical, environmental, and economic drivers aimed at improving efficiency, sustainability, and safety in shipyard and offshore operations. The increasing size and complexity of modern vessels, including liquefied natural gas (LNG) carriers, offshore platforms, and autonomous ships, demand precise and high-performance welding solutions capable of withstanding extreme marine conditions.

A key driver is automation, which addresses labor shortages, enhances productivity, and ensures consistent weld quality. The global push for decarbonization and green shipbuilding, guided by regulations such as the International Maritime Organization's (IMO's) greenhouse gas (GHG) reduction strategy and the EU Green Deal, encourages the adoption of low-emission welding processes, energy-efficient technologies, and eco-friendly consumables.

Additionally, digitalization and predictive analytics are enabling smarter decision-making through data collected from sensors, cameras, and thermal scanners. AI algorithms can now predict defects, adjust welding parameters in real time, and optimize energy use, minimizing rework and material waste. Cost competitiveness and the need for global standardization also motivate shipyards to integrate smart manufacturing systems, ensuring compliance with quality and environmental standards while maintaining operational efficiency.

Ultimately, these drivers are converging toward the creation of smart, autonomous, and sustainable shipyards, where welding operations are digitally coordinated, adaptive, and environmentally responsible.

13.1.3 Role of Emerging Technologies in Shipbuilding and Offshore Structures

Emerging technologies are redefining the fabrication, inspection, and life cycle management of ships and offshore structures. AI, ML, and digital twin systems are central to this transformation, enabling predictive control, real-time monitoring, and virtual simulation of welding processes. These technologies create a closed feedback loop where data from sensors and cameras are analyzed to continuously improve process performance and weld quality.

Additive manufacturing techniques, particularly WAAM, allow the layer-by-layer construction or repair of large metal components directly in shipyards, reducing material waste and production time. Hybrid welding systems, such as HLAW, combine the precision of lasers with the robustness of arc welding to produce high-strength, low-distortion joints in thick marine plates.

Meanwhile, robotic welding systems and collaborative robots (cobots) are revolutionizing productivity by automating complex welds in confined or hazardous environments, including underwater and offshore structures. Integration with AR and virtual reality (VR) is enhancing training and inspection efficiency, while AI-based defect recognition ensures faster, more accurate quality assessment.

In offshore engineering, these emerging technologies enable the fabrication of corrosion-resistant, fatigue-optimized, and high-performance structures, critical for energy platforms, pipelines, and subsea installations. Together, they signify a paradigm shift from traditional fabrication to intelligent, connected, and sustainable shipbuilding ecosystems, setting the stage for the next generation of marine welding innovation.

13.2 DIGITAL TWIN TECHNOLOGY IN MARINE WELDING

The advent of digital twin technology has revolutionized how shipyards and offshore industries design, monitor, and maintain welded structures. In marine welding, a digital twin serves as a dynamic, virtual replica of physical welding systems, capturing real-time data from sensors, robotics, and monitoring devices to mirror actual operations. This technology enables engineers to simulate, analyze, and optimize welding processes before implementation, enhancing accuracy, efficiency, and predictive decision-making. The digital twin thus bridges the gap between physical shipyard operations and digital engineering ecosystems, ushering in a new era of smart manufacturing in the maritime sector.

13.2.1 Concept and Principles of Digital Twin Technology

A **digital twin** is a real-time, data-driven virtual model of a physical asset, process, or system that continuously updates based on operational feedback. In marine welding, it integrates 3D modeling, process sensors, and ML algorithms to replicate the behavior of welding equipment, materials, and environmental conditions.

The core principles of digital twin technology include:

Real-Time Data Synchronization: Continuous input from sensors, cameras, and Internet of Things (IoT) devices ensures the digital model accurately mirrors the actual welding process.

Process Simulation and Prediction: Using computational models, the twin simulates heat distribution, weld penetration, distortion, and residual stress.

Closed-Loop Control: The system provides real-time feedback to adjust parameters such as current, voltage, and travel speed for optimal performance.

By integrating the physical (real-world welding) and virtual (digital simulation) domains, the digital twin enables predictive insights that improve weld quality, reduce waste, and shorten development cycles—a critical step toward Industry 4.0 and Shipyard 5.0 transformation.

Figure 13.2 illustrates the digital twin framework for marine welding, representing the integration of physical operations and virtual simulation for real-time monitoring and control. The framework is divided into four core components: the physical system, sensors and data flow, simulation model, and predictive control.

The physical system includes robotic or automated welding setups within the shipyard, where sensors such as thermal cameras, temperature probes, and current sensors continuously collect process data. This information is transmitted through data flow networks into the digital simulation model, which replicates the welding process virtually using computational algorithms. The simulation analyzes thermal

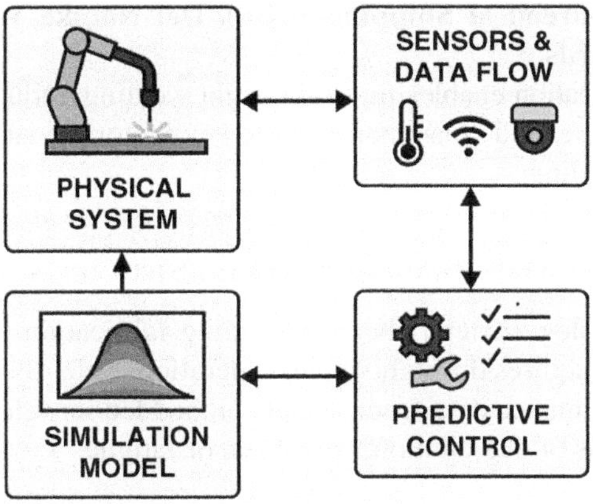

FIGURE 13.2 Digital twin framework for marine welding.

gradients, material behavior, and structural deformation to identify potential deviations or defects.

The predictive control block applies AI-based analytics to optimize process parameters, anticipate faults, and ensure weld quality. Feedback from this module is continuously sent back to the physical system, enabling self-correcting and adaptive welding operations.

Together, these interconnected components form a closed-loop system that ensures accuracy, efficiency, and proactive maintenance, marking a major step toward smart shipyard manufacturing and Industry 4.0 integration in marine welding.

13.2.2 Integration with Welding Process Simulation

One of the most powerful applications of digital twin technology is its integration with welding process simulation. In traditional shipbuilding, predicting weld quality and distortion relies heavily on trial-and-error or empirical methods. The digital twin replaces this with virtual experimentation, where process variables can be simulated, adjusted, and optimized in real time before physical welding begins.

Thermal-Mechanical Simulation: The digital twin uses computational models to visualize heat flow, phase transformation, and mechanical deformation in welded joints.

Parameter Optimization: AI algorithms fine-tune parameters such as arc energy, travel speed, and wire feed rate to achieve desired penetration and minimal distortion.

Defect Prediction: ML models embedded within the digital twin identify potential defects like porosity, cracks, or lack of fusion based on live data trends.

Virtual Qualification: Before executing critical welds (e.g., in pressure hulls or offshore risers), simulations validate the process for compliance with

American Bureau of Shipping (ABS), Det Norske Veritas (DNV), or AWS standards.

This integration enables first-time-right welding, reducing rework costs, inspection time, and consumable waste key factors in sustainable and efficient shipbuilding.

13.2.3 Predictive Maintenance of Welded Structures

Digital twin technology extends beyond welding fabrication to predictive maintenance of marine structures throughout their operational life. By continuously analyzing real-time performance data from sensors embedded in welded joints, the system identifies early signs of degradation, corrosion, or fatigue.

Condition Monitoring: Sensors measure temperature, strain, and vibration in critical weld zones. The twin compares these data with baseline models to detect deviations.

Failure Prediction: AI algorithms forecast fatigue life and corrosion progression, enabling timely maintenance before structural failure occurs.

Life Cycle Assessment (LCA): The digital twin compiles historical data for every weld, creating a traceable record of stress cycles, inspection results, and environmental exposure.

This predictive approach transforms maintenance from reactive to proactive, reducing downtime, improving safety, and extending vessel and offshore asset lifespan. It also aligns with ISO 55000 asset management principles, enhancing operational reliability and cost efficiency in marine systems.

13.2.4 Application in Shipyard Workflow Optimization

In modern smart shipyards, the digital twin acts as a central intelligence system that integrates welding operations, production planning, and quality control within a unified digital environment. By connecting design (computer-aided design [CAD]/computer-aided manufacturing [CAM]), manufacturing (robotic welding), and inspection (Non-Destructive Testing [NDT] systems), it ensures complete process transparency and optimization.

Real-Time Production Coordination: Welding robots and automated lines communicate via the digital twin to synchronize tasks and optimize resource allocation.

Data Integration with ERP Systems: The twin links production data with enterprise resource planning (ERP) platforms, improving project scheduling and traceability.

Quality Assurance and Training: VR interfaces allow supervisors to visualize weld progression, identify potential bottlenecks, and train operators interactively.

Continuous Improvement Loop: Data feedback from completed projects refines future welding models, ensuring each new ship or platform benefits from accumulated experience.

Through digital twin-driven workflow optimization, shipyards achieve higher process efficiency, reduced material waste, and enhanced safety essential traits of the next-generation digital shipyard.

13.3 SMART SENSORS AND REAL-TIME MONITORING IN MARINE WELDING

The evolution of smart sensor technologies and real-time monitoring systems has transformed modern marine welding from a manually supervised process to a data-driven, intelligent fabrication environment. In shipbuilding and offshore engineering, welding operations must maintain exceptional precision and reliability under challenging environmental conditions such as high humidity, salinity, and variable temperature. To ensure weld integrity, advanced sensors continuously collect and analyze process data such as heat input, vibration, acoustic emissions (AEs), and optical signatures, enabling real-time defect detection, quality assessment, and process optimization.

These monitoring systems, when integrated with AI and digital twin platforms, provide welders and engineers with instant feedback, allowing automatic parameter adjustments and predictive quality control. The result is a shift from reactive inspection to proactive, adaptive welding, ensuring high quality, efficiency, and sustainability across all stages of marine fabrication.

Figure 13.3 illustrates a smart sensor network for real-time monitoring and adaptive control of marine welding processes. The system integrates thermal, acoustic,

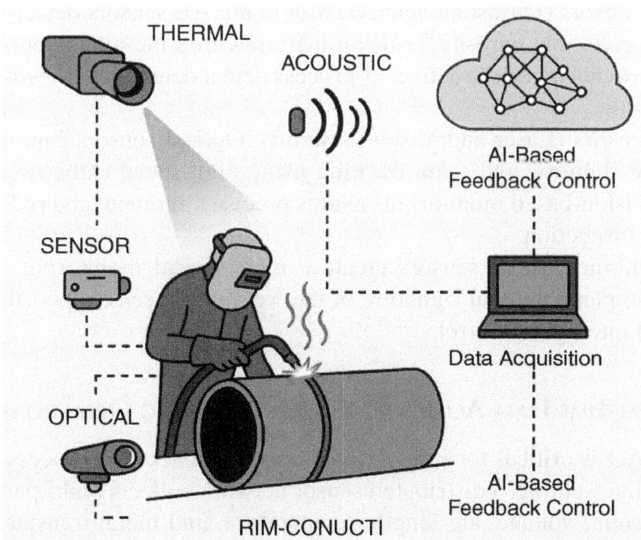

FIGURE 13.3 Smart sensor network for real-time monitoring in marine welding.

and optical sensors strategically positioned around the welding area. The thermal sensor (infrared [IR] camera) measures temperature gradients and heat input, ensuring proper penetration and preventing overheating. The acoustic sensor detects sound wave patterns to identify porosity, cracking, or arc instability, while the optical sensor (high-speed vision system) tracks weld pool geometry, arc alignment, and seam quality.

All sensor data converge into a data acquisition (DAQ) unit, which transmits information to an AI-based feedback control system through cloud or edge computing platforms. The AI system analyses real-time data, detects anomalies, and automatically adjusts parameters such as voltage, current, or travel speed, maintaining consistent weld quality.

This integrated monitoring framework transforms marine welding into a self-learning, adaptive process, reducing defects, improving efficiency, and supporting autonomous operation in smart shipyards and offshore environments.

13.3.1 Types of Sensors for Weld Quality Monitoring (Thermal, Acoustic, Optical)

Smart sensors form the foundation of intelligent welding systems, enabling precise, continuous observation of welding parameters and outcomes. The three main types of thermal, acoustic, and optical sensors capture different aspects of weld quality and process stability.

Thermal Sensors (IR and Thermocouples): These measure temperature distribution, cooling rates, and heat input profiles, essential for predicting weld penetration and preventing overheating or distortion. IR cameras generate thermal maps of the weld pool in real time, identifying inconsistencies before defects occur.

Acoustic Sensors (Ultrasonic and AE Monitors): AE sensors detect sound waves generated by cracking, porosity, or delamination within the molten pool. Ultrasonic probes also provide non-destructive, in-process defect detection, allowing immediate corrective action.

Optical Sensors (Laser and Vision Systems): Optical sensors capture weld pool geometry, arc stability, and seam tracking using high-speed cameras and laser triangulation. Vision-based monitoring ensures precise alignment and reduces reliance on post-weld inspection.

When combined, these sensors create a multi-modal monitoring network that captures a complete physical signature of the welding process, supporting advanced analytics and intelligent control.

13.3.2 Real-Time Data Acquisition during Welding Operations

Real-time DAQ is critical for integrating sensor feedback into process control systems. In marine welding, a distributed sensor network collects multi-parameter data, including current, voltage, arc length, temperature, and metal transfer characteristics, which are processed and visualized using Industrial Internet of Things (IIoT) frameworks.

Modern shipyards use edge computing devices located near welding stations to preprocess data before transmitting it to central servers or cloud systems. This ensures low latency and rapid response for process adjustments.

Key aspects of real-time DAQ systems include:

Synchronized Multi-Sensor Input: Data from thermal, optical, and acoustic sensors are time-stamped and aligned for accurate interpretation.

High-Frequency Sampling: Enables detection of transient welding events such as spatter bursts, arc instability, or microcrack initiation.

Data Fusion and Visualization: Integrates heterogeneous sensor data into intuitive dashboards for operators and engineers.

Through real-time acquisition and analysis, welding systems can transition from static control to dynamic process optimization, ensuring consistent quality even under variable environmental conditions.

13.3.3 Defect Detection and Automated Correction Systems

Defect detection is one of the most significant applications of smart sensing in marine welding. Advanced monitoring systems employ ML models and signal analysis techniques to detect irregularities in real time and initiate corrective actions automatically.

Defect Recognition: AI algorithms trained on large datasets identify characteristic patterns of porosity, incomplete fusion, or undercutting from sensor signals (thermal gradients, acoustic spikes, or visual irregularities).

Automated Correction: Once a defect is detected, the system adjusts **travel speed, current, or gas flow rate** automatically through **closed-loop control** to stabilize the weld pool and prevent defect propagation.

Post-Weld Validation: The system logs all anomalies for review, improving future parameter optimization through feedback learning.

This capability not only enhances weld quality but also reduces reliance on manual inspection, thereby increasing throughput and reliability in large-scale shipbuilding operations.

13.3.4 Remote and Underwater Weld Monitoring Solutions

In offshore fabrication and subsea repair, direct human supervision is often impractical due to safety and accessibility constraints. Remote monitoring systems using tethered sensors, remotely operated vehicles (ROVs), and autonomous underwater vehicles (AUVs) have enabled real-time inspection and control of underwater welding processes.

Underwater Sensors: Equipped with waterproof cameras, pressure-compensated acoustic detectors, and corrosion-resistant thermocouples, these sensors transmit live data to surface control centers.

Robotic Integration: ROV-mounted welding arms utilize vision and laser sensors for seam tracking and automatic adjustment under turbulent water conditions.

Data Transmission: High-bandwidth fiber-optic cables or acoustic modems ensure reliable communication between submerged systems and control stations.

This integration of smart sensing with underwater robotics allows safe, precise, and efficient welding operations for offshore pipelines, risers, and subsea structures, significantly reducing diver dependency and operational risk.

13.3.5 Integration with AI for Intelligent Weld Control

The full potential of smart sensors is realized when integrated with AI for intelligent weld control. AI systems analyze continuous data streams to understand process behavior, recognize anomalies, and make real-time decisions.

Machine Learning (ML): Trained on historical weld data, ML models predict optimal process parameters and detect early signs of instability.

Neural Networks: Deep learning algorithms interpret complex relationships between sensor inputs and weld quality outcomes, improving over time through adaptive learning.

Predictive Control Systems: AI adjusts settings proactively based on predicted thermal and mechanical responses, ensuring consistent quality with minimal human intervention.

This intelligent fusion of sensing, data, and automation forms the foundation of **next-generation smart shipyards**, where welding systems can **self-monitor, self-correct, and self-optimize**, enabling unparalleled accuracy, efficiency, and sustainability.

13.4 ARTIFICIAL INTELLIGENCE AND MACHINE LEARNING IN MARINE WELDING

The integration of AI and ML has redefined the scope and precision of marine welding, transforming it from a static, human-controlled process into a dynamic, intelligent, and data-driven operation. These technologies enable shipyards to optimize welding parameters, predict defects before they occur, and adapt to complex geometries in real time. The growing adoption of AI-assisted automation is reshaping both onshore and offshore fabrication, improving productivity, safety, and sustainability. By leveraging sensor data, simulation outputs, and digital twin feedback, AI-based systems continuously learn and evolve, ensuring consistently high weld quality across diverse marine applications.

Figure 13.4 illustrates the AI-driven framework used in intelligent marine welding systems, where real-time data from multiple sensors (thermal, acoustic, and optical) are processed through an integrated AI and ML analytics platform. The sensor array continuously monitors temperature distribution, acoustic signals, and

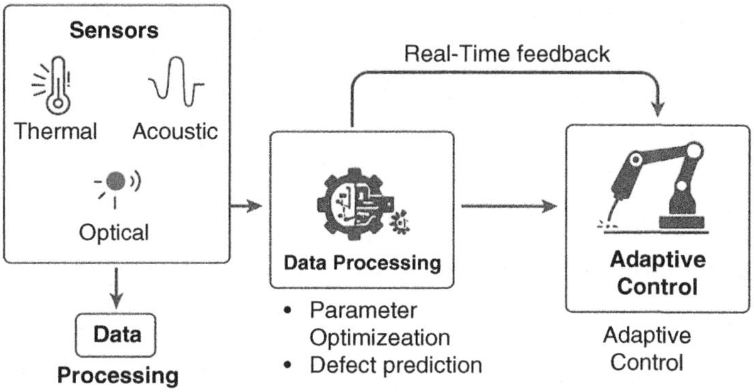

FIGURE 13.4 AI-driven framework for intelligent marine welding.

weld pool geometry during operation. This information is sent to the data processing module, where AI algorithms perform parameter optimization and defect prediction using historical datasets and real-time analysis.

The processed data are then relayed to the adaptive control system, which automatically adjusts welding parameters such as current, voltage, and torch speed to maintain optimal conditions. Feedback from the adaptive control loop ensures continuous process refinement and quality consistency. This closed-loop system exemplifies smart automation in shipyards, reducing human intervention while enhancing weld accuracy, energy efficiency, and overall productivity in marine fabrication.

13.4.1 AI-Based Weld Parameter Optimization

AI algorithms have revolutionized weld parameter selection and process optimization, traditionally performed through manual expertise or empirical testing. Using data from sensors, historical weld records, and real-time monitoring systems, ML models identify the optimal combinations of current, voltage, travel speed, wire feed rate, and shielding gas flow to achieve desired weld penetration and quality.

Neural Networks (NNs): Predict the influence of multiple parameters on weld bead geometry, mechanical strength, and defect probability.

Genetic Algorithms (GA) and Optimization Models: Evolve parameter sets over multiple iterations to reach the most efficient and stable process conditions.

Reinforcement Learning (RL): Enables welding robots to "learn" by interaction—adjusting parameters autonomously in response to changing conditions like plate thickness or joint misalignment.

By using **AI-driven optimization**, marine welders can achieve **first-pass success rates** above 95%, minimize rework, and improve **energy efficiency**, leading to lower consumable costs and reduced carbon emissions. This capability is particularly crucial for **large-scale ship hull fabrication**, where process consistency directly affects structural integrity and delivery timelines.

13.4.2 Automated Quality Assessment and Defect Prediction

AI also plays a pivotal role in automated quality assessment and defect prediction, significantly reducing the reliance on post-weld inspection and manual visual checks. By analyzing data from thermal cameras, acoustic sensors, X-ray imaging, and ultrasonic scans, AI systems can detect even minute anomalies in real time.

Computer Vision (CV): Deep learning-based image recognition identifies surface irregularities such as undercuts, spatter, or incomplete fusion with high accuracy.

Predictive Defect Modeling: ML algorithms trained on historical inspection data forecast the likelihood of internal defects like porosity, slag inclusions, or cracks during the welding process.

Adaptive Inspection Systems: Continuous feedback between sensors and AI models allows the system to flag issues and trigger corrective actions automatically, often within milliseconds.

Such automation ensures **standardized, repeatable quality**, meeting stringent classification society requirements (ABS, DNV, Lloyd's Register [LR]). Moreover, it enhances safety by minimizing the need for human exposure in hazardous welding zones such as confined ship compartments or offshore platforms.

13.4.3 Adaptive Welding Systems for Complex Marine Structures

Modern shipbuilding involves complex geometries, varying material thicknesses, and difficult-to-access areas that challenge traditional welding automation. Adaptive AI-based welding systems overcome these limitations through real-time process learning and control, enabling precise, flexible fabrication even in dynamic conditions.

AI-Guided Seam Tracking: Vision sensors combined with AI algorithms detect joint edges and contours, automatically adjusting torch position and angle during operation.

Multi-Robot Coordination: AI enables multiple robotic welding arms to collaborate in synchronized sequences, optimizing deposition rates and reducing idle time.

Structural Adaptability: For components like double-hull vessels or offshore jackets, AI systems adjust heat input and filler deposition based on geometry and stress distribution.

Additionally, when paired with **digital twin environments**, these adaptive systems can simulate welding on complex assemblies before execution, minimizing distortion and rework. This capability makes AI-enabled welding indispensable in the construction of **next-generation naval vessels, offshore rigs, and subsea infrastructure** where precision, durability, and efficiency are paramount.

13.5 AUTOMATION AND ROBOTICS IN NEXT-GENERATION MARINE WELDING

The rise of automation and robotics is transforming the landscape of marine welding, ushering in the era of smart and fully digital shipyards. Advanced robotic systems, integrated with AI-based control, sensor feedback, and digital twin platforms, are redefining how shipbuilding and offshore fabrication are executed. In modern shipyards, welding robots are no longer limited to repetitive or simple tasks, and they now operate autonomously in complex geometries, confined spaces, and high-risk environments with exceptional precision and reliability. These systems enhance productivity, reduce material waste, and ensure consistent weld quality while promoting worker safety and sustainability. The synergy between human expertise, robotic precision, and digital intelligence is laying the foundation for next-generation marine manufacturing.

Figure 13.5 depicts the automation and robotic framework that underpins smart shipyard operations. The illustration shows how CAD/CAM integration, autonomous welding robots, and human–robot collaboration (HRC) interact within a closed digital workflow.

The CAD/CAM system provides digital input derived from ship design models, automatically generating optimized weld paths and parameters. These data are transmitted to autonomous welding robots, which execute precise, high-quality welds on ship sections using sensor-guided and AI-controlled feedback. Simultaneously, HRC zones demonstrate how welders and cobots work together. Welders manage setup, inspection, and quality control, while robots perform repetitive or hazardous tasks with accuracy and endurance.

Feedback loops between the CAD/CAM system and robotic units ensure continuous monitoring, process adjustment, and traceability. This integration of digital

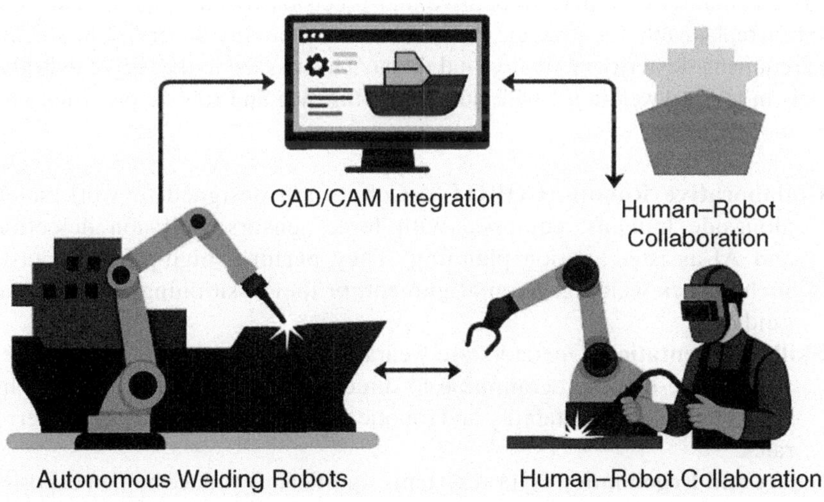

FIGURE 13.5 Automated and robotic welding framework for smart shipyards.

design, automation, and human oversight exemplifies next-generation marine welding, delivering enhanced efficiency, consistency, and safety in shipbuilding operations.

13.5.1 Autonomous Welding Robots in Shipbuilding

Autonomous welding robots are now integral to modern shipyards, capable of executing complex tasks with minimal human supervision. These robots are equipped with AI-driven vision systems, multi-axis manipulators, and real-time process monitoring tools, enabling them to adapt dynamically to varying joint configurations and welding conditions.

Capabilities: Modern robots perform multi-pass welding, seam tracking, and adaptive heat control across large ship panels, bulkheads, and hull sections. Using onboard sensors, they detect part misalignments and automatically adjust torch angles and travel speeds to maintain weld quality.

Mobility: Mobile robotic platforms, including track-mounted and crawler-type systems, navigate along curved hull surfaces or vertical structures, extending automation to areas previously restricted to manual welding.

Efficiency Gains: Robots can operate continuously, achieving up to 40% improvement in productivity and 30% reduction in material waste compared to traditional manual methods.

Autonomous robotic welding not only accelerates ship construction but also ensures precision, repeatability, and process stability, particularly for high-value components like LNG storage tanks, offshore risers, and naval vessel hulls.

13.5.2 Hybrid Human–Robot Collaboration Models

While automation dominates repetitive and precision-based tasks, human welders remain irreplaceable for strategic decision-making, quality supervision, and operations requiring dexterity or creative judgment. This has led to the rise of hybrid HRC models in marine welding, where human intelligence and robotic precision operate in harmony.

Collaborative Robots (COBOTs): Cobots are designed to work safely alongside humans, equipped with force sensors, collision detection, and AI-assisted motion planning. They perform high-precision tasks such as tack welding, seam alignment, or tool positioning under human guidance.

Skill Augmentation: Operators use wearable control systems, AR interfaces, and gesture-based programming to direct robotic arms in real time. This fusion of human adaptability and robotic accuracy reduces fatigue and error rates.

Safety and Ergonomics: Hybrid systems significantly reduce human exposure to high heat, UV radiation, and hazardous fumes while enabling real-time process validation through shared sensor feedback.

By combining human insight with robotic automation, shipyards achieve a flexible, semi-autonomous fabrication environment that enhances both safety and efficiency, particularly in custom ship design and complex assembly operations.

13.5.3 Integration with CAD/CAM for Fully Automated Welding Lines

The integration of CAD and CAM with robotic welding systems represents a major milestone toward fully automated digital shipyards. This seamless link between design and fabrication allows for direct translation of 3D ship models into executable welding programs without manual intervention.

Digital Workflow: Once a ship's structure is designed in CAD software, the system automatically generates weld paths, joint geometries, and process parameters in CAM environments, which are then transmitted to robotic controllers.

Simulation and Verification: Virtual welding simulations validate process feasibility, thermal distortions, and access constraints before execution. Integration with digital twin systems ensures that deviations detected in real time are corrected dynamically.

Data-Driven Production: Each weld operation is logged and analyzed for quality assurance, forming a traceable digital record that supports certification under **ABS, DNV, and LR standards**.

Through CAD/CAM integration, shipyards can implement fully automated welding lines, enabling continuous, high-precision manufacturing with minimal downtime. This digital continuity ensures faster production cycles, reduced rework, and enhanced cost efficiency, marking a pivotal step toward the Smart Shipyard 5.0 era.

13.6 SUSTAINABLE AND LOW-CARBON WELDING INNOVATIONS

The growing emphasis on decarbonization and environmental responsibility in the maritime industry has positioned sustainability as a central pillar of modern welding innovation. Shipyards and offshore fabrication centers are under increasing pressure to minimize GHG emissions, reduce energy consumption, and comply with international sustainability mandates such as the IMO 2050 GHG Reduction Strategy, the EU Green Deal, and the UN Sustainable Development Goals (SDGs).

In response, marine welding is undergoing a transformation toward energy-efficient, low-emission, and digitally optimized processes. Advances in process control, consumable design, and digital sustainability tracking are helping shipyards achieve cleaner operations without compromising productivity or weld integrity. These innovations not only reduce environmental impact but also enhance long-term cost efficiency and global competitiveness in ship construction and repair.

Figure 13.6 illustrates the core framework for achieving sustainability and carbon reduction in marine welding operations. The diagram integrates three primary components: energy-efficient processes, green consumables, and digital sustainability

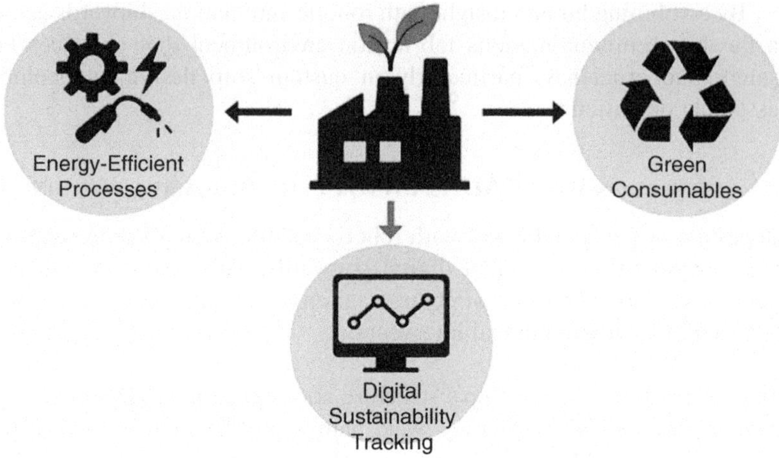

FIGURE 13.6 Sustainable and low-carbon welding framework for marine shipyards.

tracking, all converging toward a central goal of environmentally responsible shipyard operations.

The energy-efficient processes segment represents innovative methods such as HLAW, friction stir welding (FSW), and WAAM, which reduce power consumption and heat loss. The green consumables section highlights low-emission filler materials, water-based fluxes, and recycling practices that minimize waste and airborne pollutants. Finally, digital sustainability tracking embodies the use of AI, LCA, and carbon monitoring dashboards to quantify and optimize environmental performance in real time.

Collectively, these three interconnected strategies form the foundation of low-carbon, eco-conscious shipbuilding, ensuring that future marine fabrication aligns with global sustainability goals while maintaining high precision, quality, and operational efficiency.

13.6.1 Energy-Efficient Welding Processes for the Future

Energy-efficient welding represents the foundation of sustainable marine fabrication. Traditional arc-based processes consume significant electrical energy and generate high levels of heat, fumes, and CO_2 emissions. Future-oriented solutions focus on process optimization, hybrid technologies, and intelligent control to minimize waste and improve energy utilization.

Hybrid Laser-Arc Welding (HLAW): Combines deep penetration of laser welding with the filler versatility of arc welding, reducing energy consumption by up to **30%** while ensuring higher deposition efficiency.

Friction Stir Welding (FSW): A solid-state process that eliminates melting and filler metal usage, offering nearly **80% lower energy demand** and superior joint integrity.

Wire Arc Additive Manufacturing (WAAM): Enables near-net-shape fabrication and on-site component repair, drastically reducing material waste and logistics-related emissions.

Intelligent Energy Control Systems: Real-time AI monitoring adjusts voltage, current, and travel speed to optimize arc stability and prevent overheated energy losses.

The adoption of such low-energy, high-efficiency welding methods contributes significantly to achieving carbon-neutral shipbuilding while maintaining mechanical performance and compliance with structural standards.

13.6.2 Green Consumables and Environmentally Friendly Practices

Sustainability in marine welding extends beyond process efficiency to include eco-friendly consumables and waste reduction practices. New developments in welding materials and operations aim to reduce toxic emissions, conserve natural resources, and enhance workplace safety.

Low-Emission Filler Wires and Fluxes: Manufacturers are developing flux-cored wires with reduced alloying waste and low-spatter characteristics, minimizing airborne particulates and metal oxide fumes.

Water-Based and Low-VOC Shielding Agents: Replace traditional solvent-based formulations, significantly reducing volatile organic compound (VOC) emissions and improving air quality in enclosed shipyard environments.

Consumable Recycling and Reuse: Programs for reclaiming unused wire segments, slag, and metal residues are being implemented to promote circular economy principles in shipyards.

Eco-Conscious Practices: Initiatives such as waste heat recovery, fume extraction systems, and renewable energy integration (solar or wind-powered welding stations) enhance operational sustainability.

These innovations align with **MARPOL pollution prevention regulations** and emerging **ISO 14001 environmental management standards**, ensuring that modern shipyards meet global environmental compliance benchmarks.

13.6.3 Role of Digital Tools in Sustainability Tracking

Digitalization plays a transformative role in achieving and maintaining sustainability goals in marine welding operations. Digital twins, AI-based analytics, and LCA software enable real-time tracking of energy usage, emissions, and material efficiency throughout fabrication processes.

Energy and Carbon Monitoring Dashboards: Integrate data from welding machines, robots, and sensors to calculate real-time energy consumption and carbon footprint per weld joint.

LCA Integration: Digital models quantify the environmental impact of materials, processes, and logistics across a vessel's construction and operational phases.

Blockchain-Based Traceability: Ensures transparency in sourcing eco-certified consumables and recording sustainable production credentials for classification compliance.

Predictive Sustainability Optimization: AI algorithms analyze historical data to suggest process adjustments that reduce emissions and resource waste.

These digital tools not only enhance environmental accountability but also support data-driven decision-making for sustainable shipyard management. By quantifying and visualizing carbon and energy performance, shipbuilders can align operations with international sustainability metrics and corporate Environmental, Social, and Governance (ESG) frameworks.

13.7 CHALLENGES IN ADOPTING FUTURE WELDING TECHNOLOGIES

While the adoption of advanced welding technologies—such as automation, digital twins, AI-based monitoring, and low-carbon processes—offers transformative potential for the marine industry, several implementation challenges remain. Transitioning from traditional shipyard practices to intelligent, automated, and sustainable systems demands significant investment, workforce adaptation, and alignment with established maritime standards. These challenges are particularly critical for developing economies, small-scale shipyards, and maintenance yards where operational costs, legacy infrastructure, and limited digital readiness can delay modernization. Addressing these issues is vital for ensuring that future marine welding innovations deliver consistent performance, safety, and compliance across global shipbuilding operations.

Figure 13.7 presents a clear summary of the primary challenges faced in implementing next-generation marine welding systems and their corresponding mitigation

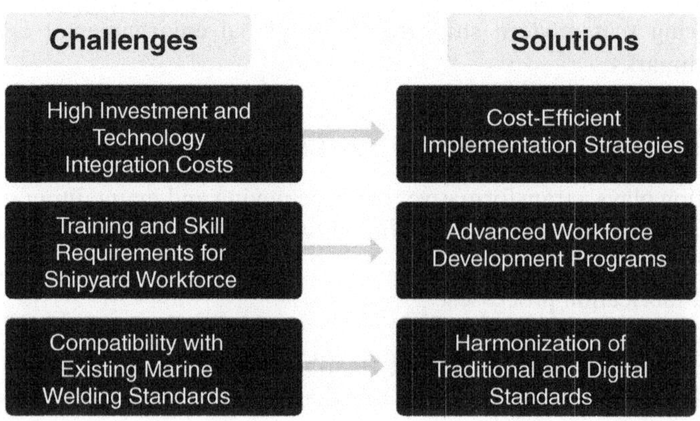

FIGURE 13.7 Challenges and solutions for adopting future marine welding technologies.

strategies. The diagram is divided into two main columns: challenges on the left and solutions on the right connected by directional arrows that illustrate the transition from problems to actionable responses.

The first challenge, high investment and technology integration costs, is addressed through cost-efficient implementation strategies, emphasizing phased adoption, shared infrastructure, and scalable digital upgrades. The second, training and skill requirements for shipyard workforce, is countered with advanced workforce development programs, including continuous training in automation, AI, and digital system management. The third, compatibility with existing marine welding standards, highlights the need for the harmonization of traditional and digital standards to ensure compliance and interoperability across shipyards.

13.7.1 High Investment and Technology Integration Costs

One of the foremost barriers to the adoption of next-generation welding systems is the high initial capital investment required for automation, robotics, and digital infrastructure. Smart shipyards need advanced robotic welding cells, sensor-integrated equipment, and AI-driven data management systems, which involve substantial procurement and setup costs.

In addition, integration costs arise from aligning new technologies with existing shipyard machinery, CAD/CAM systems, and production workflows. Many shipyards face the challenge of bridging legacy manual or semi-automated systems with modern digital platforms, requiring specialized middleware, calibration, and real-time synchronization.

Furthermore, the maintenance and software licensing of such high-tech systems add to operational expenses. Smaller shipyards or repair yards, which operate on tight margins, often find it difficult to justify the return on investment (ROI) within short project cycles.

However, as digital ecosystems mature and costs for sensors, AI computing, and robotics decrease, these technologies are expected to become more accessible, driving long-term cost savings through higher efficiency, reduced rework, and predictive maintenance.

13.7.2 Training and Skill Requirements for Shipyard Workforce

Another significant challenge lies in developing a technically skilled workforce capable of operating, maintaining, and optimizing modern welding systems. Traditional welders, who excel in manual or semi-automatic processes, must now adapt to robotic programming, data interpretation, and digital system management.

Shipyards require professionals trained in mechatronics, data analytics, and process automation, in addition to conventional welding expertise. Training programs must evolve to include:

Simulation-based learning using virtual or AR (VR/AR) to replicate complex marine welding scenarios.

Cross-disciplinary education that integrates mechanical, electrical, and software engineering concepts for effective robotic system handling.

Continuous upskilling initiatives supported by classification societies and maritime institutes to ensure safety and compliance in digital shipyards.

Without targeted skill development, there is a risk of a digital skill gap, which could hinder the effective use of advanced welding systems. Building a digitally literate, adaptable workforce is therefore crucial for the successful deployment of next-generation marine welding technologies.

13.7.3 Compatibility with Existing Marine Welding Standards

A persistent concern in adopting innovative welding technologies is ensuring compatibility with existing marine standards and certification frameworks. Bodies such as **ABS**, **DNV**, **LR**, and **ISO** maintain strict requirements for welding procedures, materials, and quality assurance.

Emerging techniques such as WAAM, HLAW, and AI-based adaptive control systems often outpace the rate at which new standards are developed or updated. This creates uncertainty for shipyards regarding approval, classification, and acceptance of novel processes in commercial ship construction.

To overcome this, international collaboration among standardization bodies, research institutions, and shipbuilders is essential to establish testing protocols, validation methods, and certification pathways for emerging technologies. Harmonizing traditional and digital standards will facilitate a smoother transition toward Industry 4.0-compliant shipyards while ensuring safety, reliability, and regulatory consistency.

13.8 OUTLOOK AND LONG-TERM TRENDS

The future of marine welding engineering is being redefined by the convergence of autonomous systems, digital intelligence, and sustainable manufacturing paradigms. As global shipyards embrace automation, AI, and cyber-physical systems, welding will no longer be a manual craft but a self-optimizing, data-driven discipline embedded within the broader framework of smart and green shipbuilding. The coming decades will see the rise of fully autonomous welding ecosystems, greater human-machine collaboration through Industry 5.0 principles, and international cooperation driving technological standardization and innovation. Together, these trends will transform shipyards into integrated digital environments capable of achieving superior quality, safety, and environmental performance.

Figure 13.8 presents a conceptual view of the future marine welding ecosystem, illustrating how autonomous systems, digital intelligence, and human collaboration will coexist in next-generation shipyards. The figure highlights four key elements: the autonomous welding system, HRC, digital twin integration, and global digital connectivity, all operating within a sustainable smart shipyard powered by renewable energy.

In this framework, autonomous robots execute precision welds guided by AI-based feedback, while digital twins provide real-time virtual simulations and predictive analytics. Human operators, enhanced with wearable technology and AR interfaces,

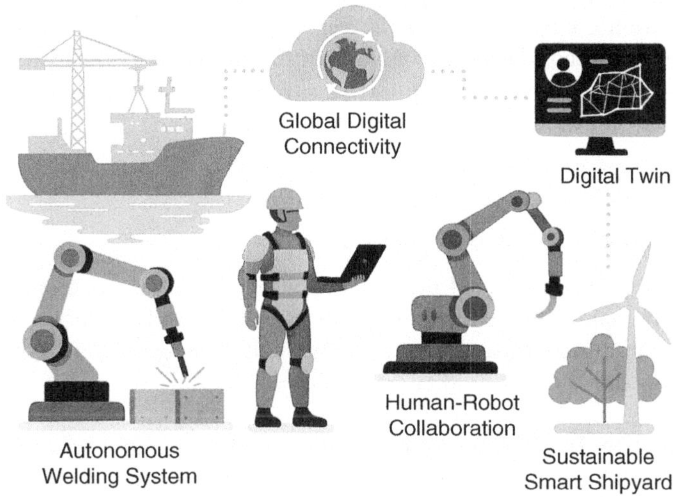

FIGURE 13.8 Vision of future marine welding ecosystem.

supervise complex tasks and decision-making through seamless human-machine collaboration. Data flow continuously between on-site equipment and global digital networks, enabling cross-shipyard optimization and standardization.

The inclusion of sustainability indicators such as wind turbines and green energy systems emphasizes the maritime industry's shift toward carbon-neutral manufacturing. Overall, the figure encapsulates the long-term vision of integrated, intelligent, and eco-efficient shipyards, where advanced automation, digital ecosystems, and global collaboration define the future of marine welding engineering.

13.8.1 Fully Autonomous Shipyard Welding Systems

The next frontier in marine fabrication lies in the realization of fully autonomous welding systems, where robotic and AI technologies operate with minimal human intervention. These systems integrate AI-driven path planning, multi-sensor feedback, and digital twin coordination to execute precise welds, adjust parameters dynamically, and validate quality in real time.

- **Self-Optimizing Welding Cells**: Robots equipped with machine vision and thermal imaging can identify joint configurations, compensate for distortion, and ensure uniform bead geometry without external input.
- **AI-Based Decision-Making**: Predictive analytics enable welding systems to forecast material responses and adapt to environmental variations such as humidity, vibration, or surface contamination.
- **Autonomous Maintenance**: Digital twins linked with onboard diagnostics predict tool wear, schedule servicing, and update operational parameters automatically.

Such systems mark the transition toward lights-out shipyards, where fabrication continues around the clock with minimal supervision. These autonomous ecosystems not only maximize productivity but also minimize energy waste, rework, and emissions, advancing the maritime sector toward zero-defect and zero-carbon operations.

13.8.2 Integration with Industry 5.0 and Human-Centric Manufacturing

While Industry 4.0 established the foundation for digital automation, Industry 5.0 emphasizes human-centric collaboration, sustainability, and resilience. In the marine sector, this paradigm ensures that technology complements human expertise rather than replacing it.

Collaborative Intelligence: AI systems augment human decision-making by offering predictive insights and adaptive recommendations, while welders oversee complex judgments requiring creativity and contextual understanding.

Enhanced Safety and Ergonomics: Exoskeleton-assisted operators, wearable sensors, and AR/VR interfaces reduce fatigue and risk, especially in confined or hazardous marine environments.

Ethical and Sustainable Design: Industry 5.0 encourages integrating social and environmental goals into manufacturing such as minimizing the carbon footprint of welding consumables and optimizing energy use through intelligent controls.

By harmonizing digital precision with human adaptability, Industry 5.0 fosters resilient, inclusive, and sustainable shipyards, where both machines and humans collaborate seamlessly to achieve high-quality fabrication.

13.8.3 Global Collaboration on Smart Marine Welding Technologies

The transformation of marine welding into an intelligent and sustainable discipline requires global cooperation among shipyards, regulatory bodies, research institutions, and technology developers. Unified frameworks for innovation, standardization, and knowledge exchange will ensure that advanced welding methods are implemented safely and consistently worldwide.

Standard Harmonization: Organizations like **IMO, ISO, ABS, DNV, and LR** will play pivotal roles in defining standards for AI-driven, robotic, and additive welding processes.

Joint Research Initiatives: Cross-border projects will promote the development of AI algorithms, sustainable consumables, and digital twin validation tools tailored for the maritime sector.

Knowledge and Skill Networks: Global training alliances and maritime technology hubs will bridge the gap between developed and emerging shipbuilding nations, ensuring equitable technological adoption.

By fostering international collaboration, the maritime industry can accelerate innovation while ensuring safety, interoperability, and environmental stewardship, creating a unified roadmap toward smart, sustainable shipyards of the future.

13.9 SUMMARY

The chapter explores the technological evolution shaping the next generation of shipbuilding and offshore fabrication, emphasizing automation, digital intelligence, and sustainability. It highlights how innovations such as digital twin technology, AI-driven process optimization, smart sensors, and autonomous robotic welding systems are revolutionizing weld precision, quality, and efficiency. The integration of CAD/CAM systems, HRC, and Industry 5.0 principles fosters a human-centric yet highly automated shipyard environment. Additionally, the chapter underscores the emergence of low-carbon welding processes, eco-friendly consumables, and data-driven sustainability tracking, aligning with global green manufacturing goals. Despite challenges such as high investment costs, skill adaptation, and standardization gaps, the long-term outlook envisions fully autonomous, interconnected, and environmentally responsible shipyards driven by global collaboration, ensuring safer, smarter, and more sustainable marine fabrication in the decades ahead.

13.10 REVIEW QUESTIONS

1. What are the key technological trends influencing the future of marine welding engineering?
2. Explain how digital transformation is reshaping welding practices in modern shipyards.
3. What are the main drivers behind innovation in marine welding processes and equipment?
4. Discuss the role of emerging technologies in enhancing efficiency and safety in offshore structures.
5. How does the shift from Industry 4.0 to Industry 5.0 impact the marine welding sector?
6. Define the concept of a digital twin and describe its application in marine welding.
7. How does digital twin technology contribute to predictive maintenance in welded ship structures?
8. Explain how simulation and real-time data integration improve shipyard workflow optimization.
9. What are the primary components of a digital twin-based welding system?
10. Discuss the advantages and potential challenges of adopting digital twin technology in shipbuilding.
11. Identify the main types of sensors used for weld quality monitoring in marine fabrication.
12. How does real-time data acquisition improve weld quality control?

13. Explain the role of AI-based systems in detecting and correcting welding defects automatically.
14. Describe how remote and underwater monitoring systems are used in marine welding operations.
15. Discuss how intelligent sensing contributes to the development of fully automated welding systems.
16. How does AI optimize welding parameters in complex marine structures?
17. What are the benefits of using machine learning algorithms for automated quality assessment?
18. Explain the function of adaptive welding systems and their role in next-generation shipyards.
19. Discuss how AI-based predictive defect models enhance process reliability.
20. What are the limitations of current AI applications in marine welding environments?
21. Compare autonomous welding robots and collaborative robots (cobots) used in shipbuilding.
22. How does CAD/CAM integration facilitate fully automated welding lines in shipyards?
23. Explain how human–robot collaboration models enhance safety and productivity.
24. What are the major challenges in deploying robotic systems in large ship structures?
25. Discuss how robotic automation contributes to digital continuity in marine manufacturing.
26. What are the key features of energy-efficient welding processes such as FSW, HLAW, and WAAM?
27. How do green welding consumables reduce environmental impact in shipyards?
28. Explain the importance of digital tools in tracking sustainability performance and carbon footprint.
29. Discuss the role of renewable energy and eco-friendly materials in future marine welding.
30. How can sustainable welding technologies contribute to meeting IMO and global emission reduction goals?

Bibliography

Adil, G. K., & Bhole, S. D. (1992). HAZ hardness and microstructure predictions of arc welded steels—I: Review of predictive models. *Canadian Metallurgical Quarterly*, 31(2), 151–157. https://doi.org/10.1179/cmq.1992.31.2.151.

Adrian, H., Drodten, P., Engell, H. J., Grimme, D., Herbsleb, G., Horstmann, D., … & Schwenk, W. (1977). Effect of metallurgical development on the corrosion of carbon and low-alloy construction steels for ship building and harbour construction. *Corrosion Science*, 17(2), 143–153. https://doi.org/10.1016/0010-938X(77)90014-2.

Alkahla, I., & Pervaiz, S. (2017, September). Sustainability assessment of shielded metal arc welding (SMAW) process. In *IOP Conference Series: Materials Science and Engineering* (vol. 244, no. 1, p. 012001). https://doi.org/10.1088/1757-899X/244/1/012001.

Anand, A., & Khajuria, A. (2015). Welding processes in marine application: A review. *International Journal of Mechanical Engineering and Robotics Research*, 2(1), 215–225 https://www.ijmerr.com/uploadfile/2015/0409/20150409110634573.pdf.

Bhuvanesh Kumar, M., Sathiya, P., & Senthil, S. M. (2023). A critical review of wire arc additive manufacturing of nickel-based alloys: Principles, process parameters, microstructure, mechanical properties, heat treatment effects, and defects. *Journal of the Brazilian Society of Mechanical Sciences and Engineering*, 45(3), 164. https://doi.org/10.1016/j.mtcomm.2022.103739.

Billingham, J., Sharp, J. V., & Spurrier, J. (2003). Review of the performance of high strength steels used offshore. Technical Report, Report Number RR-105, Cranfield University, United Kingdom. TRN: GB0351006, HSE Books, Suffolk.

Chernyshov, E. A., Romanov, A. D., & Romanova, E. A. (2016). High-strength shipbuilding steels and alloys. *Metallurgist*, 60(1), 186–190. https://doi.org/10.1007/s11015-016-0271-1.

Di Bella, G., Alderucci, T., Salmeri, F., & Cucinotta, F. (2022). Integrating the sustainability aspects into the risk analysis for the manufacturing of dissimilar aluminium/steel friction stir welded single lap joints used in marine applications through a Life Cycle Assessment. *Sustainable Futures*, 4, 100101. https://doi.org/10.1016/j.sftr.2022.100101.

Dickinson, D. W., Franklin, J. E., & Stanya, A. (1980). Characterization of spot-welding behavior by dynamic electrical parameter monitoring. *Welding Journal*, 59(6), 170.

Fleury, G., & Schofield, R. (1979, February). Dry habitat connections and repairs of subsea pipe-line and structures. In *SPE Middle East Oil and Gas Show and Conference*, Bahrain (p. SPE-7770). SPE. https://doi.org/10.2118/7770-MS.

Ganguly, S. (2023). Welding of dissimilar metals. In Khoshnaw, F. (ed.), Welding of Metallic Materials (pp. 317–365). Elsevier. https://doi.org/10.1016/B978-0-323-90552-7.00014-6.

GeeksforGeeks. (n.d.). https://www.geeksforgeeks.org/electrical-engineering/electric-resistance-welding/ (Accessed on September 15, 2025).

Ghazanlou, S. I., Amini, A. M., Carrier, F. A., & Javidani, M. (2024, August). Advances in materials selection, processing, and welding techniques for marine steels in polar icebreakers: A review. In: *Conference of Metallurgists*, Halifax, Nova Scotia, Canada (pp. 119–123). Springer. https://doi.org/10.1007/978-3-031-67398-6_25.

Gould, J. E. (1993). Projection welding. In Olson, D. L., Siewert, T. A., Liu, S., & Edwards, G. R. (eds.), *ASM Handbook: Welding, Brazing, and Soldering* (vol. 6). ASM International. https://doi.org/10.31399/asm.hb.v06.a0001364.

Gyasi, E. A. (2019). Welding processes of metals for offshore environment: Underwater welding. Research Report No. 103. LUT Scientific and Expertise Publications, LUT University. https://urn.fi/URN:ISBN:978-952-335-433-3.

Haldar, N., Anand, S., Datta, S., & Das, A. (2024). Microstructure and mechanical property characterization of wire arc additively manufactured SS308L built part: Study of heat interaction phenomena. *Journal of Materials Engineering and Performance*, 33(23), 13178–13202. https://doi.org/10.1080/01694243.2025.2499055.

Ibarra, S., Grubbs, E. C., & Olson, D. L. (1988). Metallurgical aspect of underwater welding. *JOM*, 40(12), 8–10. https://doi.org/10.1080/2374068X.2023.2184582.

Ichiyama, Y., & Kodama, S. (2006). Flash-butt welding of high strength steels. *Shinnittetsu Giho*, 385, 74. https://www.nipponsteel.com/en/tech/report/nsc/pdf/n9514.pdf.

igus®. (n.d.). https://toolbox.igus.com/143/underwater-welding-robotics (Accessed on October 1, 2025).

Inrotech A/S. (n.d.). https://www.inrotech.com/inrotech-solve-well-known-challenges-in-shipbuilding/ (Accessed on September 25, 2025).

Kah, P., & Martikainen, J. (2012). Current trends in welding processes and materials: Improve in effectiveness. *Reviews on Advanced Materials Science*, 30(2), 189–200. https://www.ipme.ru/e-journals/RAMS/no_23012/08_kah.pdf.

Krause, G., Filgueira, R., Ahmed, N., Alexander, K. A., Asif, F., Fanning, L., … & Villasante, S. (2025). The local turn in a global sea: Identifying sustainability trade-offs in regionalized marine aquaculture systems. *Reviews in Aquaculture*, 17(4), e70071. https://doi.org/10.1111/raq.70071.

Lee, D., Ku, N., Kim, T. W., Kim, J., Lee, K. Y., & Son, Y. S. (2011). Development and application of an intelligent welding robot system for shipbuilding. *Robotics and Computer-Integrated Manufacturing*, 27(2), 377–388. https://doi.org/10.1016/j.rcim.2010.08.006.

Mandal, N. R. (2017). *Ship* Construction and Welding (pp. 413–417). Springer. https://doi.org/10.1007/978-981-10-2955-4.

Meyer, D. W. (1993) Flux-cored arc welding. In LeRoy Olson, D., Siewert, T. A., Liu, S., & Edwards, G. R. (eds.), *ASM Handbook: Welding, Brazing, and Soldering* (vol. 6, pp. 186–189). ASM International. https://doi.org/10.31399/asm.hb.v06.a0001355.

Mishra, R. S., & Ma, D. Z. (2005). Friction stir welding and processing. *Materials Science and Engineering: R: Reports*, 50(1–2), 1–78. https://doi.org/10.1016/j.mser.2005.07.001.

Miwa, Y., Sato, S., Kojima, T., Owaki, K., & Hirose, N. (2002). Underwater welding using remote controlled robots: Development of remote underwater welding technology with a high power YAG laser. *Kensa Gijutsu*, 7. https://www.osti.gov/etdeweb/biblio/20255103.

Moinuddin, S. Q., Saheb, S. H., Dewangan, A. K., Cheepu, M. M., & Balamurugan, S. (Eds.). (2024). Automation in the Welding Industry: Incorporating Artificial Intelligence, Machine Learning and Other Technologies. Wiley.

Moore, P. L. (2009). The importance of welding quality in ship construction. In Guedes Soares, C., & Das, P. K. (eds.), Analysis and Design of Marine Structures (pp. 381–388). CRC Press. https://www.taylorfrancis.com/chapters/edit/10.1201/9780203874981-49/importance-welding-quality-ship-construction-philippa-moore.

Norrish, J. (2017). Recent gas metal arc welding (GMAW) process developments: The implications related to international fabrication standards. *Welding in the World*, 61(4), 755–767. https://doi.org/10.1007/s40194-017-0463-8.

Olalerea, B. I., Gidiagbab, J. O., Fawolec, A. A., Egbokhaebhod, B. A., Ninduwezuor-Ehiobue, N., & Okparaekef, J. I. (2023). Review of advanced welding and testing for safety in offshore oil and gas. *Materials & Corrosion*, 4(2), 37–43. https://doi.org/10.26480/macem.02.2023.37.43.

Pemberton, H. N. (1953). Welding in marine engineering. *Proceedings of the Institution of Mechanical Engineers*, 167(1), 48–61. https://doi.org/10.1243/PIME_PROC_1953_167_013_02.

Pogrebisky, D. M. (2016). *Welding* of Metals: Classification, Brief History, Development. Insty Prints, Paton Publishing House.

Rowe, M., & Liu, S. (2001). Recent developments in underwater wet welding. *Science and Technology of Welding and Joining*, 6(6), 387–396. https://doi.org/10.1179/stw.2001.6.6.387.

Short, A. B. (2009). Gas tungsten arc welding of α+ β titanium alloys: A review. *Materials Science and Technology*, 25(3), 309–324. https://doi.org/10.1179/174328408X389463.

Singh, B., Khan, Z. A., & Siddiquee, A. N. (2013). Review on effect of flux composition on its behavior and bead geometry in submerged arc welding (SAW). *Journal of Mechanical Engineering Research*, 5(7), 123–127. https://doi.org/10.5897/JMER2013.0284.

Stavrov, D., & Bersee, H. E. N. (2005). Resistance welding of thermoplastic composites-an overview. *Composites Part A: Applied Science and Manufacturing*, 36(1), 39–54. https://doi.org/10.1016/j.compositesa.2004.06.030.

Tang, F., & Yu, Y. (2020). Nondestructive testing method for welding quality in key parts of ocean-going ships. *Journal of Coastal Research*, 110(SI), 91–94. https://doi.org/10.2112/JCR-SI110-022.1.

TheFabricator. (n.d.). https://www.thefabricator.com/thefabricator/article/laserwelding/hybrid-laser-arc-welding-takes-on-heavy-transportation (Accessed on October 1, 2025).

Tzeng, Y. F. (2000). Parametric analysis of the pulsed Nd: YAG laser seam-welding process. *Journal of Materials Processing Technology*, 102(1–3), 40–47. https://doi.org/10.1016/S0924-0136(00)00447-7.

Vijay, A., & Radhika, N. (2024). A comprehensive review on underwater welding: Methods, property evaluations and challenges. *Advances in Materials and Processing Technologies*, 10(2), 923–959. https://doi.org/10.1080/2374068X.2023.2184582.

Vimeo.com, Inc. (n.d.). https://vimeo.com/528392166/d6e4eb581d (Accessed on October 1, 2025).

Vishnukumar, M., Pramod, R., & Kannan, A. R. (2021). Wire arc additive manufacturing for repairing aluminium structures in marine applications. *Materials Letters*, 299, 130112. https://doi.org/10.1016/j.matlet.2021.130112.

Walz, C., Stiebe-Springer, I., El Rayes, M., Seefeld, T., & Sepold, G. (2001, June). Hybrid welding of steel for offshore applications. In Paper presented at *The Eleventh International Offshore and Polar Engineering Conference (IOPEC)*, Stavanger, Norway.

WEIHUA. (n.d.). https://www.weihuagantrycrane.com/product/Gantry-Crane-for-Shipyard-Shipbuilding.html (Accessed on September 25, 2025).

Wert, David & DiSabella, Ryan. (2006). Strong, corrosion-resistant stainless steel. *Advanced Materials and Processes*. 164(8), 34; EID: 2-s2.0-33845363204.

Wikipcdia. (n.d.). https://cn.wikipcdia.org/wiki/Flux-corcd_arc_wclding (AcccsscdonAugust8, 2025).

Wikipedia. (n.d.). https://en.wikipedia.org/wiki/Friction_stir_welding (Accessed on September 25, 2025).

Wikipedia. (n.d.). https://en.wikipedia.org/wiki/Gas_tungsten_arc_welding (Accessed on August 8, 2025).

Wikipedia. (n.d.). https://en.wikipedia.org/wiki/Hyperbaric_welding (Accessed on October 1, 2025).

Wikipedia. (n.d.). https://en.wikipedia.org/wiki/Submerged_arc_welding (Accessed on August 8, 2025).

Index

A

ABS Rules for Materials and Welding Procedures, 198
Acoustic Emission Testing (AET), 166, 168
AI-Based Defect Recognition, 188
American Bureau of Shipping (ABS), 21, 101, 162, 184, 192, 216, 238, 268
Arc Welding Processes, 3, 31, 41, 42, 108
Austenitic Stainless Steels, 24, 62, 65, 118, 227, 231
Automation and Robotics in Shipyard Welding, 11
Autonomous Welding Robots in Shipbuilding, 276
Auxiliary and Support Structures, 133, 134

C

Certification Processes in Marine Projects, 210
Challenges in Implementing Robotic Welding, 85
Collaborative Robots, 70, 77, 87, 265, 276
Corrosion Prevention Techniques, 225, 226
Corrosion Resistance and Material Selection, 9
Criteria for Choosing a Welding Process, 54

D

Digital Twin Technology, 103, 122, 138, 155, 157, 266–268
Dry Hyperbaric Welding, 31, 98, 145, 147–149
Dry Welding (Habitat Welding), 97
Duplex Stainless Steels, 6, 24, 57, 59, 61, 62, 65, 66, 135, 138, 140, 141, 142, 158, 227
Dye Penetrant Testing (DPT), 13

E

Eddy Current Testing (ECT), 155, 166, 167, 179
Emerging Technologies in Shipbuilding, 265
Evolution of Welding Automation in Shipbuilding, 69

F

Fatigue in Marine Welds, 216, 229
Fixed Offshore Platforms, 130, 131, 207, 240
Fixed Welding Robots, 73–75
Flash Butt Welding, 48–51
Floating Offshore Structures (FOS), 131
Friction stir welding (FSW), 5, 23, 31, 51, 58, 192, 251, 254, 278
Future Trends in Marine Welding Automation, 87

G

Gas Metal Arc Welding (GMAW), 21, 33, 73, 89, 111, 131, 199, 246, 264

H

Hardness and Toughness Testing in Marine Welds, 66
Historical Development of Marine Welding, 2
Hybrid and Laser-Based Welding, 158
Hybrid welding techniques, 92
Hyperbaric Welding, 95, 96, 98, 102, 128, 132, 137, 141, 143–147

I

Importance of Welding in Shipbuilding, 6
Indian Register of Shipping (IRS), 21
Infrared Thermography (IRT), 166, 168

L

Laser and Optical Scanning, 166, 168
Lightweight Marine Vessels, 5
LR Inspection and Audit Practices, 205

M

Machine Learning in Marine Welding, 272
Magnetic Particle Testing (MPT), 13
Metal Arc Welding (GMAW), 199, 246, 264
Mobile and Modular Welding Units, 103
Mobile Welding Robots, 74, 75

P

Phased Array Ultrasonic Testing (PAUT), 155, 161, 166, 170, 237
Procedure and Production Audits, 199
Projection Welding, 47, 48, 50, 51

R

Radiographic Testing (RT), 101, 155, 161, 165, 167, 171–173, 182
Remotely Operated Welding Systems, 129, 136, 150

Resistance Welding, 41–49
Robotic and Autonomous Welding Systems, 156

S

Seam Welding, 46, 47, 50, 51, 53
Shielded Metal Arc Welding (SMAW), 21, 31, 42, 69, 96, 97, 131, 246
Spot Welding, 44–47, 50, 51
Stress Corrosion Cracking (SCC), 57, 64, 67, 141, 163, 218, 224, 225, 236
Submerged Arc Welding (SAW), 31, 39, 41, 42, 64, 69, 131, 252, 264
Subsea and Offshore Pipelines, 132, 133
Subsea Production Systems, 132
Sustainability and Green Manufacturing, 103

T

Testing and Quality Control Methods, 102
Thermomechanically controlled processed (TMCP) steels, 60
Tungsten arc welding (GTAW), 31, 37, 38, 42, 90, 111
Types of Corrosion in Marine Welds, 219

U

Ultrasonic Testing (UT), 13, 101, 134, 155, 161, 165, 167, 168, 170, 182, 198, 205, 234
Underwater Repair Welding, 5
Underwater Weld Monitoring Solutions, 271
Underwater welding, 94

W

Welder Qualification, 199
Welding Consumable Approval, 199
Welding dissimilar metals, 10, 25
Wet Hyperbaric Welding, 145–147
Wet Welding, 8, 9, 96
Wire Arc Additive Manufacturing (WAAM), 107, 109, 114, 254

For Product Safety Concerns and Information please contact our EU representative GPSR@taylorandfrancis.com Taylor & Francis Verlag GmbH, Kaufingerstraße 24, 80331 München, Germany

Printed and bound by CPI Group (UK) Ltd, Croydon, CR0 4YY
14/09/2026
02216256-0013